Rapid Assessment Program

Smithsonian Institution/Monitoring and Assessment of Biodiversity Program

12

RAP Working Papers

SI/MAB Series 6

Biological and Social Assessments of the Cordillera de Vilcabamba, Peru

Leeanne E. Alonso, Alfonso Alonso,
Thomas S. Schulenberg, and
Francisco Dallmeier, Editors

CENTER FOR APPLIED BIODIVERSITY SCIENCE (CABS)
CONSERVATION INTERNATIONAL (CI)
CONSERVATION INTERNATIONAL – PERU (CI-PERU)
THE FIELD MUSEUM
SMITHSONIAN INSTITUTION / MONITORING AND ASSESSMENT OF BIODIVERSITY PROGRAM (SI/MAB)
ASOCIACION PARA LA CONSERVACION DEL PATRIMONIO DEL CUTIVIRENI (ACPC)
SHELL PROSPECTING AND DEVELOPMENT, (PERU) B.V. (SPDP)
MUSEO DE HISTORIA NATURAL, UNIVERSIDAD NACIONAL MAYOR DE SAN MARCOS

RAP Working Papers are published by:
Conservation International
Center for Applied Biodiversity Science
Department of Conservation Biology
1919 M Street NW, Suite 600
Washington, DC 20036
USA
202-912-1000 tel
202-912-1030 fax
www.conservation.org

Editors: Leeanne E. Alonso, Alfonso Alonso,
Thomas S. Schulenberg, and Francisco Dallmeier
Assistant Editors: Jed Murdoch and Mariana Varese
Design: Glenda P. Fábregas
Maps: Jed Murdoch, Alejandro Queral-Regil, and Mark Denil
Cover photographs: Thomas S. Schulenberg
Translations: Ruth Nogeron

Conservation International is a private, non-profit organization exempt from federal income tax under section 501 c(3) of the Internal Revenue Code.

ISBN 1-881173-51-8

Library of Congress Card Catalog Number 98-75555

RAP Working Papers has been changed to RAP Bulletin of Biological Assessment. Numbers 14 and higher will be published under the new series title.

Suggested citation:
L. E. Alonso, A. Alonso, T. S. Schulenberg, and F. Dallmeier (eds). 2001. Biological and social assessments of the Cordillera de Vilcabamba, Peru. RAP Working Papers 12 and SI/MAB Series 6, Conservation International, Washington, D.C.

Printed on recycled paper.

These studies and publication have been funded in part by CI-USAID Cooperative Agreement #PCE-5554-A-00-4028-00. Publication was also funded by Shell International, The Guiliani Family, and CEPF.

TABLE OF CONTENTS

PARTICIPANTS AND AUTHORS

Conservation International – RAP Expeditions

Leeanne E. Alonso (Editor)
Director, Rapid Assessment Program
Center for Applied Biodiversity Science
Conservation International
1919 M Street NW, Suite 600
Washington, D.C. 20036
l.alonso@conservation.org

Mónica Arakaki (Botany)
Museo de Historia Natural
Universidad Nacional Mayor de San Marcos
Av. Arenales 1256 - Lima 14, Perú
marakaki@mail.utexas.edu
marakaki@lycos.com

Hamilton Beltrán (Botany)
Museo de Historia Natural
Universidad Nacional Mayor de San Marcos
Av. Arenales 1256 - Lima 14, Perú
Hamilton@musm.edu.pe

Brad Boyle (Botany)
Organization for Tropical Studies
Box 90630
Durham, NC 27708-0630
Mailing address;
13810 Long Lake Road
Ladysmith, BC V9G 1G5
Canada
Abboyle2@aol.com

Avecita Chicchón (Anthropology)
Program Officer
Program on Global Security and Sustainability
The John D. and Catherine T. MacArthur Foundation
140 S. Dearborn St., Suite 1100
Chicago IL 60603-5285
achiccho@macfound.org

Fonchii Chang (Fishes; deceased)
Museo de Historia Natural
Universidad Nacional Mayor de San Marcos
Av. Arenales 1256 - Lima 14, Perú

Louise Emmons (Mammals)
Division of Mammals, MRC-108
U.S. National Museum of Natural History
Washington, D.C. USA 20560
emmons.louise@nmnh.si.edu

William Evans (Coordination)
Asociación para la Conservación del Patrimonio del Cutivireni (ACPC)
Av. Javier Prado Este 255- Of. 302 B
Lima 27 – Perú
ACPC@terra.com.pe

Robin Foster (Botany, Overflight)
Environmental and Conservation Program
Field Museum of Natural History
1400 S. Lakeshore Dr.
Chicago, IL 60605
rfoster@fmnh.org

Juan Grados (Butterflies)
Museo de Historia Natural
Universidad Nacional Mayor de San Marcos
Av. Arenales 1256 - Lima 14, Perú
Grados@musm.edu.pe

Bruce K. Holst (Botany)
Marie Selby Botanical Gardens
811 South Palm Ave.
Sarasota, FL USA 34236
bholst@virtu.sar.usf.edu

Gerardo Lamas (Butterflies)
Museo de Historia Natural
Universidad Nacional Mayor de San Marcos
Av. Arenales 1256 - Lima 14, Perú
gerardo@musm.edu.pe

Lawrence López (Birds)
Department of Biological Sciences
Florida International University
Miami, FL USA 33199
pipra@usa.net

Lucia Luna (Mammals)
Division of Mammals
The Field Museum
1400 S. Lakeshore Drive
Chicago, IL 60605-2496, USA
lluna@fmnh.org
Museo de Historia Natural
Departamento de Mastozoología
Aptdo. 14-0434
Lima 14, Perú

Carlos Rivera (Amphibians and Reptiles)
Alferez Tavara 352
Iquitos, Perú
Junglexmail@cosapidata.comp.pe

Lily Rodríguez (Amphibians and Reptiles)
Asociación Peruana para la Conservación de la Naturaleza (APECO)
Parque Jose Acosta 187 - Lima 1, Perú
lrodriguez@datos.limaperu.net

Mónica Romo (Coordination, Mammals)
University of Turku
Department of Biology
Amazon Research Team
20014 Turku, Finland
romomonica@hotmail.com

Thomas S. Schulenberg (Birds, Team Leader)
Environmental and Conservation Program
Field Museum of Natural History
1400 S. Lakeshore Dr.
Chicago, IL 60605
tschulenberg@fmnh.org

Grace Servat (Birds)
University of Missouri - St. Louis
Biology Department
8001 Natural Bridge Rd
St Louis, MO USA 63121-4499
gservat@aol.com

Glenn H. Shepard, Jr. (Anthropology)
Instituto Nacional de Pesquisas da Amazônia (INPA)
c/o da Silva
Conj. Villar Camara
Rua 3, Casa 105
Bairro do Aleixo
Manaus, AM 69083-000
BRAZIL
ghs@inpa.gov.br
GShepardJr@aol.com

Armando Valdes (Birds)
Alexander Koenig Zoological Research Institut and Museum of Zoology
Museum Koenig, Adenauerallee 160,
53113 Bonn, Germany
uzspu8@ibm.rhrz.uni-bonn.de

Smithsonian Institution – SI/MAB Expedition

Raul Acosta (Aquatic Systems)
Departamento de Entomología
Universidad Nacional Agraria "La Molina"
Lima, Perú
racorivas@hotmail.com

Alfonso Alonso (Editor, Team Leader)
Smithsonian Institution/MAB Program
1100 Jefferson Drive, S.W., Suite 3123
Washington, D.C. 20560-0705
aalonso@ic.si.edu

Jessica Amanzo (Mammals)
Museo de Historia Natural
Universidad Nacional Mayor de San Marcos
Av. Arenales 1256 - Lima 14, Perú
jessica_amanzo@yahoo.com

Constantino Aucca (Birds)
Urbanizacion Ttio Q-1-13
Pasaje "Uriel Garcia"
Wanchac
Cusco, Perú

Severo Baldeon (Botany)
Museo de Historia Natural
Universidad Nacional Mayor de San Marcos
Av. Arenales 1256 - Lima 14, Perú

Patrick Campbell (Botany)
Smithsonian Institution/MAB Program
1100 Jefferson Drive, S.W., Suite 3123
Washington, D.C. 20560-0705
pcampbell@ic.si.edu

Edgardo Castro (Aquatic Systems)
Museo de Historia Natural
Universidad Nacional Mayor de San Marcos
Av. Arenales 1256 - Lima 14, Perú

James Comiskey (Botany)
Smithsonian Institution/MAB Program
1100 Jefferson Drive, S.W., Suite 3123
Washington, D.C. 20560-0705
jac@ic.si.edu

Saida Córdova (Invertebrates)
Museo de Historia Natural
Universidad Nacional Mayor de San Marcos
Av. Arenales 1256 - Lima 14, Perú
aracno@musm.edu.pe

Francisco Dallmeier (Editor)
Smithsonian Institution/MAB Program
1100 Jefferson Drive, S.W., Suite 3123
Washington, D.C. 20560-0705
fdallmeier@ic.si.edu

Rafael De la Colina (Botany)
Departmento de Botánica
Facultad de Ciencias Biológicas
Universidad San Antonio
Abad del Cusco
Cusco, Perú

Alicia de la Cruz (Invertebrates)
Departamento de Entomología
Universidad Nacional Agraria "La Molina"
Lima, Perú
adecruza@hotmail.com

Lupe Guinand (Coordination)
Smithsonian Institution
Las Tunas 125
Urbanización la Fontana
La Molina
Lima, Perú
cparedes@amauta.rcp.net.pe

Max Hidalgo (Aquatic Systems)
Museo de Historia Natural
Universidad Nacional Mayor de San Marcos
Av. Arenales 1256 - Lima 14, Perú

Javier Icochea (Amphibians and Reptiles)
Apartado 14-219, Lima 14, Perú
javiericochea@hotmail.com

Manuel Laime (Invertebrates)
Museo de Historia Natural
Universidad Nacional Mayor de San Marcos
Av. Arenales 1256 - Lima 14, Perú

William Nauray (Botany)
Departmento de Botánica
Facultad de Ciencias Biológicas
Universidad San Antonio
Abad del Cusco
Cusco, Perú
wnauray@hotmail.com

Percy Núñez (Botany)
Departmento de Botánica
Facultad de Ciencias Biológicas
Universidad San Antonio Abad del Cusco
Cusco, Perú

Tatiana Pequeño (Birds)
Museo de Historia Natural
Universidad Nacional Mayor de San Marcos
Av. Arenales 1256 - Lima 14, Perú

Roberto Polo (Invertebrates)
Departamento de Entomología
Universidad Nacional de la Libertad (Trujillo)
Trujillo, Perú

Elias Ponce (Amphibians and Reptiles)
Facultad de Ciencias Biológicas
Universidad Nacional de San Agustin
Arequipa, Perú

Alfredo Portilla (Amphibians and Reptiles)
Maestria en Gestión Ambiental
Facultad de Ingenieria Ambiental
Universidad Nacional de Ingeneria, Lima, Perú

Eliana Quispitupac (Amphibians and Reptiles)
Maestria en Zoología
Facultad de Ciencias Biológicas
Universidad Nacional Mayor de San Marcos
Av. Arenales 1256 - Lima 14, Perú

Daisy Reyes (Aquatic Systems)
Departamento de Entomología
Universidad Nacional de la Libertad
Trujillo, Perú

Juan José Rodríguez (Mammals)
Proyecto BIODAMAZ - IIAP
Ap. Postal 454
Iquitos - Perú
Tel. 51 94 264060
jjrodriguez@iiap.org.pe

Edwin Salazar (Birds)
Departamento de Ornitología
Universidad San Agustin
Arequipa, Perú
esazap@hotmail.com

Norma Salcedo (Aquatic Systems)
Museo de Historia Natural
Universidad Nacional Mayor de San Marcos
Av. Arenales 1256 - Lima 14, Perú
norma@musm.edu.pe

Jose Santisteban (Coordination, Invertebrates)
Museo de Historia Natural
Universidad Nacional Mayor de San Marcos
Av. Arenales 1256 - Lima 14, Perú
santisteban@musm.edu.pe

Sergio Solari (Mammals)
Museo de Historia Natural
Universidad Nacional Mayor de San Marcos
Av. Arenales 1256 - Lima 14, Perú
ssolari@musm.edu.pe

Shana Udvardy (Coordination, Botany)
Smithsonian Institution/MAB Program
1100 Jefferson Drive, S.W., Suite 3123
Washington, D.C. 20560-705
sudvardy@simab.si.edu

Gorky Valencia (Invertebrates)
Universidad San Antonio
Abad del Cusco
Urbanizacion Mateo Pumacahua
Segunda Etapa H11
Wanchaq, Cusco, Perú
gorkyv@yahoo.com

Paul Velazco (Mammals)
Museo de Historia Natural
Universidad Nacional Mayor de San Marcos
Av. Arenales 1256 - Lima 14, Perú

Elena Vivar (Mammals)
Museo de Historia Natural
Universidad Nacional Mayor de San Marcos
Av. Arenales 1256 - Lima 14, Perú
vivar@musm.edu.pe

ORGANIZATIONAL PROFILES

CONSERVATION INTERNATIONAL

Conservation International (CI) is an international, non-profit organization based in Washington, D.C. CI acts on the belief that the Earth's natural heritage must be maintained if future generations are to thrive spiritually, culturally, and economically. Our mission is to conserve biological diversity and the ecological processes that support life on Earth, and to demonstrate that human societies are able to live harmoniously with nature.

Conservation International
1919 M Street NW, Suite 600
Washington, DC 20036 USA
202-912-1000 (telephone)
202-912-1030 (fax)
http://www.conservation.org

FIELD MUSEUM

The Field Museum (FM) is an educational institution concerned with the diversity and relationships in nature and among cultures. Combining the fields of Anthropology, Botany, Geology, Paleontology and Zoology, the Museum uses an interdisciplinary approach to increasing knowledge about the past, present, and future of the physical earth, its plants, animals, people, and their cultures. In doing so, it seeks to uncover the extent and character of biological and cultural diversity, similarities and interdependencies so that we may better understand, respect, and celebrate nature and other people. Its collections, public learning programs, and research are inseperably linked to serve a diverse public of varied ages, background and knowledge.

Field Museum
1400 South Lakeshore Dr.
Chicago, IL 60605
312-922-9410
312-665-7932 (fax)

CONSERVACION INTERNACIONAL - PERU

CI-Perú was created in 1989 and in 1992, the program office moved to Perú. CI-Perú is a decentralized office, responsible for the development of its programs and activities in key ecosystems in Perú. Presently, CI-Perú works in three extremely important regions for biodiversity conservation: Tambopata, Vilcabamba and Condor. Additionally, the program works in thematic areas with broader geographical impacts including: protected area policy, hydrocarbon policy, training in biodiversity rapid appraisal, conservation-based enterprise development and communications. Program growth has been rapid and now counts with a full time staff of 55 natural resource professionals, technicians, community promoters and administrative personnel. CI-Perú uses an interdisciplinary approach to work in high biodiversity areas with local populations in association with other organizations and institutions to generate economic development based on the conservation of natural resources. This effort has the goal of contributing to the construction of a society that has as one of its fundamental bases the conservation of natural ecosystems.

Conservación Internacional - Perú
Chinchón 858, Dept. A
San Isidro
Lima 27 PERÚ
(51-1) 4408-967 (telephone/fax)
ci-peru@conservation.org
http://www.fieldmuseum.org

SMITHSONIAN INSTITUTION/ MONITORING AND ASSESSMENT OF BIODIVERSITY PROGRAM (SI/MAB)

Since 1986, The Smithsonian Institution/Monitoring and Assessment of Biodiversity Program (SI/MAB) has promoted biodiversity conservation. We maintain that biological diversity is one of the most valuable assets on earth, yet few human, technological and financial resources are targeted to assure its conservation. We thus conduct research projects, biodiversity assessments and monitoring programs, as well as education and training courses, to understand ecosystems and to reach the most diverse audience as we build in-country capacities to carry out this important work. Biodiversity assessments provide the first picture of the species abundance and distribution. Long-term monitoring is the movie that shows the dynamics of the system. Through an integrated approach of research and training, we provide scientific information and build in-country capacity to foster sustainable use of natural resources. We focus on tropical and temperate forested regions within a growing network of long-term biodiversity monitoring sites in Latin America and the Caribbean, North America, Africa, and Asia.

Smithsonian Institution/MAB Program
1100 Jefferson Drive, S.W., Suite 3123
Washington, DC 20560-0705
202-357-4793 (telephone)
202-786-2557 (fax)
http://www.si.edu/simab

SHELL PROSPECTING AND DEVELOPMENT (PERU) B.V.

On May 17,1996 Shell Prospecting and Development (Perú) B.V. (SPDP) and its partner Mobil Exploration and Producing Perú Inc, Sucursal Peruana signed a 40 year licence agreement with the Peruvian authorities. This agreement included a 2-year appraisal phase in which Shell, the project operator, would drill three appraisal wells and conduct market studies before deciding whether or not to proceed with the main phase of the project. In March 1997, Shell and Mobil also signed a licence agreement for exploration activity in Block 75. This area covers some 7,900 square kilometers adjacent to the Camisea blocks and extending to the North. This agreement included at least one exploration well and a limited amount of seismic activity.

The key role of SPDP was to develop the gas and condensate resources of the Camisea fields. The mission in the first two years was to undertake the necessary studies to enable a clear decision on how to proceed. This required the evaluation of technical issues, costs, development of markets, and gaining of the environmental and social 'licence to operate' from the authorities, the communities and other key stakeholders. To ensure the latter company objectives included 'attaining the highest standards' and implementing 'full consultation'.

Shell Prospecting and Development (Perú) B.V.
Please see: www.shell.com

ASOCIACION PARA LA CONSERVACION DEL PATRIMONIO DEL CUTIVIRENI (ACPC)

The Asociación para la Conservación del Patrimonio del Cutivireni is a non-governmental organization established on September 3, 1987. Its main objective is the preservation of the cultural and natural heritage of the area of the Vilcabamba Mountain Range, which is the source of water for important tributaries of the Ene-Apurimac and Urubamba river systems.

ACPC members include foreigners, Asháninka, Peruvians, and others that work in close coordination with the local native communities (primarily Asháninka and Matsiguenga). ACPC's key work has been located around the Cutivireni region, which is a special area of Amazonian rainforest, encompassing several different ecosystems from the valleys where communities are located to approximately 4000 meters above sea level.

Asociación para la Conservación del Patrimonio del Cutivireni
Hector Vega, President
Av. Javier Prado Este 255 of 302-B
Lima 14, PERÚ
(51-1) 421-0946 (Telephone/fax)
ACPC@terra.com.pe

MUSEO DE HISTORIA NATURAL DE LA UNIVERSIDAD NACIONAL MAYOR DE SAN MARCOS

The Museo de Historia Natural is a branch of the Biology Department of the University of San Marcos in Lima, a government institution, reporting directly to the Office of the Rector. Since its creation in 1918, the museum has contributed greatly to the scientific knowledge of the fauna, flora and geology of Peru. The museum's main goal is to develop scientific collections to conduct systematic research, to provide the data, expertise, and human resources necessary for understanding Peru's biogeography, and to promote the conservation of Peru's many ecosystems.
The museum conducts field studies in the areas of Botany, Zoology, Ecology, and Geology-Paleontology. Each of the fifteen departments has its own curator, associated researchers, and students. Over the past 15 years the museum staff have conducted intensive field work in protected areas such as Manu National Park, Abiseo National Park, and Pacaya-Samiria National Reserve.

Museo de Historia Natural
Universidad Nacional Mayor de San Marcos
Apartado 14-0434
Lima-14
Lima PERÚ
(51-1) 4710-117 (telephone)
(51-1) 4265-6819 (fax)
postmaster@musm.edu.pe

ACKNOWLEDGMENTS

CI - RAP Expeditions

Expeditions into remote, undeveloped areas are made with the support and assistance of many organizations and individuals. Our 1997 expedition was possible only with the assistance of the Aviación del Ejercito Peruano; we are particularly indebted to Mayor Javier Lazabara Alay and to Comandante Luis Rojas Merion. Alas de Esperanza, and in particular pilot Enrique Tante, provided excellent logistic support for an initial aerial reconnaissance in May 1997, and for all portions of the 1998 expedition. Our expeditions also benefited greatly from the logistical support provided by the Asociación para la Conservación del Patrimonio del Cutivireni, especially Billy Evans, Javier Narvaez (Tino), David Llanos, Percy Peralta, Antonio Sánchez, and Michel Saenz.

We are indebted to Lic. Luis Alfaro, Director General de Areas Naturales Protegidas y Fauna Silvestre from INRENA, who has supported RAP efforts in Peru and who has committed his office to the continuity of conservation efforts in the Zona Reservada Apurimac.

Our expeditions also would not have been possible without the support and collaboration of the members of communities living on the lower slopes of the Cordillera Vilcabamba. We are very grateful for the generous hospitality that we received from the community of Tangoshiari, especially from Nicolás Sagastizabal, Vice-President of the Community of Tangoshiari, who assisted in all aspects of coordinating our expedition, and to Daniel Zapata, President of the Community of Tangoshiari, and a first class camp assistant.

The 1997 expedition up the Río Picha was facilitated by Mateo Italiano, a Matsigenka field assistant; Julio Urquia, Yine motorist; and all the members of the Río Picha Matsigenka native communities of Camaná, Mayapo and Puerto Huallana, particularly Rani Ríos, Julio Pacaya, and Oswaldo Bernaldes. Luis Dávalos prepared a baseline assessment of the region, for which we are grateful. Richard Piland visited the Ene region and provided valuable insights about the Asháninka organizations. Anthropologists Carlos Mora (Chevron) and Lelis Rivera (CEDIA) also provided initial key guidance about the Urubamba region.

The following botanists kindly helped in identifying plant specimens or photographs: John T. Atwood (Orchidaceae), Thomas Croat (Araceae), Harry Luther (Bromeliaceae), and Hans Wiehler (Gesneriaceae). Emmons, Luna, and Romo thank James L. Patton for generously processing tissue samples, which greatly aided in the identification of small mammals. Victor Pacheco helped with the identification of murid rodents, especially the genus *Thomasomys*, and Alfred Gardner and Michael Carleton made helpful suggestions about identification. For hosting the study of specimens at the American Museum of Natural History, we thank Rob Voss and Guy Musser.

The editors thank Mariana Varese, Monica Romo, and Ana Maria Chonati for providing background materials and for helpful comments on this report.

Financial support for these investigations was provided under USAID cooperative agreement #PCE-5554-A-00-4028-00. Publication of the report was funded by Shell International, The Guiliani Family, and The Critical Ecosystem Partnership Fund (CEPF).

SI/MAB Expedition

Our sincere appreciation goes to Alan Hunt, general manager of Shell Prospecting and Development (Perú) B.V. (SPDP), and to Tony Lancione, general manager of the Bechtel- Cosapi-Odebrecht (BCO) consortium, for their commitment and support for the biodiversity assessment along the proposed pipeline route. Murray Jones, manager of SPDP Health, Safety and Environment, was instrumental in conducting this project. SPDP's Gert van der Horst, Peter Dushe, Craig Schenk, Ted Murray, Allan Sayers, Mark Hammerton, Patricia Zavala, Miguel Ruiz-Larrea, Mary Malca, Augusto Baldoceda, Lincoln Williamson, and Alonso Zarzar, as well as BCO's Jack Laurijssen, Jacob Teerink, Bruce Skinner, Steve Macklin, Curtis Meininger, James Streeter, Manuel Durazo, David Barry, John Buckley, Jorge Juarez, and Alejandro Camino, surmounted significant logistical, operational, medical, and safety obstacles. Victor Grande, Carlos Guillen, Alejandro Alvarez, Lissette Giudice, Marco Marticorena, and Julio Cesar Postigo were invaluable in their roles as liaisons with Native communities and in finding excellent field assistants.

Alan Dabbs and Marcelo Andrade of Pro-Natura USA continue to make helpful suggestions concerning sustainable use of natural resources. SPDP/BCO provided transportation to the study sites, through the helicopter services of Canadian Helicopters, Ltd. and Fuerza Aerea Peruana. We thank Nuevo Mundo managers Aarnoud Smit and Robert Smit, and the Kimbiri manager Ing. Jorge Juarez for logistic/supply arrangements. Field work progressed more smoothly through the help of local guides Celestino Gutierrez Mauri, Hugo Gutierrez Mauri, Roger Rodriguez Aguilar, and Santiago Lopez Otiari from Manitinkiari and Crisostomo Yucra Cardenas, Gaudencio Yucra Cardenas, Raúl Martinez Ramos, and Joel Rodas Martinez from Comunidad Campesina Pueblo Libre. The cooperation of Native communities has been immensely helpful. Concession and Catering provided facilities for our field operations and very good meals.

Asociación Peruana para la Conservacion de la Naturaleza (APECO) was proficient in handling administrative matters and some of the logistics from Lima. The Instituto Nacional de Recursos Naturales (INRENA) granted permits for sampling scientific specimens. The Museo de Historia Natural, Universidad Nacional Mayor de San Marcos, directed by Dr. Neils Valencia, provided support. Many of our counterpart researchers were from that institution, and it is one of the depositories for the specimens sampled, along with the Universidad de San Antonio Abad del Cusco and the Universidad Nacional Agraria "La Molina."

We are indebted to Lupe Guinand for coordinating the SI/MAB program in Lima, Tatiana Pacheco for her adminstrative/ management skills, and Sandy Thomas for her legal advice. Thanks to Shana Udvardy and José Santisteban for coordinating field camp operations, Antonio Salas for assisting with translations, and Deanne Kloepfer for reviewing all of the papers.

We thank Ruth Nogeron for translating the Executive Summary into Spanish.

PREFACE

The Cordillera de Vilcabamba is an extraordinary place. It contains some of the world's most diverse and unique ecosystems, plants, animals, and human cultures. It is considered by Conservation International and other organizations as one of the "hottest" of biodiversity hotspots. It is one of the last remaining intact tropical montane areas in South America. It is also the home of four ancestral ethnic groups. However, it may not remain this way for long. Despite its remoteness and rugged terrain, development in the form of human settlements, oil exploration, and logging are entering the region.

The protection and preservation of the Cordillera de Vilcabamba - its forests, watersheds, biodiversity, and human culture- deserve a great amount of effort from all of us. The health and survival of local communities and natural ecosystems depend on the future condition of the forests of the Vilcabamba mountains.

To this end, many diverse groups have been involved in conservation efforts in the Cordillera de Vilcabamba. Recognizing that collaborative efforts will take us further toward our common goal of protecting this unique region, we have formed coalitions and have been working together to study the region, to develop consensus among groups, local communities, and governmental agencies, and to propose protected areas within the Cordillera.

This report is the result of landmark collaborations between diverse groups to study the biological and cultural diversity of the Cordillera de Vilcabamba for conservation. Conservation International- Perú (CI-Perú) and the Asociación para la Conservación del Patrimonio de Cutivireni (ACPC), two Peru based non-governmental organizations working in the Vilcabamba region, collaborated with CI's Rapid Assessment Program (RAP) to organize and conduct scientific expeditions in 1997 and 1998 in the Northern Cordillera de Vilcabamba. In a similar fashion, the Smithsonian Institution/Monitoring and Assessment of Biodiversity Program (SI/MAB), a U.S. based scientific institution, collaborated with Shell Prospecting and Development (Perú) B.V. (SPDP) to survey the Southern Cordillera de Vilcabamba.

We present information from three scientific surveys together in this report in order to provide a comprehensive overview of the biodiversity of the Cordillera de Vilcabamba. We also include an overview of other collaborative efforts to conserve the region, including a current proposal for the establishment of two communal reserves and a national park.

We trust that this document will not only promote further partnerships, but will provide ample justification for the official protection of the Cordillera de Vilcabamba's unique and globally important natural and cultural resources.

Carlos F. Ponce
Field Vice-President for the Andean Countries
CI-Peru Director
Conservation International

Francisco Dallmeier
Director
Smithsonian Institution/
Monitoring and Assessment of Biodiversity Program

REPORT AT A GLANCE

BIOLOGICAL AND SOCIAL ASSESSMENTS OF THE CORDILLERA DE VILCABAMBA, PERU

1) Dates of Studies

CI-RAP Expedition I (Camps One and Two):
6 June – 4 July, 1997
CI-RAP Expedition II (Camp Three):
28 April – 23 May, 1998
SI/MAB Expedition:
11 July – 11 August, 1998

2) Description of Location

The Cordillera de Vilcabamba, comprising more than three million hectares (approximately 2.4% of Peru's land area), is located in the Andes Mountain chain in south-eastern Peru. In 1988, 1,699,300 hectares of the Cordillera de Vilcabamba was set aside by the Peruvian government as the Zona Reservada del Apurímac, a designation that provides it transitory protected status until a permanent classification is made. The Cordillera exhibits wide variation in altitude and rainfall, from the highest peaks above 4,300 m of altitude, to the lower river basins at around 400 - 500 m. Rainfall ranges from 1,200 mm/year (47 inches/year) to a high of 5,600 mm/ year (220 inches/year). The Cordillera de Vilcabamba contains seven life and two transition vegetation zones and is primarily covered by Pre-montane Moist Tropical Forest and Moist Tropical Forest. The steep slopes and isolation of the Cordillera de Vilcabamba have created the setting for the evolution of a high number of plant, amphibian, insect, and mammal species, many of which are found no where else on earth and are therefore endemic to the Cordillera.

The Cordillera de Vilcabamba is recognized nationally and internationally as an area of high conservation importance due to its great indigenous diversity. The region is home to several native Amazonian groups: Asháninka, Matsigenka (or Matsiguenga), Nomatsiguenga, and Yine. The steep mountains and forests protect the headwaters of many important South American rivers, thereby maintaining a clean supply of freshwater for the fertile soils in the lowlands below. The Cordillera contains 55 waterfalls that range from 80 to 260 meters tall, the world's longest natural bridge, and innumerable unexplored lagoons, canyons, forests, and subterranean rivers.

3) Reason for Expeditions

Remote and difficult to access, the Cordillera de Vilcabamba is still poorly known to biologists. The range of elevations and resulting high variety of habitat types of the Cordillera have long led biologists to suspect that it harbors an incredible diversity of animals and plants, many of which will be new to science. However, the area is increasingly threatened by encroaching natural gas and oil exploration, logging, and changing land-use patterns of the native indigenous groups.

In order to increase our knowledge of the biological diversity of the region, three scientific expeditions were carried out in 1997 and 1998. Two expeditions were organized by Conservation International's Rapid Assessment Program (RAP) and CI-Perú, and one by the Smithsonian Institution/Monitoring and Assessment of Biodiversity Program (SI/MAB). These expeditions were designed to provide an initial survey of selected taxa and to provide biodiversity data to support conservation efforts in the region. The biological data collected and presented here will contribute to formulating management plans for the sustainable use and conservation of this unique area.

4) Major Results

The results from the biodiversity surveys reported here confirm that the Cordillera de Vilcabamba is indeed rich in biodiversity and that conservation of the area is highly recommended. All sites contained a remarkable substrate heterogeneity, which contributed to high levels of habitat heterogeneity, from humid canyons to high-elevation pajonales. Species richness of vertebrates was high, with many endemic and geographically restricted species of mammals and birds, including the discovery of a new mammal genus. Most of the amphibians observed are endemic to the Cordillera de Vilcabamba. A high diversity of all taxonomic groups was recorded.

Many of the species recorded are new to science, were found in unexpected habitats or in surprising abundance, further indication that this area is a special place in relation to biodiversity. The higher elevation sites were distinct from each other and from lower altitudes. The high habitat heterogeneity observed at all survey sites strongly indicates that the Cordillera de Vilcabamba contains a tremendous floral and faunal diversity, which we have only begun to document and understand.

5) Conservation Recommendations

The results of the five biodiversity surveys reported herein highlight the Cordillera de Vilcabamba as a region of high global biodiversity importance, deserving of conservation efforts and protection, and support the proposal put forward by CI-Perú, ACPC, and CEDIA to establish two communal reserves and a national park within the Cordillera de Vilcabamba (see Proposed Protected Areas in the Cordillera de Vilcabamba, this volume). The Cordillera de Vilcabamba should also be linked to nearby protected areas through biodiversity corridors, which are essential for the long-term survival of animals, plants and people.

Conservation priorities within the Cordillera include protecting the outlying ridges to sustain rare and local bird species, protecting areas above 1400 m to ensure conservation of the high habitat heterogeneity and to provide a refuge for mammals, protecting the watershed services of the Cordillera, protecting valleys that reach as low as 1000 m, continuing to include indigenous communities living in and near the Cordillera in determining future conservation and development plans, routing gas pipelines outside of the Cordillera de Vilcabamba, prohibiting road construction, controlling human immigration to the area, and establishing monitoring stations within and around the Cordillera de Vilcabamba.

Research priorities include further studies of high elevations areas, exploration of the valleys (800-1500 m), sampling during the rainy season, study of the role of the underlying substrates and fire in maintaining and determining habitat heterogeneity, establishment of a long-term monitoring program, and study of the effects of the proximity of human communities on game animals and selected plant species to aid management and conservation programs.

Number of species recorded:

	CI-RAP Camp 1 (3350 m)	CI-RAP Camp 2 (2050 m)	CI-RAP Camp 3 (1000 m)	ALL CI-RAP Sites Combined	SI/MAB Wayrapata (2445 m)	SI/MAB Llactahuaman (1710 m)	ALL SI/MAB Sites Combined
Vascular Plants*	247	428	220	--	130/0.1 ha	121/0.1 ha	--
Birds	43	115	150	290	92	111	176
Mammals	12	28	58	87	31	31	46
Reptiles	1	2	9	12	7	9	13
Amphibians	3	11	29	43	11	8	16
Invertebrates							
Butterflies and Large Moths	58	19	--	76	--	--	--
Spiders	--	--	--	--	45	60	92
Beetles	--	--	--	--	--	166	--
Crickets	--	--	--	--	17	22	34
Bees & Wasps	--	--	--	--	67	102	152
Aquatic Invertebrates	--	--	--	--	37/creek	47/creek	191

* Most plant species have yet to be identified.

New species recorded:

	CI-RAP Camp 1 (3350 m)	CI-RAP Camp 2 (2050 m)	CI-RAP Camp 3 (1000 m)	SI/MAB Wayrapata (2445 m)	SI/MAB Llactahuaman (1710 m)	Genera
Vascular Plants*	1	1	--	--	--	*Greigia raporum* *Greigia vilcabambae*
Birds	--	1?	--	--	--	*Grallaria* nr. *erythroleuca*
Mammals	3	1?	--	--	1	*Thomasomys* spp. *Cuscomys ashaninka* *Carollia* sp.
Reptiles	1	1	--	--	--	*Proctoporus* sp. *Dipsas* sp.
Amphibians	3	7	2	1	--	*Eleuthrodactylus* spp. *Gastotheca* spp. *Phrynopus* sp. *Centrolene* sp. *Telmatobius* sp. *Colostethus* sp. *Atelopus* sp.
Fishes	--	--	2-3	--	--	*Scopaeocharax* sp. *Tyttocharax* sp. *Corydoras* sp. ?
Butterflies	11	1	--	--	--	*Pedaliodes* sp.

REPORTE EN BREVE

EVALUACION BIOLOGICA Y SOCIAL EN LA CORDILLERA DE VILCABAMBA, PERÚ

1) Fechas de Estudio

Expedición I de CI-RAP (Campamentos Uno y Dos)
6 junio – 4 julio, 1997
Expedición II de CI-RAP (Campamento Tres)
28 abril – 23 mayo, 1998
Expedición del SI/MAB
11 julio – 11 agosto, 1998

2) Descripción de la Zona

La Cordillera de Vilcabamba, con más de tres millones de hectáreas (aproximadamente 2.4% del área terrestre del Perú), esta localizada en las montañas de los Andes en el Sur-Este del Perú. En 1988, un área de 1,669,300 hectáreas de la Cordillera de Vilcabamba fue establecida como la Zona Reservada de Apurímac, una categoría que protege el área en forma temporal, al tiempo que promueve estudios para el establecimiento de categorías de protección permanente. La Cordillera muestra una gran variación altitudinal, desde los picos más altos a 4,300 m.s.n.m., a la cuenca de ríos a 400-500 m.s.n.m. La precipitación también varía de 1,200 mm/año (47 pulgadas/año) a 5,600 mm/año (220 pulgadas/año). Con base en la vegetación se han descrito siete zonas de vida y dos transicionales, siendo las principales el bosque tropical húmedo y el bosque tropical húmedo pre-montano. Las pendientes marcadas y el aislamiento en la Cordillera de Vilcabamba han creado un ambiente propicio para la evolución de muchas especies de plantas y animales, incluyendo varios anfibios, insectos, y mamíferos. Muchas de estas especies son endémicas de la Cordillera, y por tanto sólo se encuentran en este lugar.

La Cordillera de Vilcabamba es también reconocida como una de las áreas de prioridad para la conservación en el Perú y en el mundo por su gran diversidad de grupos indígenas. En esta región viven Los Asháninka, Los Matsigenka (o Matsiguenga), Los Nomatsiguenga, y Los Yine. La Cordillera protege las cuencas de muchos ríos, constituyendo una fuente limpia y abundante de agua que alimenta a ríos, bosques y asentamientos humanos en las áreas más bajas. El área contiene 55 cascadas de 80 a 260 metros de caída libre, el puente natural más grande del mundo, y muchas lagunas, cañones, bosques y ríos subterraneos que todavía no han sido explorados.

3) Metas de las Expediciones

Remota y de difícil acceso, la Cordillera de Vilcabamba ha permanecido poco explorada por biólogos. La gran variación en altitud y habitats han hecho que los investigadores sospechen que tiene una diversidad muy alta de plantas y animales, muchas de las cuales son nuevas para la ciencia. Sin embargo, el área esta amenazada por varias actividades humanas, incluyendo la exploración para el desarrollo de reservas de gas natural, el aprovechamiento de los recursos forestales, y los cambios en los patrones del uso del suelo por los grupos indígenas.

Con el fin de ampliar el conocimiento de la diversidad biológica de la Cordillera de Vilcabamba, se llevaron a cabo tres expediciones científicas entre 1997 y 1998. Dos expediciones fueron organizadas por el Programa de Evaluación Rápida (RAP) de Conservación Internacional (CI) y Conservación Internacional-Perú (CI-Perú), y una por el Programa de Evaluación y Monitoreo de la Biodiversidad de la Institución Smithsonian (SI/MAB). Estas expediciones fueron diseñadas para obtener un muestreo inicial de grupos taxonómicos selectos para apoyar los planes de conservación del área. La información biológica que se presenta en este reporte contribuirá a formular planes de manejo para el uso sostenible y de conservación de esta área única en el mundo.

4) Resultados Principales

Los estudios de biodiversidad que se reportan en este trabajo confirman que la Cordillera de Vilcabamba es ciertamente muy rica en biodiversidad. Los cinco sitios tienen una alta heterogeneidad en el substrato, que contribuye a altos niveles de heterogeneidad de los hábitats, que van de quebradas profundas y húmedas hasta pajonales de altura. La riqueza de especies de vertebrados fue alta, con muchas especies de mamíferos y aves endémicas y geográficamente restringidas, incluyendo una nueva especie de roedor. Se confirmó que muchos de los anfibios observados son endémicos de la Cordillera de Vilcabamba. En breve, se registró una alta diversidad en todos los grupos taxonómicos estudiados.

Muchas de las especies registradas son nuevas para la ciencia o representan nuevos registros para el área. La ocurrencia de ciertas especies en hábitats inesperados o en gran abundancia, es un indicador adicional de que Vilcabamba es una área muy especial en términos de biodiversidad. Las áreas de estudio en las elevaciones más altas son distintas entre sí y también de las áreas de menor elevación. La gran heterogeneidad de hábitats observada en todas las áreas de estudio indica con gran claridad que la Cordillera de Vilcabamba alberga una tremenda diversidad de flora y fauna, algo que sólo ahora se ha comenzado a documentar y entender.

5) Recomendaciones para la Conservación

Los cinco estudios de biodiversidad llevados a cabo por CI-RAP y SI/MAB demuestran que la Cordillera de Vilcabamba es un área biológicamente rica y única que bien merece ser conservada y protegida. Se recomienda que la preservación de los recursos naturales de la Cordillera de Vilcabamba sea una prioridad nacional y global. Los resultados apoyan la propuesta recomendada por CI-Perú, ACPC y CEDIA de establecer dos reservas comunales y un parque nacional en la Cordillera de Vilcabamba (véase la sección de Areas Protegidas Propuestas en la Cordillera de Vilcabamba en este reporte). La Cordillera de Vilcabamba también deberá conectarse a otras áreas protegidas cercanas a través de corredores biológicos, los cuales son esenciales para la supervivencia a largo plazo de animales, plantas y poblaciones humanas.

Las prioridades de conservación de La Cordillera incluyen proteger las crestas de las montañas que mantienen especies de aves locales y raras, proteger áreas por encima de los 1400 m.s.n.m. para asegurar la conservación de la mayor parte de la heterogeneidad de los hábitats y para proveer refugio para mamíferos, proteger la Cordillera de Vilcabamba como una cuenca, proteger algunos de los valles que llegan hasta los 1000 m.s.n.m., continuar involucrando a las comunidades indígenas que viven dentro de y en los alrededores de la Cordillera en la determinación de futuros planes de conservación y desarrollo, mantener la posible ruta del gasoducto fuera de zona protegida de la Cordillera de Vilcabamba, prohibir la construcción de caminos en la Cordillera de Vilcabamba, controlar la migración al área, y establecer estaciones de monitoreo dentro de y en los alrededores de la Cordillera de Vilcabamba.

Prioridades de investigación incluyen más estudios biológicos en los valles y montañas entre los 800 y 1500 m.s.n.m., más muestreos durante la época de lluvias, estudios del rol del substrato y el fuego en el sostenimiento y la delimitación de la heterogeneidad de hábitats, establicimiento de un programa de monitoreo a largo plazo, y estudios de los efectos de la proximidad de asentamientos humanos en los animales sujetos a cacería y en especies de plantas seleccionadas para ayudar a los programas de manejo y conservación.

Numeros de especies registrados:

	CI-RAP Camp 1 (3350 m)	CI-RAP Camp 2 (2050 m)	CI-RAP Camp 3 (1000 m)	Todos los Sitios del CI-RAP Combinados	SI/MAB Wayrapata (2445 m)	SI/MAB Llactahuaman (1710 m)	Todos los Sitios del SI/MAB Combinados
Plantas Vasculares*	247	428	220	--	130/0.1 ha	121/0.1 ha	--
Aves	43	115	150	290	92	111	176
Mamíferos	12	28	58	87	31	31	46
Reptiles	1	2	9	12	7	9	13
Anfibios	3	11	29	43	11	8	16
Invertebrados							
Mariposas	58	19	--	76	--	--	--
Arañas	--	--	--	--	45	60	92
Escarabajos	--	--	--	--	--	166	--
Grillos	--	--	--	--	17	22	34
Abejas y Avispas	--	--	--	--	67	102	152
Invertebrados Acuáticos	--	--	--	--	37/arroyo	47/arroyo	191

* La mayoria de las plantas todavia no han sido identificadas.

Nuevas especies registradas:

	CI-RAP Camp 1 (3350 m)	CI-RAP Camp 2 (2050 m)	CI-RAP Camp 3 (1000 m)	SI/MAB Wayrapata (2445 m)	SI/MAB Llactahuaman (1710 m)	Genera
Plantas Vasculares	1	1	--	--	--	*Greigia raporum* *Greigia vilcabambae*
Aves	--	1?	--	--	--	*Grallaria* nr. *erythroleuca*
Mamíferos	3	1?	--	--	1	*Thomasomys* spp. *Cuscomys ashaninka* *Carollia* sp.
Reptiles	1	1	--	--	--	*Proctoporus* sp. *Dipsas* sp.
Anfibios	3	7	2	1	--	*Eleuthrodactylus* spp. *Gastotheca* spp. *Phrynopus* sp. *Centrolene* sp. *Telmatobius* sp. *Colostethus* sp. *Atelopus* sp.
Peces	--	--	2-3	--	--	*Scopaeocharax* sp. *Tyttocharax* sp. *Corydoras* sp. ?
Mariposas	11	1	--	--	--	*Pedaliodes* sp.

EXECUTIVE SUMMARY

INTRODUCTION

Although "tropical biodiversity" is associated primarily with the richness of tropical rainforest, the Andes Mountains of western South America are every bit as important in species diversity and in numbers of endemic species as is the adjacent Amazon basin (e.g., Stotz et al. 1996). The Tropical Andes region has been designated as one of Conservation International's 25 global biodiversity "hotspots," areas of exceptional concentrations of endemic species that are experiencing high rates of habitat loss (Myers et al. 2000). Of the 25 global hotspots, the Tropical Andes contains the highest percentage of the world's endemic plant (7%) and vertebrate (6%) species.

In many ways the once pristine natural habitats of the Andes face threats even greater than those confronting other tropical regions. The Andes hotspot contains only 25% of its original forest cover and has only 25% of its current forests under protection, making it one of the most important regions for biodiversity conservation in the world (Myers et al. 2000). The Andes Mountains have long been targeted for human settlement and conversion to agriculture, a trend that is accelerated whenever roads open up new opportunities for colonization. Montane areas increasingly are subject to additional threats from mineral exploration and exploitation. At the same time, these species-rich habitats occupy relatively small surface areas of the Earth, squeezed by constraints of elevation and topography.

Within the Andes Mountains of Peru, the Cordillera de Vilcabamba has an almost mythical hold on the imagination (e.g., Baekeland 1964). Remote and difficult to access, no other area of even remotely comparable size anywhere in the Andes has remained so poorly known to biologists, causing the region to be a place of wonder and fascination for years. The forests of the Cordillera de Vilcabamba are critical to the healthy functioning of both highland and lowland regions of not only Peru, but a large part of South America. The Cordillera de Vilcabamba protects the headwaters of many rivers, maintaining a clean and abundant source of water to feed the rivers, the forests, and the human communities at lower elevations. The forests, the water, and the local people are closely linked.

The potential opportunities represented by such a large area of wilderness also have made this mountain range a national priority for conservationists (Rodríguez 1996). No area, no matter how large or how difficult to access, remains immune to human encroachment; already the western slope of the Cordillera de Vilcabamba is heavily settled, with concomitant habitat loss (Terborgh 1985). As early as the 1960s, the Organization of American States proposed the establishment of the Cutivereni National Park within the Cordillera de Vilcabamba. In April 1988, a new regulation established an area of 1,669,300 ha as the Zona Reservada de Apurímac, a category that temporarily preserves an area while research is conducted to determine the status and boundaries of permanent protected areas within the zone. Despite the fact that the Cordillera de Vilcabamba has remained unknown biologically, this mountain range continues to be recognized as one of the priority areas for conservation in Peru (Rodríguez 1996).

THE CI-RAP AND SI/MAB EXPEDITIONS

In order to increase the knowledge and understanding of the biological diversity of the Cordillera de Vilcabamba to support conservation plans for the area, three scientific expeditions were carried out in 1997 and 1998. These expeditions were designed to provide an initial survey of selected taxa and to obtain a preliminary assessment of the region's biodiversity and importance to conservation. The Rapid Assessment Program (RAP) of Conservation International (CI) and Conservation International-Peru (CI-Peru), in collaboration with the Asociación para la Conservación del Patrimonio del Cutivireni (ACPC) and biologists from the Museo de Historia Natural de la Universidad Nacional Mayor de San Marcos (Lima, Peru), conducted two biological surveys at three sites within the northern part of the Cordillera de Vilcabamba. A third survey along the southern edge of the Cordillera was completed by the Smithsonian Institution's Monitoring and Assessment of Biodiversity Program (SI/MAB) and their Peruvian counterparts as part of a larger biodiversity study that included portions of lowland forest in the adjacent Lower Urubamba region.

CI-RAP Expeditions

Since 1990, CI's Rapid Assessment Program (RAP) has conducted over 30 terrestrial, freshwater aquatic, and marine biodiversity surveys in 14 countries around the world, including Bolivia, Ecuador and Peru. Previous RAP activities in Perú surveyed the Cordillera del Condor (Schulenburg and Awbrey 1997) and the Tambopata/Candamo Reserved Zone (Foster et. al. 1994), which was recently granted permanent protected status as Bahuaja-Sonene National Park and Tambopata National Reserve. RAP expeditions are designed to quickly gather biological data useful for conservation planning.

The RAP surveys in the Cordillera de Vilcabamba were carried out by a team of twenty scientists from Peru (12) and the United States (8). The team surveyed vegetation (trees and understory vegetation), birds, mammals, amphibians and reptiles, fishes, and butterflies and moths. The assessments also included a rapid field study of ethnoecology and resource use among the lowland native indigenous Matsigenka populations of the Río Picha. The area that could be quickly studied by the CI-RAP surveys was relatively limited. Therefore, study sites were chosen to sample the greatest possible variety of elevational zones and habitats within the Cordillera. Three sites were selected for study (see Gazetteer and Map 2):

Camp One, 3350 m, in the headwaters of the Río Pomureni, 1997.
Camp Two, 2050 m, in the headwaters of the Río Poyeni, 1997.
Camp Three (Ridge Camp), 1000 m, in the upper Río Picha drainage, 1998.

Sampling methods included standard techniques adapted to the rapid time frame of the surveys. To sample vegetation, a modified version of the rapid transect methodology developed by Robin Foster was used. Each transect consisted of separate sub-transects designed to apportion the sampling effort equally across a continuum of vascular plant size classes and growth patterns. Mammals were sampled with Sherman, Tomahawk, and Victor snap traps, Conibear traps, and pitfall trap lines. Sampling of amphibians, reptiles, and birds was completed primarily through visual contact and by listening, and the recording of calls. Dragnets, canting nets, and hooks and lines were used to sample fish specimens, while some of the larger fish species were photographed in the field. Butterflies were sampled with entomological nets and baited traps; moths were attracted to a white sheet adjacent to a 250-volt mercury vapor light source.

SI/MAB Expedition

From 1996 through 1999, the Smithsonian Institution's Monitoring and Assessment of Biodiversity Program (SI/MAB) and Shell Prospecting and Development (Perú) B.V. (SPDP) worked in concert in the lower Urubamba region of the Peruvian Amazon to achieve environmentally sensitive development of natural gas resources. SI/MAB was responsible for conducting a biodiversity assessment in the study area and for establishing a long-term monitoring program at the well sites, along a proposed pipeline route, and in the general area of influence.

This project was carried out under guidelines devised by SI/MAB for long-term, multi-taxa forest biodiversity monitoring at permanent research sites. Specific emphasis was placed on capacity building of Peruvian scientists and guides from local Matsigenka communities. Five phases of the project were completed before SPDP terminated its operations because the company and the Peruvian government could not reach an agreement. Results from the first four phases of the project have been published elsewhere (Dallmeier and Alonso 1997; Alonso and Dallmeier 1998, 1999) and those from Phase V are presented in this report.

Phase V focused on assessments conducted in July and August of 1998 at sites along the southern border of the Cordillera de Vilcabamba where SPDP proposed a route for a pipeline to carry gas and gas condensates from the lower Urubamba region west across the Andes to the Peruvian coast. The assessments covered six biological groups, vegetation, aquatic systems, invertebrates, amphibians and reptiles, birds, and mammals.

The pipeline assessment offered a unique opportunity to conduct biodiversity surveys along elevation gradients in a portion of the Andes that has not been extensively studied. Given the length—more than 700 kilometers (km)—of the proposed pipeline, plans called for constructing it in three spreads. SI/MAB studies focused on Spread I. This area was approximately 150 km in length from the Río Camisea across the Cordillera to the Río Apurimac. It included a number of forest types, which ranged from lowland tropical rainforest (some 400 m in elevation at Las Malvinas, where forest types are characteristic of animal and plant communities in the highly diverse Amazon Basin) to the Apurimac drainage system, where vegetation is representative of that from higher elevations. The proposed route crossed two mountain ranges that rise to 2600 m and contain diverse cloud forests typical of the Andean highlands.

Because SPDP terminated its operations in Peru, SI/MAB was unable to study the ten habitat types originally planned and agreed to set up two field camps (see Gazetteer and Map 2):

Llactahuaman, 1710 m, "land of Huaman," named after the local guide who found the site, located about 4 km northeast of the community of Pueblo Libre.
Wayrapata, 2445 m, "house of winds," located 10 km northeast of the community of Pueblo Libre.

Both camps were located on the western slope of the Vilcabamba mountain range about 40 km south of Kimbiri, in the Department of Cusco. Kimbiri was the logistics base camp for this phase of field work. Access to the field camps followed SPDP's "off-shore" policy of using helicopters for transporting all personnel and equipment.

SI/MAB survey teams consisted of 32 people, 16 researchers assisted by eight guides from the local communities and eight logistical support personnel. To sample vegetation, the SI/MAB team established seven modified Whittaker study plots at Llactahuaman in a montane forest zone, which is within the range of the cloud forest. At Wayrapata, the team established 10 modified Whittaker study plots in the area's primarily elfin forest (also known as subalpine tropical forest), again within the range of the cloud forest.

CONSERVATION RECOMMENDATIONS

The CI-RAP and SI/MAB biodiversity surveys of five sites within the Cordillera de Vilcabamba, each in a different elevational zone, demonstrated that the Cordillera holds every promise of being as biologically rich as expected. The diversity of habitats and species provides a picture of a biologically rich and unique area, well deserving of protection and conservation efforts. We recommend that preservation of the Cordillera de Vilcabamba's natural resources become a national and global priority.

Regional Conservation Strategy

The editors of this report fully endorse the proposal put forward by CI-Peru, ACPC, and CEDIA to establish two communal reserves and a national park within the Cordillera de Vilcabamba (see Proposed Protected Areas in the Cordillera de Vilcabamba, this volume). The biological data presented here will contribute to formulating management plans for the sustainable use and conservation of this unique area. The Cordillera de Vilcabamba should also be linked to nearby protected areas through biodiversity corridors. Biodiversity corridors do not require strict protection of all connected habitat, but integrate protected areas with multi-use areas that are compatible with biodiversity protection.

Biodiversity corridors maintain connectivity between habitats and populations of animals and plants, thereby enhancing genetic diversity and the functioning of ecosystem processes such as pollination, migration, and nutrient cycling. This connectivity is essential for the long-term survival of the animals, plants, and people of the Cordillera de Vilcabamba.

Long-term monitoring of the biological diversity of the Cordillera is also essential to its conservation and maintenance. Appendix 1 outlines proposed guidelines for a biodiversity monitoring program.

Conservation Priorities

- **Protect the outlying ridges to sustain rare and local bird species.** Many bird species documented during the surveys are restricted to low outlying ridges along the base of the Andes. Consequently, their distributions are exceptionally narrow and naturally fragmented, making these species of particular interest for conservation. The outlying ridges are a zone rich in bird species diversity and rare species, but relatively poorly protected.
- **Protect areas above 1400 m to ensure conservation** of most of the habitat heterogeneity and biological diversity in the Cordillera, including mammal (and other) species that are either endemic to the Cordillera de Vilcabamba or have small Andean geographic ranges. Because lowland mammals, including primates and other game species, are also found above 1000 m, a reserve protecting the fauna above this elevation would also function as a reservoir and refuge for species under hunting pressures at lower elevations.
- **Protect the Cordillera de Vilcabamba as a watershed.** The Cordillera is the source of the quantity and quality of the water that flows from clear highland streams into the rivers that are vital for maintaining the human communities around the base of the Cordillera. By protecting the mountains and the forests, we also protect the backbone that regulates climate in the region.
- **Protect at least a few valleys that reach as low as 1000 m,** either by incorporating them into strict reserves or protected areas, or managing them in such a way that the overall integrity of the forests are maintained.
- **Continue including indigenous communities living in and near the Cordillera de Vilcabamba** in determining future conservation and development

plans. The enhancement of local capacities for land-use management should be considered a conservation and development priority.

- **Route gas pipelines outside of the Cordillera de Vilcabamba.** Should a pipeline from the Camisea gas fields be developed, we recommend that it be routed outside the limits of the Cordillera de Vilcabamba and not through it. Any deforestation scars created by pipeline construction should be immediately reforested so that the scar does not provide a route for colonization of the Cordillera.
- **Prohibit road construction in the Cordillera de Vilcabamba.** No roads should be constructed in or near the Cordillera in relation to pipeline development or any other development activity. This would encourage colonization by humans and allow invasive plant and animal species to become established.
- **Control human immigration to the area** by colonists to prevent further deforestation for agriculture or logging.
- **Establish monitoring stations within and around the Cordillera de Vilcabamba.** We recommend the establishment of several biological stations throughout the Cordillera de Vilcabamba to facilitate a long-term monitoring program for the region which will provide continuous monitoring of species, particularly endemic species that are likely to be the most sensitive to changes in their habitat. If a gas pipeline is constructed near the Cordillera, also establish monitoring stations along its route to monitor any possible environmental impacts (see Appendix 1).

Research Priorities

The studies reported here are only among the first to be conducted in the region. Much more must be done to document the variety of life forms of the Cordillera de Vilcabamba, their interrelationships, and the functioning of the ecosystems of which they are a part. Further studies will also go far to inform future management decisions for this region.

- **Studies of high elevations areas** will undoubtedly reveal yet more unknown species, new records, and high endemism due to the high habitat heterogeneity.
- **Further biological studies of the valleys and mountains** (800-1500 m) behind the outer hill series, and mid-elevation areas (1500-3000 m) on the eastern slopes of the main massif undoubtedly will yield many interesting and new species for science.
- **Further sampling should be conducted during the rainy season** to obtain a more accurate assessment of diversity, particularly for amphibians and reptiles. Since many of the amphibian species documented during these studies are new to science, many more likely remain to be discovered.
- **Study the role of the underlying substrates and fire** in maintaining and determining habitat heterogeneity.
- **Establish a long-term monitoring program** to provide continuous and more in-depth biodiversity and habitat data (see Appendix 1).
- **Study the effects of the proximity of human communities on game animals and selected plant species** to aid management and conservation programs.

SUMMARY OF RESULTS

The results from the biodiversity surveys reported here confirm that the Cordillera de Vilcabamba is indeed rich in biodiversity and that conservation of the area is highly recommended. The principal observation by the CI-RAP team at their three study sites was the remarkable heterogeneity of the substrate, which contributed to high levels of habitat heterogeneity, ranging from humid canyons to high-elevation pajonales. Over 700 species of vascular plants were collected from the two higher elevation sites, with twice the species richness at 2050 m than at 3350 m. Little overlap in plant species richness was observed between habitats within an elevation. This is a clear indication of the tremendous diversity in an area with the varied geology and topography of the Cordillera de Vilcabamba. The lower elevation site (1000 m) contained a mix of Amazonian and Andean species. Species richness of vertebrates was high at the three sites visited, with many endemic and geographically restricted species of mammals and birds, including a new genus of abrocomid rodent (*Cuscomys ashaninka*). Most of the amphibians observed are endemic to the Cordillera de Vilcabamba, with at least 12 new species of frogs recorded. Twelve new species of butterflies were also discovered.

The two field sites selected for study by the SI/MAB team represented two distinct habitats that varied both in floristic composition and structure. A higher proportion of larger trees were recorded at the lower elevation. In all, 191 morphospecies of aquatic and semi-aquatic macroinvertebrates, distributed among 13 orders and 69 families, were recorded at Llactahuaman and Wayrapata. A combined total of 28 small mammal species were recorded from both camps; eleven of these species were common to both localities, while eight were restricted to the montane forests of Llactahuaman and nine to the elfin and transitional forests of Wayrapata. Sixteen species of amphibians, all of them anurans, and 13 reptile species, four lizards and nine snakes, were collected. Nineteen species of large mammals, or 68% of known existing species in the area, and 176 species of birds were also recorded. A new species of the bat genus, *Carollia*, and a new species of the toad genus, *Atelopus*, were discovered.

Many of the species recorded during the surveys of the Cordillera de Vilcabamba clearly are either new to science, likely to be proven so following further comparative studies, or represent new records for the area. The occurrence of a number of species in unexpected habitats or in surprising abundance is further indication that this area is a special place in relation to biodiversity. The high levels of biodiversity observed were similar to those of other Peruvian sites, including Machu Picchu and Manu National Park, despite the absence of some representative groups.

The biogeographic importance of this study lies in its use in delimiting zones of high endemism and diversity. The higher elevation sites were distinct from each other and from lower altitudes. The high habitat heterogeneity observed at all survey sites strongly indicates that the Cordillera de Vilcabamba contains a tremendous floral and faunal diversity, which we have only begun to document and understand. The results of the surveys reported herein highlight the Cordillera de Vilcabamba as a region of high global biodiversity importance, deserving of conservation efforts and protection.

RESULTS BY SITE

Integrated results from all taxonomic groups surveyed are presented here by site, from the highest elevation on down.

CI-RAP Survey, Camp One, Elevation 3350 m (1997)
Although the three principal vegetation types visible from an overflight (pajonales, mixed-species forest, and *Polylepis* forest) had led RAP botanists to expect considerable heterogeneity, the great differences in species composition at this site were surprising. Two survey transects conducted in mixed-species forest differed greatly in structure and species composition, with only 26% of their woody floras in common.

Some of the habitat variation may relate directly to heterogeneity in the underlying substrate, as in the case of *Polylepis* forest, which was always found on blocky limestone substrate. Distributions of other vegetation types, such as the dwarf pajonales, can be explained by differences in soil drainage. Other distributional patterns are harder to explain. For example, elfin forest and tallgrass pajonales occurred as a mosaic on well-drained slopes and ridge crests. While such patterns may reflect unobserved properties of the underlying substrates, the role of fire in maintaining the grassland vegetation should also be considered. Whatever the causes of this habitat heterogeneity, it implies that diversity at the landscape scale is likely to be high. The total number of plant species collected in the vicinity of Camp One, roughly 300, was exceptional for such a high elevation site, and implies much higher diversity in the area as a whole.

A total of 43 bird species was recorded at Camp One, which is lower than expected but understandable given that the preferred bird habitat at this altitude, tall humid forests, was not fully surveyed. One of the bird species (*Schizoeaca vilcabambae*) discovered by Terborgh and Weske (1969, 1972, 1975) during their surveys of the Cordillera de Vilcabamba in the 1960s-70s, was common in forest edge and in bushy areas on the pajonal. At least two species (*Buteo polysoma* and *Notiochelidon murina*) not reported by Terborgh and Weske were found at this site, although both species are widespread and were expected to occur in this area. By contrast, several expected species (e.g., *Metallura aeneocauda, Diglossa brunneiventris*) were not found in pajonal or forest edge habitats.

Surprisingly, despite the extensive *Polylepis* forests around Camp One, none of the bird species often associated with *Polylepis* elsewhere in the Andes were observed. Possible explanations include particular features of the altitude, isolation, or age of the *Polylepis* forests or the composition and accessibility of resources needed by these bird species.

Mammal surveys were conducted in the open pajonal and *Sphagnum* bogs, the *Polylepis/Weinmannia/Chusquea* forest and the sparsely wooded *Weinmannia* open woodlands. The mammal fauna was dominated taxonomically by rodents, with 10 species recorded, and numerically by the grass mouse, *Akodon torques*, which were captured in high numbers in all habitats. In the forest, three species of Thomas' paramo mice, *Thomasomys* spp. were also captured, two of which may be new to science. The open pajonal had good populations of guinea pigs (*Cavia tschudii*). The one individual marsupial collected, *Gracilinanus* cf. *aceramarcae*, is only the second record of this species for Peru and represents a considerable northward range extension. There were a few old signs of Andean bears (*Tremarctos ornatus*), but they were evidently uncommon. There were also many signs, including hair and bones, of a small deer which matched those of *Mazama chunyi*, which is a rare species known from only a few specimens near Cusco.

The most exciting mammalian find from this site was a large arboreal rodent of the family Abrocomidae, which bears close resemblance to *Abrocoma oblativa*, known only from two skulls and some other bones excavated from pre-Columbian burial sites at Machu Picchu. Our specimen, externally very unlike other *Abrocoma*, shows that these specimens, collectively, represent a new genus and species that has been described as *Cuscomys ashaninka* (Emmons 1999).

Species diversity of amphibians and reptiles was slightly lower than expected as compared to similar elevations elsewhere in Peru. However, all four species found at this site, three frogs and one lizard, appear to be new to science, and belong to genera (*Gastrotheca,*

Phrynopus, Eleutherodactylus and *Proctoporus*) that are either restricted to or are most diverse at higher elevations.

A comprehensive survey of butterflies and moths revealed 29 species of butterflies and 29 species of larger moths at Camp One, which was lower than expected. However, a proportionately high number of new taxa were found at Camp One, including 11 new species of butterflies. The remaining 18 butterfly species were a mixture of widespread species (*Dione glycera, Vanessa braziliensis, Tatochila xanthodice*) and species found mostly in the upper Andes of central and southern Peru and Bolivia. Among the moths, most of the sphingids collected are widespread species, although both species of *Euryglottis* collected are restricted to areas above 1000 m. Rare species found at this site include the saturniid *Bathyphlebia aglia*, known only from a handful of specimens, and the cercophanid *Janiodes bethulia*, previously known only from the type-locality. Biogeographically, the butterflies found at Camp One are related to the fauna known from Machu Picchu to the south-east.

SI/MAB Survey, Wayrapata, 2445 m (1998)

Structurally, the Wayrapata vegetation biodiversity plots were dominated by trees between one and five cm dbh. Only a few individuals were in the larger size classes. Overall, species richness (the number of species per a given sized area) at Wayrapata was low compared to other montane sites in Peru at similar elevations. The more southerly location of this site increased its susceptibility to strong climatic oscillations originating in Patagonia from May to October. These "southerlies" are noted for their ability to lower temperatures up to 8° C from normal levels. In addition, the transition from lower, more species-rich forest types to higher, less species-rich forest types in this study area has been reported to occur at lower elevations than in montane regions further to the north. Both of these factors may play a role in the relatively low plant species richness at Wayrapata when compared to other sites.

A total of 92 bird species belonging to 26 families were observed at this site. Of these, the most speciose family was Emberezidae with 22 species (23.7% of the total), followed by Trochilidae with 12 species (12.9%) and Tyrannidae with 10 species (10.8%). With the exceptions of parrots, swifts, swallows, and vultures, the majority of species records at Wayrapata were obtained by mist net captures. Relatively few records were based on direct observation because of the low visibility in the first hours of the day, which is precisely the time of greatest bird activity in this area.

Twenty species of small mammals were recorded at Wayrapata, nine of which were only found in the elfin and transitional forests at this site. Most of the small mammals species were typical of montane forests, but some were from pre-montane forests or had a wide altitude range that included several types of montane forests. Noteworthy records from this site included the sympatry (i.e., species from the same systematic group that live in the same location) within several genera. These genera included marsupials such as *Monodelphis* (two species), rodents such as *Akodon* (two species), and bats such as *Anoura* (three species) and *Platyrrhinus* and *Sturnira* (three species of each).

Another important finding was the absence of the highly diverse native genus *Thomasomys*, which is generally present in eastern-slope montane forests in Peru. It could be the genus had highly diminished populations during the time of the year that our study was conducted. Whatever the cause, this phenomenon has not been manifested in other studies of cloud and montane forests.

Twelve species of large mammals, or 68% of species reported to exist in the area, were recorded at Wayrapata. The most effective method of recording large mammals was by observing tracks (footprints and feces). Six species were recorded using this method, while the scent station yielded only five species.

High values of abundance and diversity of aquatic macroinvertebrates were recorded at the Wayrapata sampling stations, with 737 individuals/m^2 in the lower portion of the Cascada Creek ravine. The Wayrapata ravines presented relatively large index values for family richness but a low uniformity value in relation to abundance distribution.

Eleven species of amphibians, all of them anurans, and seven reptile species—two lizards and five snakes, were collected from Wayrapata. The greatest amphibian diversity was found in the family Leptodactylidae (10 species), which is typical of Andean montane forests. A new species of toad of the Bufonidae family belonging to the genus *Atelopus* was recorded. The first record for Peru of the snake *Liophis andinus* was also made at this site. Previously, this snake was known only from Incachaca in Bolivia. Further sampling for amphibians and reptiles should be conducted during the rainy season to obtain a more accurate assessment of diversity. In addition, the researchers noted that the visual sampling method yielded the most information and recommended that this technique be incorporated into future studies.

Terrestrial invertebrate diversity was also high at this site. Forty-five species of spiders from 18 families and 17 cricket species from four families were recorded. Only three species of dung beetles were sampled, possibly due to the low incidence of large mammals that provide dung. Sixty-seven species of bees and wasps (Hymenoptera-Aculeata, not including ants) from nine families were collected. The difficult conditions at Wayrapata (exposed ridge tops, strong winds, stunted and densely compacted vegetation, semi-permanent cloudiness) may be favorable for nest-forming social bee and wasp genera such as *Trigona*, *Partamona*, *Nannotrigona*, and *Melipona*. These species, which appear to be better adapted to the sudden environmental changes that are likely common in the Wayrapata habitats, were more abundant at Wayrapata than at the lower elevation site, Llactahuaman.

CI-RAP Survey, Camp Two, 2050 m (1997)

Most of the vegetation at 2000 m is tall cloud forest on moderate to steep slopes; open pajonal vegetation is relatively rare below the summit crest of the Cordillera. The distribution of vegetation types appears to be governed primarily by differences in drainage. As in Camp One, overlap in species composition between transects was low, with an overlap of only 7% between two transects through tall cloud forest. In general, diversity was similar between tall humid forest and ridge crest forest, but ridge forest and dwarf *Clusia* forest shared only 12% of their species. Only 2% of species were shared between the elfin swamp forest and tall humid forest.

The open vegetation in the center of the plateau was strikingly similar to the tall pajonal at Camp One, dominated by *Chusquea* over a thick carpet of *Sphagnum* moss. A second type of open vegetation, lacking *Chusquea* and with a much lower abundance of *Sphagnum* moss, occurred on wetter soils adjacent to the margins of small lagoons. Species diversity in this vegetation type was lower (about 13 species per 50 individuals sampled) than in any open vegetation type sampled at Camp One (about 19 species per 50 individuals). However, these open pajonales were the exception in terms of species diversity; comparable vegetation types at Camp Two were approximately twice as rich in species than at Camp One at 3300 m. The high alpha diversity at Camp Two, combined with what appears to be higher beta (between-habitat) diversity, highlights the potentially high regional diversity at this elevation.

Bird species richness (115 species) at Camp Two was consistent with what would be expected at this elevation (1800-2100 m). Another nine species were recorded here that were not reported for the Cordillera de Vilcabamba by Terborgh and Weske. Most of these species are widespread Andean birds and their presence here was expected. Several species, such as *Otus albogularis* (White-throated Screech-Owl) and *Basileuterus luteoviridis* (Citrine Warbler), are typically found at much higher elevations elsewhere in Peru. The Red-and-White Antpitta (*Grallaria erythroleuca*), was fairly common at Camp Two; this species has a very restricted distribution, known only from the Vilcabamba and Vilcanota mountain ranges. Although detailed comparisons have not yet been made, the population in the Cordillera de Vilcabamba site may represent an undescribed taxon.

Mammals surveys were concentrated on the variety of forested habitats, with only a few traps and a line of pitfalls in the *Sphagnum* bog. The same rodent species, *Akodon torques*, that was numerically dominant at Camp One also was dominant at this site. Five marsupials were captured at 2050 m, including two species of short-tailed opossums (*Monodelphis* spp.). For one of these species, our specimens are only the third and fourth ever collected in Peru. Dense stands of bamboo thickets were occupied by montane bamboo rats (*Dactylomys peruanus*); this site may be a new range extension for this rarely collected species. At least two species of primates, spider and night monkeys, and possibly also capuchin monkeys were observed at this site. Because primates have been severely reduced by hunting in the inhabited lowlands, the higher elevations of the Cordillera de Vilcabamba are an important local refuge.

The herpetofauna sampled at Camp Two was composed of 13 species (11 frogs and 2 snakes). The frog fauna at this site was basically similar to what would be found at comparable elevations in Manu National Park, but was unusual in that we found no representatives of *Hyla* or *Phrynops*, which are normally present at similar elevations. It is possible that the absence of *Hyla* is due to a lack of suitable habitat. Up to two thirds of the frog species found at this site may be new to science.

Due to limited sampling efforts, the butterflies collected at Camp Two (19 species) represent only a small sample of the fauna that ought to occur there. Most of the species collected are widespread in Andean montane forests, but a new record for *Pedaliodes*, otherwise known only from the Vilcanota and Santa María valleys in Cusco, was made. Butterflies from Camp Two are biogeographically related to the Chanchamayo fauna.

SI/MAB Survey, Llactahuaman, 1710 m (1998)

Vegetation plots at this site contained a higher proportion of larger trees than at Wayrapata (2445 m) but far fewer than the number of larger trees at the Pagoreni gas well site in the lowlands, studied during Phase IV of the SI/MAB project.

A total of 111 bird species belonging to 28 families were observed at this site. The family with the largest representation was Emberezidae with 27 species (23.9% of the total), followed by Tyrannidae with 17 species (15.0%), Formicaridae with 12 species (10.6%), and Trochilidae (8.0%).

Nineteen small mammal species were found at Llactahuaman. Most of the small mammals species were typical of montane forests, but some were from pre-montane forests or had a wide altitude range that included several types of montane forests. The diverse genus, *Thomasomys*, which is generally present in eastern-slope montane forests in Peru was not recorded at this site. Twelve species of large mammals were also recorded. The most effective method of recording large mammals at Llactahuaman was the scent station, where five species were noted. Three large mammals species at that site were visually observed.

Eight species of amphibians (all anurans) and nine species of reptiles (four lizards and five snakes) were recorded. The greatest diversity was found within the Leptodactylidae family of frogs, with five species.

Two fish species (*Trichomycterus* sp. and *Astroblepus* sp.) were recorded at three Llactahuaman sampling stations, both belong to the suborder Suliformes. Relatively low density of aquatic macroinvertebrates, 96 individuals/m^2, was recorded along the upper portion of the Bagre Creek ravine.

A total of 60 species of spiders from 16 families and 22 species of crickets from four families was collected. The team also recorded high beetle diversity, 166 beetle species from 21 families, which was much higher than at Wayrapata, perhaps due to the higher diversity of specialized habitats at this site. A total of 102 species of bees and wasps (Hymenoptera-Aculeata, not including ants) from 10 families was collected.

CI-RAP Survey, Camp Three, 1000 m (1998)

Camp Three contained a wide range habitat types, from humid, deep shaded canyons with high herbaceous plant diversity, to sharp limestone ridges dominated by palms, to mid-montane cloud forests with high tree species diversity. While the majority of the plant specimens collected have yet to be identified to species, RAP botanists recorded approximately 220 morphospecies of trees > 10 cm dbh out of 500 trees sampled, and identified nearly 90 families of flowering plants from this site. The number of pteridophyte (fern) families is not known, but will likely be significant.

From the transect data and general observations, it appears that there is very little overlap in plant species occurrence at the different elevation levels sampled at this site. Major differences in soil type and depth, as well as climatological factors are likely responsible causes for this high turnover in plant species. The most common or speciose herbaceous plant families in the area were: Acanthaceae, Araceae, Bromeliaceae, Cyclanthaceae, Gesneriaceae, Marantaceae, Orchidaceae, and Piperaceae.

The area surveyed around Camp Three represents the upper limit of the Amazonian avifauna, and the lower limit of Andean bird species. We found at least 36 bird species from this camp that were not recorded by Terborgh and Weske in the upper Apurimac valley or on the western slopes of the Vilcabamba. The majority of these are widespread Amazonian species, and their apparent absence in the upper Apurimac possibly may be due to a "filter" effect of the drier, deciduous forests of the Río Ene, downstream from the Río Apurimac, on the movements of Amazonian species up the Apurimac valley.

A number of characteristic "upper tropical" bird species were observed at this site, including *Phaethornis koepckeae* (Koepcke's Hermit), *Heliodoxa branickii* (Rufous-webbed Brilliant), *Hemitriccus rufigularis* (Buff-throated Tody-Tyrant), *Phylloscartes parkeri* (Cinnamon-faced Tyrannulet), *Pipreola chlorolepidota* (Fiery-throated Fruiteater), *Ampeliodes tschudii* (Scaled Fruiteater), *Lipaugus subularis* (Gray-tailed Piha), and *Oxyrunus cristatus* (Sharpbill). Most of these species are known in Peru from only a few localities and several species apparently are restricted to low outlying ridges along the base of the Andes. Consequently, their distributions are exceptionally narrow and naturally fragmented making these species of particular interest for conservation. Protection of the outlying ridges which sustain rare and local species should be a priority for conservation.

Fifty-eight species of mammals were recorded at this site. As expected, this mammal fauna was composed largely of lowland genera and species. The spiny rat *Proechimys simonsi* numerically dominated the small mammal fauna, especially in bamboo forests, which also harbor dense populations of the bamboo rat, *Dactylomys boliviensis*. The common murid species of the lowlands, *Oryzomys megacephalus* was rare at our camp, while *Oryzomys macconnelli*, which is rare in the lowlands, was exceedingly common. Within the elevation range surveyed, the upper regions are increasingly isolated from faunas at similar elevations within the Andean chain. Thus, the mammal fauna of the Cordillera de Vilcabamba becomes more distinctive and divergent with increasing elevation.

Relatively few amphibian and reptile species (five toads, 24 frogs, five lizards, and four snake species) were recorded at this site. Moreover, two of these records are of great interest. A marsupial frog, *Gastrotheca* sp., probably is an undescribed species, and may represent one of the lowest elevation localities for members of this genus in Peru. The other species of interest is an undescribed *Colostethus* sp., belonging to the family of poison arrow frogs (Dendrobatidae). The genus *Colostethus* has a few large species in the Andean foothills, but its taxonomy is poorly known and systematic studies remain to be done.

CI-RAP Survey, Tributaries of the Río Picha, 500 m (1997)

Fish samples were collected from various tributaries of the Río Picha, near the native communities of Camaná, Mayapo, and Puerto Huallana. During the survey, 86 fish species from 19 different families were recorded. The fish species composition is dominated by characids, pimelodids, and loricariids. The ichthyofauna here is quite similar to that of the Río Camisea, a tributary of the Río Urubamba, and tends to share various species with the Río Manu watershed (Madre de Dios). Several fish species, including two in the genera *Scopaeocharax* and *Tyttocharax* may be new to science. The former appears to be related to *S. atopodus*, a species known from the Río Huallaga in Tingo María. The undescribed *Tyttocharax* species is closely related to *T. tambopatensis*, a species described from the Madre de Dios. Many of the species described, especially those of the Characidae family, live in a watershed similar to that of the Río Urubamba and were previously described from research conducted during the beginning of the past half century.

In addition to biological surveys, CI-RAP anthropologists also conducted a rapid field study of ethnoecology and resource use among the lowland Matsigenka populations of the Río Picha, adjacent to the proposed Communal Reserve. CI-RAP anthropologists spent four to six days each in the Matsigenka communities of Camaná, Mayapo and Puerto Huallana during May 1997. In addition, the communities of Nuevo Mundo, Nueva Luz and the Catholic Mission of Kirigueti (all along the Río Urubamba) were visited in passing. The three principal study communities had populations of more than 300 people each, which is much denser than traditional Matsigenka settlements of 30-80 people. The population of native communities of the Río Picha and tributaries was close to 1500. In order to take advantage of the diverse and scattered resources of the tropical forest, the Matsigenka of the modern communities of the Río Picha maintain a strong tradition of seasonal treks and frequent migrations.

Matsigenka swidden agriculture, like that of other groups in the Amazon (Posey and Balée 1989, Boster 1984), is characterized by a relatively small area of forest disturbance, multicropping, great genetic diversity of crop cultivars and a rapid process of forest regeneration. Most of the protein requirements of the Matsigenka come not from agriculture but from wild resources such as fish, game, fruits, nuts and palm hearts. A map was drawn in each community with the help of indigenous community members detailing the names and relative locations of rivers, mountains, past and present human settlements, gardens, trails, animal and plant resources and other landmarks surrounding each community. A preliminary study of Matsigenka forest classification was carried out; the Matsigenka are able to distinguish a great diversity of habitat types, including high-elevation cloud forest, pajonal and high-Andean vegetation, despite the fact that the Matsigenka currently do not live in these regions.

LITERATURE CITED

Alonso, A. and F. Dallmeier (Eds.). 1998. Biodiversity assessment and monitoring of the Lower Urubamba Region, Perú: Cashiriari-3 well site and the Camisea and Urubamba Rivers. SI/MAB Series #2. Washington, DC. Smithsonian Institution/MAB Biodiversity Program.

Alonso, A. and F. Dallmeier (Eds.). 1999. Biodiversity assessment and monitoring of the Lower Urubamba Region, Perú: Pagoreni well site assessment and training. SI/MAB Series #3. Washington, DC. Smithsonian Institution/MAB Biodiversity Program.

Baekeland, G. B. 1964. By parachute into Peru's lost world. National Geographic 126: 268-296.

Boster, J. S. 1984. Classification, cultivation, and selection of Aguaruna cultivars of *Manihot esculenta* (Euphorbiaceae). Advances in Economic Botany 1: 34-47.

Dallmeier, F. and A. Alonso (Eds.). 1997. Biodiversity assessment of the Lower Urubamba Region, Perú: San Martin-3 and Cashiriari-2 well sites. SI/MAB Series #1. Washington, DC. Smithsonian Institution/MAB Biodiversity Program.

Emmons, L. H. 1999. A new genus and species of a brocomid rodent from Peru (Rodentia: Abrocomidae). American Museum Novitates, Number 3279.

Foster, R. B., J. L. Carr, and A. B. Forsyth (eds.) 1994. The Tambopata-Candamo Reserved Zone of southeastern Perú: a biological assessment. RAP Working Papers Number 6. Washington, DC. Conservation International.

Myers, N., R. A. Mittermeier, C. G. Mittermeier, G. A. B. da Fonseca, and J. Kent. 2000. Biodiversity hotspots for conservation priorities. Nature. 403: 853-858.

Posey, D. and W. Balée (eds.). 1989. Resource Management in Amazonia: Indigenous and Folk Strategies. Advances in Economic Botany, Vol. 7. New York Botanical Gardens. New York.

Rodríguez, L. O. 1996. Diversidad biológica del Perú: zonas prioritarias para su conservación. Lima: Proyecto Fanpe (GTZ and INRENA). Deutsche Gesellschaft für Technische Zusammenarbeit (GTZ) and Ministerio de Agricultura, Instituto Natural de Recursos Naturales (INRENA).

Schulenberg, T. and K. Awbrey (eds.). 1997. The Cordillera del Condor region of Ecuador and Peru: A biological assessment. RAP Working Papers Number 7. Washington, DC. Conservation International.

Stotz, D. F., J. W. Fitzpatrick, T. A. Parker III, and D. K. Moskovits. 1996. Neotropical birds: ecology and conservation. Chicago, University of Chicago Press.

Terborgh, J. 1985. The role of ecotones in the distribution of Andean birds. Ecology 66: 1237-1246.

Terborgh, J., and J. S. Weske. 1969. Colonization of secondary habitats by Peruvian birds. Ecology 50: 765-782.

Terborgh, J., and J. S. Weske. 1972. Rediscovery of the Imperial Snipe in Peru. Auk 89: 497-505.

Terborgh, J., and J. S. Weske. 1975. The role of competition in the distribution of Andean birds. Ecology 56: 562-576.

RESUMEN EJECUTIVO

INTRODUCCION

Aunque la "biodiversidad tropical" se relaciona principalmente con la riqueza de los bosques tropicales, las montañas de los Andes en el Oeste Sudamericano tiene tanta importancia en diversidad biológica y en número de especies endémicas como la cuenca del Amazonas adyacente a los Andes (e.g. Stotz et al. 1996). La región tropical de los Andes ha sido reconocida por Conservación Internacional como uno de los 25 "hotspots" de biodiversidad global. Son áreas con concentraciones excepcionales de especies endémicas que experimentan altas tasas de destrucción de hábitats (Myers et al. 2000). De los 25 "hotspots" globales, la región de los Andes contiene el porcentaje más alto en el mundo de especies de plantas (7%) y vertebrados (6%) endémicos.

En muchos sentidos, el una vez pristino hábitat de los Andes se enfrenta a amenazas aún mayores que las que enfrentan otras regiones tropicales. En la región de los Andes orientales sólo queda el 25% de la cubierta forestal original y sólo el 25% de sus bosques actuales cuenta con algún tipo de protección. Esta situación hace que la Cordillera de los Andes sea una de las regiones más importantes para la conservación de la biodiversidad en el mundo (Myers et al. 2000). Por mucho tiempo, en la Cordillera de los Andes se han establecido asentamientos humanos y campos agrícolas, una tendencia acelerada por la construcción de caminos que abren oportunidades para la colonización. Las áreas montañosas, a su vez, son objeto de amenazas continuas provenientes de la exploración y explotación de recursos mineros. Al mismo tiempo, estos hábitats ricos en especies ocupan una superficie relativamente pequeña de la tierra, encontrándose restringidos por limitaciones topográficas y de elevación.

En los Andes peruanos, la Cordillera de Vilcabamba ocupa un lugar casi mítico en la imaginación (e.g., Baekeland 1964). Remota y de difícil acceso, ninguna otra área de dimensiones comparables en los Andes ha permanecido tan poco conocida para los biólogos, por lo que esta región ha causado admiración y fascinación por años. Los bosques de la Cordillera de Vilcabamba son críticos para el funcionamiento saludable de regiones altas y bajas de Perú y también de una porción grande de Sudamérica. La Cordillera de Vilcabamba protege las cuencas de muchos ríos, constituyendo una fuente limpia y abundante de agua que alimenta a ríos, bosques y asentamientos humanos en las áreas más bajas. Por lo tanto, los bosques, el agua y la población local están íntimamente ligados.

Las oportunidades potenciales de un área silvestre tan grande también contribuyen para que esta cadena montañosa tenga prioridad nacional para los conservacionistas (Rodríguez 1996). Ningún área, no importa qué tan grande sea o qué tan difícil acceso tenga, permanece inmune al avance humano. La ladera occidental de la Cordillera de Vilcabamba está densamente ocupada por asentamientos humanos, con la consecuente pérdida de hábitat que éstos generan (Terborgh 1985). Ya desde los años sesenta, la Organización de Estados Americanos propuso el establecimiento del Parque Nacional Cutivereni en la Cordillera de Vilcabamba. En abril de 1988, una nueva ley estableció un área de 1,669,300 hectáreas como Zona Reservada de Apurímac, una categoría que protege áreas en forma temporal, al tiempo que promueve estudios para el establecimiento de categorías de protección permanente. A pesar de que la Cordillera de Vilcabamba ha permanecido desconocida en términos biológicos, esta cadena montañosa continúa siendo reconocida como una de las áreas de prioridad para la conservación en el Perú (Rodríguez 1996).

LAS EXPEDICIONES DEL CI-RAP Y DEL SI/MAB

Con el fin de ampliar el conocimiento de la diversidad biológica de la Cordillera de Vilcabamba y para apoyar los planes de conservación para el área, se llevaron a cabo tres expediciones científicas entre 1997 y 1998. Estas expediciones fueron diseñadas para obtener muestreas iniciales de los grupos taxonómicos selectos y para hacer una evaluación preliminar de la biodiversidad de la región y de su importancia para la conservación. El Programa de Evaluación Rápida (RAP) de Conservación Internacional (CI) y Conservación Internacional- Perú (CI- Perú), en colaboración con la Asociación para la Conservación del Patrimonio del Cutivireni (ACPC) y biólogos del Museo de Historia Natural de la Universidad Nacional Mayor de San Marcos (Lima, Perú) condujeron dos inventarios biológicos en tres áreas del Norte de la Cordillera de Vilcabamba. Un tercer inventario fue conducido por el Programa de Evaluación y Monitoreo de la Biodiversidad de la Institución Smithsonian (SI/MAB) y sus contrapartes peruanas a lo largo del borde Sur de la Cordillera, como parte de un estudio de biodiversidad que incluyó porciones de los bosques bajos en la adyacente región del bajo Río Urubamba.

Las Expediciones de CI-RAP

Desde 1990, el Programa de Evaluacion Rápida de CI (CI-RAP) ha conducido más de 30 evaluaciones de biodiversidad en ecosistemas terrestres, aguas marinas y aguas dulces, en 14 países alrededor del mundo incluyendo Bolivia, Ecuador y Perú. Actividades previas de CI-RAP en Perú incluyen la evaluación de la Cordillera del Cóndor (Schulenburg y Awbrey 1997) y lo que constituyo la Zona Reservada Tampobata-Candamo, actualmente Parque Nacional Bahuaja-Sonene y Reserva Nacional Tambopata (Foster et.al. 1994). Las expediciones de CI-RAP han sido diseñadas para recolectar en forma rápida información biológica útil para la planificación de la conservación.

Un equipo de 20 científicos de Perú (12) y Estados Unidos (8) llevó a cabo las evaluaciones de CI-RAP en la Cordillera de Vilcabamba. El equipo estudió la vegetación (árboles y sotobosque), aves, mamíferos, anfibios y reptiles, así como peces y mariposas diurnas y nocturnas. Además se hizo un estudio rápido de campo sobre la etnoecología y el uso de recursos naturales en las tierras bajas por los grupos indígenas Matsigenka en el Río Picha.

El área que podría ser estudiada rápidamente a través de los sondeos del CI-RAP fue relativamente pequeña. Por ello, las áreas de estudio fueron seleccionadas de tal forma que incluyeran la más alta variedad posible de zonas de elevación y hábitats de la Cordillera. Como resultado, se eligieron las siguientes áreas de estudio, llamadas "campamentos" (véase el Gazetteer y Mapa 2):

Campamento Uno, 3350 m.s.n.m., en la base de la cuenca del Río Pomureni, 1997.
Campamento Dos, 2050 m.s.n.m., en la base de la cuenca del Río Poyeni, 1997.
Campamento Tres (Ridge Camp), 1000 m.s.n.m., en la parte alta del área de drenaje del Río Picha, 1998.

Los métodos de muestreo incluyeron técnicas estándar adaptadas al corto tiempo en que se efectuaron los sondeos. De esta manera, para las muestras de vegetación se utilizó una versión modificada del método de transecto rápido desarrollado por Robin Foster. Cada transecto está conformado por sub-transectos separados, diseñados para que cada uno tenga un esfuerzo de muestreo equitativo a las diferentes clases, tamaños y formas de crecimiento de las plantas vasculares. Los mamíferos fueron muestreados utilizando trampas Sherman, Tomohawk y ratoneras Victor, trampas Conibear y trampas de caída libre en línea. El muestreo de anfibios, reptiles y aves se hizo principalmente a través de observaciones visuales y auditivas, así como grabando los cantos de cada especie. Se utilizaron redes de arrastre, anzuelos e hilo de pescar para hacer un muestreo de las comunidades de peces. En el caso de especies grandes, se fotografiaron los especimenes en el campo. Las mariposas se capturaron con redes entomológicas y trampas cebadas y se utilizó una sábana blanca cerca de una luz de vapor de mercurio de 250 voltios para atraer a las mariposas nocturnas.

La Expedición del SI/MAB

Entre 1996 y 1999, SI/MAB y Shell Prospecting and Development (Perú) B.V. (SDPD) trabajaron concertadamente en la región del bajo Río Urubamba en la Amazonía peruana, con el objetivo de lograr el desarrollo de recurso gas natural teniendo en cuenta el ambiente. SI/MAB se responsabilizó en hacer evaluaciones de la biodiversidad en el área de estudio y establecer un programa de monitoreo a largo plazo en las locaciones de los pozos, a lo largo de la ruta propuesta para el gasoducto y en el área de influencia.

Este proyecto se llevó a cabo bajo los lineamientos diseñados por SI/MAB para monitorear la biodiversidad de varios grupos taxonómicos en lugares permanentes de investigación. Se hizo especial énfasis en capacitar a jóvenes científicos peruanos y a los guías de las comunidades locales Matsigenka. Se completaron cinco fases del proyecto antes de que SPDP terminara sus operaciones en el área, debido a que la compañía y el gobierno no llegaron a un acuerdo. Los resultados de las primeras cuatro fases han sido presentados en otras publicaciones (Dallmeier and Alonso 1997; Alonso and Dallmeier 1998, 1999). Este reporte presenta los resultados de la quinta fase.

Los estudios de la quinta fase se llevaron a cabo en julio y agosto de 1998 en áreas de estudio a lo largo del borde Sur de la Cordillera de Vilcabamba, siguiendo la ruta propuesta por SPDP para el gasoducto que transportaría gas natural y condensados de gas desde la región del Bajo Urubamba hacia la costa peruana, cruzando los Andes en dirección Oeste. La evaluación cubrió seis grupos biológicos: vegetación, sistemas acuáticos, invertebrados, anfibios y reptiles, aves y mamíferos.

La evaluación de la ruta del gasoducto ofreció una oportunidad única para realizar muestreos a lo largo de gradientes de elevación en una parte de los Andes que previamente no había sido estudiada con intensidad. Dada la longitud de la ruta propuesta para el gasoducto - más de 700 km - la evaluación se planificó en tres secciones. Los estudios del SI/MAB se concentraron en la Sección I. Esta sección, de aproximadamente 150 km de longitud, se extiende desde el Río Camisea cruzando la Cordillera hasta el Río Apurímac. Incluye tipos de bosques que varían desde bosques tropicales en tierras bajas (aproximadamente 400 m de elevación en Las Malvinas, donde los tipos de bosque son característicos de las comunidades de flora y fauna de la altamente diversa cuenca amazónica), hasta la cuenca del Río Apurímac, donde la vegetación es representativa de elevaciones más altas. La ruta propuesta para el gasoducto cruza dos cadenas montañosas que se elevan hasta los 2600 m y donde se encuentra diversos bosques de neblina típicos de las montañas andinas.

Debido a que SPDP cerró sus operaciones en Perú, no fue posible que SI/MAB terminara de estudiar los diez tipos de hábitats que se planeó originalmente y se acordó establecer dos campamentos base (véase el Gazetteer y Mapa 2):

> **Llactahuaman** (1710 m.s.n.m.), la "tierra de Huaman", llamado así en honor del guía local que fundó el lugar, está localizado aproximadamente a 4 km al Noreste de la comunidad de Pueblo Libre.
> **Wayrapata** (2445 m.s.n.m.), la "casa de los vientos", localizado a 10 km al Noreste de la comunidad de Pueblo Libre.

Ambos campamentos están ubicados en la ladera occidental de la cadena montañosa de Vilcabamba, a unos 40 km al Sur de Kimbiri, en el Departamento de Cusco. Kimbiri fue el campamento base para operaciones logísticas durante esta fase de trabajo de campo. El acceso a los campamentos siguió la política "off-shore" establecida por SPDP de utilizar helicópteros para el transporte de todo el equipo y personal.

El equipo de investigación de SI/MAB estuvo compuesto por 32 personas, 16 investigadores asistidos por ocho guías de las comunidades locales y ocho personas de apoyo logístico. Para estudiar la vegetación en Llactahuaman, el equipo del SI/MAB estableció siete parcelas con el método modificado de Whittaker, en una zona de bosque montañoso, dentro del rango de bosque de neblina. Del mismo modo, se establecieron en Wayrapata 10 parcelas con el método modificado de Whittaker en el bosque enano predominante en el área (también conocido como bosque sub-alpino tropical de baja altura).

RECOMENDACIONES PARA LA CONSERVACION

Los estudios sobre biodiversidad llevados a cabo por CI-RAP y SI/MAB en cinco áreas con diferentes elevaciones en la Cordillera de Vilcabamba demostraron que la Cordillera es biológicamente tan rica como se creía. La diversidad de hábitats y especies representan un área biológicamente rica y única, que bien merece ser conservada y protegida. Se recomienda que la preservación de los recursos naturales de la Cordillera de Vilcabamba sea una prioridad nacional y global.

Estrategia de Conservación Regional

Los editores de esta publicación apoyan totalmente la propuesta recomendada por CI-Perú, ACPC y CEDIA de establecer dos reservas comunales y un parque nacional en la Cordillera de Vilcabamba (véase la sección de Areas Protegidas Propuestas en la Cordillera de Vilcabamba en este reporte). La información biológica presentada aquí contribuirá a formular planes de manejo para uso sostenible y conservación de esta área única.

La Cordillera de Vilcabamba también deberá conectarse a otras áreas protegidas cercanas a través de corredores biológicos. Los corredores biológicos no requieren la protección estricta de todos los hábitats conectados, sino que integran las áreas protegidas con áreas de uso múltiple que son compatibles con la protección de la biodiversidad. Los corredores biológicos mantienen la conexión entre hábitats y poblaciones animales y vegetales, mejorando de esta manera la diversidad genética y el funcionamiento de los procesos de los ecosistemas, tales como la polinización, migración y el ciclo de nutrientes. Esta conexión es esencial para la supervivencia a largo plazo de animales, plantas y poblaciones humanas de la Cordillera de Vilcabamba.

El monitoreo a largo plazo de la biodiversidad biológica de la Cordillera es tambien de suma importancia para su conservación y mantenimiento. El Apéndice 1 presenta los protocolos propuestos para un programa de monitoreo de la biodiversidad.

Prioridades de Conservación

- **Proteger las crestas de las montañas para sostener especies de aves locales y raras.** Muchas especies de aves documentadas durante los estudios están restringidas a las estribaciones de los Andes. En consecuencia, sus distribuciones son excepcionalmente limitadas y fragmentadas, convirtiéndolas en especies de particular interés para la conservación. Las estribaciones son zonas ricas en especies de aves y especies raras, pero cuentan con relativamente poca protección.

- **Proteger áreas por encima de los 1400 m.s.n.m.** para asegurar la conservación de la mayor parte de la heterogeneidad de hábitats y la diversidad biológica, incluyendo especies de mamíferos (y otras) endémicas de la Cordillera de Vilcabamba o que tienen reducidos rangos geográficos en los Andes. Debido a que los mamíferos de tierras bajas, incluyendo primates y otras especies de caza, fueron hallados por encima de los 1000 m, una reserva que proteja la fauna a esta altura funcionaría también como un reservorio y refugio para especies bajo presión de caza a menores altitudes.
- **Proteger la Cordillera de Vilcabamba como una cuenca.** La Cordillera de Vilcabamba es la fuente de la cantidad y calidad de agua que fluye desde claros arroyos de altura hasta los ríos que son vitales para las comunidades humanas en la base de la Cordillera. Al proteger las montañas y los bosques, también se protege la columna vertebral que regula el clima de la región.
- **Proteger por lo menos algunos de los valles que llegan hasta los 1000 m.s.n.m.**, ya sea incorporándolos en áreas de protección estricta u otro tipo de área protegida, o manejándolos de tal manera que se mantenga la integridad general de los bosques.
- **Continuar involucrando a las comunidades indígenas que viven dentro y en los alrededores de la Cordillera de Vilcabamba** en la determinación de futuros planes de conservación y desarrollo. La mejora de las capacidades locales para la gestión del uso de la tierra debe ser considerada una prioridad para la conservación y el desarrollo.
- **Establecer las rutas posibles de gasoductos fuera de la Cordillera de Vilcabamba.** En el caso en que se decida construir un gasoducto desde los yacimientos de hidrocarburos de Camisea, se recomienda que éste sea construido fuera de los límites de la Cordillera de Vilcabamba y no a través de ella. Cualquier tipo de rastro de deforestación producida por la construcción del gasoducto debe ser inmediatamente reforestado, de tal manera que los rastros deforestados no sean utilizados después como rutas para la colonización de la Cordillera.
- **Prohibir la construcción de caminos en la Cordillera de Vilcabamba.** No es recomendable que se construyan caminos dentro de y en los alrededores de la Cordillera de Vilcabamba, ya sea como parte del establecimiento del gasoducto o de cualquier otro tipo de actividad de desarrollo. La construcción de caminos promovería la colonización y permitiría el establecimiento de especies invasoras de plantas y animales.
- **Controlar la migración al área** por parte de colonos a fin de prevenir una mayor deforestación por el establecimiento de parcelas agrícolas o la extracción maderera no controlada.
- **Establecer estaciones de monitoreo dentro y en los alrededores de la Cordillera de Vilcabamba.** Se recomienda el establecimiento de varias estaciones biológicas a lo largo de la Cordillera de Vilcabamba para facilitar un programa de monitoreo a largo plazo. Este esfuerzo regional ofrecerá un monitoreo continuo de especies, particularmente de las especies endémicas, que son potencialmente más sensibles a los cambios en su hábitat. Si se construye un gasoducto cerca de la Cordillera, también es necesario establecer estaciones de monitoreo a lo largo de la ruta del mismo para hacer seguimiento de posibles impactos ambientales (véase el Apéndice 1).

Prioridades de investigación

Los estudios que se reportan en este documento son de los primeros que han sido llevados a cabo en esta región. Falta mucho por hacer para documentar la variedad de formas de vida en la Cordillera de Vilcabamba, sus interrelaciones y el funcionamiento de los ecosistemas de los cuales son parte. Estudios más profundos serán importantes para futuras decisiones de manejo en la región:

- **Estudios de las zonas altas** revelarán sin duda la existencia de muchas especies hasta ahora desconocidas, nuevos registros y un alto número de endemismos como resultado de la gran heterogeneidad de hábitats.
- **Más estudios biológicos en los valles y montañas** entre los 800 y 1500 m.s.n.m. y áreas de elevación media (1500-3000 m.s.n.m.) sobre las laderas orientales del macizo principal proporcionarían indudablemente muchas y nuevas especies para la ciencia.
- **Se deben conducir más muestreos durante la época de lluvias** para obtener una evaluación adecuada de la diversidad, particularmente de anfibios y reptiles. Como muchas de las especies de anfibios documentadas durante estos estudios son nuevas para la ciencia, es probable que nuevos estudios descubran más especies.

- **Estudiar el rol del substrato y el fuego** en el sostenimiento y la delimitación de la heterogeneidad de hábitats.
- **Establecer un programa de monitoreo a largo plazo** para proporcionar información continua y profunda sobre la biodiversidad y los hábitats (véase el Apéndice 1).
- **Estudiar los efectos de la proximidad de asentamientos humanos en los animales sujetos a cacería y en especies de plantas seleccionadas**, para ayudar a los programas de manejo y conservación.

RESUMEN DE RESULTADOS

Los resultados de los estudios de biodiversidad que se reportan en este trabajo confirman que la Cordillera de Vilcabamba es ciertamente rica en biodiversidad y se recomienda enfáticamente su conservación. La observación principal del equipo CI-RAP en sus tres áreas de estudio fue la alta heterogeneidad del substrato, que contribuye a altos niveles de heterogeneidad de hábitat, que a su vez va desde quebradas profundas y húmedas hasta pajonales de altura. Se recolectaron más de 700 especies de plantas vasculares en las dos áreas de estudio de mayor elevación, encontrándose el doble de riqueza de especies a 2050 m.s.n.m. que a 3350 m.s.n.m. Se observó poca sobreposición de riqueza de especies vegetales entre los diferentes hábitats de una misma elevación. Esto es un indicador claro de la inmensa diversidad de un área con una geología y topografía tan variada como la de la Cordillera de Vilcabamba. El área de estudio de menor elevación (1000 m) contiene una mezcla de especies amazónicas y andinas. La riqueza de especies de vertebrados fue alta en las tres áreas estudiadas, con muchas especies de mamíferos y aves endémicas y geográficamente restringidas, incluyendo un nuevo género de roedor (*Cuscomys ashaninka*). Muchos de los anfibios observados son endémicos de la Cordillera de Vilcabamba, con al menos 12 nuevos registros de ranas. También se descubrieron 12 nuevas especies de mariposas.

Las dos áreas de estudio seleccionadas por el equipo del SI/MAB representan dos hábitats distintos que varían tanto en la estructura como en la composición de la flora. Se registró una mayor proporción de árboles grandes en el área de menor altitud. En total, se registraron en Llactahuaman y Wayrapata 191 morfoespecies de macroinvertebrados acuáticos y semi-acuáticos. De igual manera, se registró un total de 28 especies de pequeños mamíferos en ambos campamentos. Once de éstas especies fueron comunes en ambas áreas de estudio, mientras que ocho estuvieron restringidas a los bosques montañosos de Llactahuaman y nueve a los bosques de Wayrapata. Se muestrearon 16 especies de anfibios, todos ellos anuros, 13 especies de reptiles, cuatro lagartijas y nueve serpientes. También se encontraron 19 especies de mamíferos grandes distribuidas en 12 familias, que representan el 68% del total de especies conocidas en el área y 176 especies de aves. Se descubrió una nueva especie de murciélago del género *Carollia*, así como también una nueva especie de sapo del género *Atelopus*.

Muchas de las especies registradas durante las investigaciones en la Cordillera de Vilcabamba son evidentemente nuevas para la ciencia o representan nuevos registros para el área. Esta información será confirmada en estudios comparativos posteriores. La ocurrencia de ciertas especies en hábitats inesperados o en sorprendente abundancia, es un indicador adicional de que Vilcabamba es una área especial en términos de biodiversidad. Los altos niveles de biodiversidad observados en la Cordillera de Vilcabamba fueron similares a aquéllos obtenidos en otros sitios del Perú, como Machu Picchu y el Parque Nacional Manú, a pesar de la ausencia de algunos grupos representativos.

La importancia biogeográfica de este estudio radica en el uso que se dé a la información obtenida para delimitar zonas de alto endemismo y biodiversidad. Las áreas de estudio en las elevaciones más altas son distintas entre sí y también de las áreas de menor elevación. La gran heterogeneidad de hábitats observada en todas las áreas de estudio indica con gran claridad que la Cordillera de Vilcabamba alberga una tremenda diversidad de flora y fauna, algo que sólo ahora se ha comenzado a documentar y entender. Los resultados de los muestreos reportados en este documento destacan la Cordillera de Vilcabamba como una región de elevada importancia con relación a la biodiversidad global que merece ser conservada y protegida.

RESULTADOS SITIO POR SITIO

Los resultados integrados de todos los grupos taxonómicos estudiados se presentan a continuación para cada área de estudio, de mayor a menor elevación.

Estudio de CI-RAP, Campamento Uno, 3350 m.s.n.m. (1997)

A pesar de que los tres principales tipos de vegetación visibles desde la avioneta (pajonales, bosques heterogéneos y bosques de *Polylepis*) llevaron a que los botánicos de CI-RAP esperaran encontrar una heterogeneidad considerable, las grandes diferencias en composición de especies en esta área de estudio fueron sorprendentes. Los dos transectos de evaluación en bosques heterogéneos difirieron enormemente en estructura y en composición de especies, presentando sólo un 26% de especies vegetales leñosas en común.

Algunas de las variaciones de hábitats pueden estar directamente relacionadas con la heterogeneidad del substrato, como es el caso de los bosques de *Polylepis*, que siempre se encontraron sobre un substrato de suelo calizo.

Las distribuciones de otros tipos de vegetación, tales como los pajonales enanos, pueden ser explicadas por las diferencias en el drenaje del suelo, mientras que otros patrones de distribución son difíciles de explicar. Por ejemplo, los bosques enanos y los pajonales de pastos altos aparecen en forma de mosaico sobre laderas bien drenadas y crestas de los cerros. Mientras que estos patrones pueden reflejar propiedades de los substratos que no habían sido observadas, también se debe considerar el papel del fuego en el mantenimiento del pastizal. Independientemente de cuáles sean las causas de la heterogeneidad de hábitats, ésta implica que la diversidad a escala de paisaje es probablemente alta. El número total de especies vegetales colectadas en los alrededores del Campamento Uno, aproximadamente 300, fue excepcional para un lugar tan elevado e implica una diversidad mucho más grande para el área en su conjunto.

Se registraron 43 especies de aves en el Campamento Uno, menos de lo que se esperaba pero explicable debido a que el hábitat preferido por las aves a esta altitud, bosques húmedos altos, no fue estudiado completamente. Una de las especies de aves (*Schizoeaca vilcabambae*) descubierta por Terborgh y Weske (1969, 1972, 1975) durante los muestreos realizados en la Cordillera de Vilcabamba en los años sesenta y setenta, es común en los bordes de bosque y en las áreas arbustivas del pajonal. Por lo menos dos especies (*Buteo polysoma* y *Notiochelidon murina*), que no fueron reportadas por Terborgh y Weske se encontraron en esta área de estudio; ambas son de amplia distribución y era de esperar que se encontraran en el área. En cambio, varias especies esperadas (como *Metallura aeneocauda* y *Diglossa brunneiventris*) no fueron halladas en los hábitats de pajonal ni en los bordes del bosque.

Sorprendentemente, a pesar de los extensos bosques de *Polylepsis* del Campamento Uno, no se encontró ninguna de las especies de aves asociadas frecuentemente con *Polylepsis* en otras regiones de los Andes. Como posibles explicaciones de esta observación se discute que puede ser con base a la altitud, el aislamiento de estos bosques de *Polylepsis* o por la composición y accesibilidad de los recursos que esas especies de aves necesitan.

Los estudios de mamíferos se llevaron a cabo en el pajonal abierto, en cenegales de *Sphagnum*, en bosques de *Polylepsis*/*Weinmannia*/*Chusquea* y en los claros de bosque de *Weinmannia*. La mastofauna estuvo dominada taxonómicamente por roedores, registrándose 10 especies, y numéricamente por el ratón campestre, *Akodon torques*, que fue colectado en grandes cantidades en todos los hábitats. También se capturaron en el bosque tres especies de *Thomasomys* spp., dos de las cuales son nuevas para la ciencia. El pajonal abierto también presentó poblaciones del cuy silvestre, *Cavia tschudii*, en buenas condiciones. El único individuo marsupial recolectado, *Gracilinanus* cf. *aceramarcae* constituye el segundo registro de esta especie en el Perú y representa una ampliación considerable de su distribución hacia el Norte. Hubo algunos rastros viejos del oso andino (*Tremarctos ornatus*), pero fueron evidentemente poco común. Se encontraron muchos rastros, incluyendo pelo y huesos, de un venado pequeño, que coinciden con los de *Mazama chunyi*, una especie rara conocida sólo por unos pocos especímenes encontrados cerca de Cusco.

El hallazgo más impresionante entre los mamíferos de este lugar, fue un roedor grande arbóreo de la familia Abrocomidae muy similar al *Abrocoma oblativa*, especie conocida solamente a partir de dos cráneos y otros huesos encontrados en excavaciones de entierros precolombinos en Machu Picchu. Este especimen, que por su apariencia externa sea probable *Abrocoma*, muestra que estos especímenes, colectivamente, representan un nuevo género y especie, que han sido descritos como *Cuscomys ashaninka* (Emmons 1999).

La diversidad de anfibios y reptiles fue ligeramente menor a lo esperado, comparada con la diversidad encontrada a altitudes similares en otras partes del Perú. Sin embargo, las cuatro especies halladas en este estudio, tres ranas y una lagartija, parecen ser nuevas para la ciencia y pertenecen a géneros (*Gastrotheca*, *Phrynopus*, *Eleutherodactylus* y *Proctoporus*) que son restringidos o más diversos en mayores altitudes.

El estudio de mariposas en el Campamento Uno identificó 29 especies diurnas y 29 especies nocturnas grandes. Estos números fueron inferiores al esperado. Sin embargo, se encontró un número proporcionalmente alto de nuevos taxa, incluyendo 11 especies nuevas de mariposas diurnas. Las 18 especies restantes son una mezcla de especies de distribución amplia (*Dione glycera*, *Vanessa braziliensis*, *Tatochila xanthodice*) y de especies encontradas principalmente en las partes altas de los Andes del Centro y Sur de Perú y Bolivia. Entre las mariposas nocturnas, la mayoría de los esfingidos colectados fueron especies de distribución amplia, aunque dos de las especies de *Euryglottis* colectadas están restringidas a áreas sobre los 1000 m.s.n.m. Otras especies raras encontradas en esta área de estudio incluyen al saturnido *Bathyphlebia aglia*, conocido solamente a través de unos pocos especímenes y al cercofanido *Janiodes bethulia*, previamente conocido solamente por su localidad tipo. Biogeográficamente, las mariposas encontradas en el Campamento Uno están relacionadas a la fauna conocida en Machu Picchu, al Sureste de esta área.

Estudio del SI/MAB, Wayrapata, 2445 m.s.n.m. (1998)

Estructuralmente, la vegetación de las parcelas de estudio de biodiversidad de Wayrapata están dominadas por árboles de entre uno y cinco centímetros de diámetro a la altura del pecho (dap). Sólo unos pocos individuos formaron parte de clases diamétricas más grandes. En general, la riqueza de especies (el número de especies por unidad de área) en Wayrapata fue menor en comparación a otros lugares montañosos de similar elevación en el Perú. Entre más al Sur estan ubicadas las áreas de estudio, mayor es su susceptibilidad a las fuertes oscilaciones climáticas que se originan en la Patagonia entre mayo y octubre.

Estos frentes del Sur ("friajes" o "surazos") son conocidos por su capacidad de bajar las temperaturas hasta 8°C por debajo de los niveles normales. Además, se ha reportado que la transición de bosques de partes más bajas y con mayor riqueza de especies a bosques de partes más altas y con menor riqueza de especies ocurre en elevaciones más bajas en las regiones montañosas más al Norte. En Wayrapata ambos factores pueden jugar un papel importante en la riqueza de especies vegetales relativamente baja, cuando se comparan con otros sitios.

En Wayrapata se observaron 92 especies de aves de 26 familias. De estas, la familia con mayor número de especies fue *Emberezidae*, que incluye 22 especies (23.7% del total), seguida por *Trochilidae* con 12 especies (12.9%) y *Tyrannidae* con diez especies (10.8%). Con la excepción de loros, vencejos, golondrinas y buitres, la mayoría de las especies registradas en Wayrapata fueron capturadas con redes de niebla. Relativamente pocos registros estuvieron basados en observaciones directas debido a la poca visibilidad durante las primeras horas del día, que son precisamente las horas de mayor actividad de las aves en esta área.

En Wayrapata se registraron 20 especies de pequeños mamíferos; nueve sólo se hallaron en los bosques de baja altura y en los bosques de transición. La mayoría de las especies de mamíferos pequeños son típicas de bosques montañosos, aunque algunas lo son también de bosques premontanos o tienen un amplio rango altitudinal que incluye diferentes tipos de bosques montañosos. Entre los aspectos resaltantes de este sitio de muestreo es la simpatría (especies del mismo grupo sistemático que viven en la misma localidad) en varios géneros. Estos géneros incluyen marsupiales tales como *Monodelphis* (dos especies), roedores como *Akodon* (dos especies), y murciélagos como *Anoura* (tres especies) y *Platyrrhinus* y *Sturnira* (tres especies).

Otro hallazgo importante fue la ausencia del género nativo *Thomasomys*, que es altamente diverso y está generalmente presente en los bosques montañosos de las laderas orientales del Perú. Es posible que las poblaciones de este género estuvieron fuertemente disminuidas durante la temporada del año en que se realizó el estudio. Cualquiera sea la causa, este fenómeno no se ha detectado en otros estudios en bosques de neblina o bosques montañosos.

Se registraron en Wayrapata 12 especies de mamíferos grandes, el equivalente al 68% de las especies que se ha reportado para el área. El método más efectivo para registrar mamíferos de gran tamaño fue la observación de rastros (huellas y heces). A través de este método, se registraron seis especies, mientras que en las estaciones de olor se registraron solamente cinco especies.

Los valores más altos de abundancia y diversidad de macroinvertebrados acuáticos fueron registrados en las estaciones de muestreo de Wayrapata, con 737 individuos/m^2 encontrados en la parte baja del arroyo Cascada. Las quebradas de Wayrapata presentaron índices relativamente altos de riqueza de familias pero un valor bajo de uniformidad en relación con la distribución de abundancia.

Se colectaron 11 especies de anfibios, todos ellos anuros, y siete especies de reptiles –dos lagartijas y cinco serpientes. Se encontró la mayor diversidad de anfibios en la familia Leptodactylidae (10 especies), típico de los bosques de montaña andinos. Se registró una nueva especie de sapo de la familia Bufonidae del género *Atelopus*. El primer registro de la serpiente *Liophis andinus* también se realizó en esta área de estudio. Hasta ahora, esta serpiente era sólo conocida en la región de Incachaca, Bolivia. Es necesario llevar a cabo una evaluación más profunda de los reptiles y anfibios durante la estación lluviosa, para obtener una evaluación más precisa de la diversidad de este grupo. Por otro lado, los investigadores observaron que el método de muestreo visual produjo la mayor información y se recomienda que esta técnica se aplique en estudios futuros.

La diversidad de invertebrados terrestres también fue alta en Wayrapata. Se registraron 45 especies de arañas de 18 familias y 17 especies de grillos distribuidas en cuatro familias. Solamente se muestrearon tres especies de escarabajos de estiércol posiblemente por la baja incidencia de mamíferos grandes que sirvieran como fuente de heces. También se colectaron 67 especies de abejas y avispas (Hymenoptera-Aculeata, sin incluir las hormigas) distribuidas en nueve familias. Las difíciles condiciones en Wayrapata (crestas desnudas, fuertes vientos, vegetación achaparrada y densamente compacta, neblina semi-permanente) podrían ser favorables para especies de abejas y avispas sociales que construyen nidos, como *Trigona*, *Partamona*, *Nannotrigona* y *Melipona*. Estos géneros, que parecen estar mejor adaptados a los bruscos cambios ambientales que probablemente son comunes en los hábitats de Wayrapata, fueron más abundantes en esta área de estudio que en Llactahuaman, de menor altitud.

Evaluación de CI-RAP, Campamento Dos, 2050 m.s.n.m. (1997)

A los 2000 m.s.n.m., la mayor parte de la vegetación fue bosque alto de neblina en las laderas de pendiente moderada a alta. La vegetación de pajonal abierto es relativamente rara por debajo de la cresta de la Cordillera. La distribución de los tipos de vegetación parece estar gobernada principalmente por diferencias en drenaje. Al igual que en el Campamento Uno, existe poca sobreposición de especies entre los transectos: sólo se observó 7% de sobreposición entre los dos transectos ubicados en el bosque alto de neblina. En general la diversidad fue similar entre los bosques húmedos altos y los bosques de crestas de montaña, aunque los bosques de crestas de montaña y los bosques enanos de *Clusia* compartieron solamente el 12% de sus especies. Por otro lado, solamente el 2% de las especies fueron compartidas entre los bosques enanos de pantano y los bosques altos húmedos.

La vegetación abierta del centro de la meseta del Campamento Dos fue muy similar a la del pajonal alto del Campamento Uno, dominada por *Chusquea* sobre una gruesa alfombra de musgo *Sphagnum*. Un segundo tipo de vegetación, carente de *Chusquea* y con una abundancia mucho menor de musgo *Sphagnum*, se encontró en los suelos más húmedos adyacentes a las márgenes de pequeñas lagunas. La diversidad de especies en este tipo de vegetación fue más baja (cerca de 13 especies por cada 50 individuos muestreados), que en todos los tipos de vegetación abierta muestreados en el Campamento Uno (aproximadamente 19 especies por cada 50 individuos).

Sin embargo, estos pajonales abiertos fueron la excepción en términos de diversidad de especies. Los tipos de vegetación comparables fueron aproximadamente dos veces más ricos en el Campamento Dos que en el Campamento Uno a 3300 m.s.n.m. La alta diversidad alfa del Campamento Dos, combinada con lo que parece ser una alta diversidad beta (entre hábitats), indica una diversidad regional potencial alta a esta elevación.

La riqueza de especies de aves (115 especies) del Campamento Dos es consistente con lo que se esperaba encontrar a esta altitud (1800-2100 m). Nueve especies, que no habían sido reportadas para la Cordillera de Vilcabamba por Terborgh y Weske, fueron registradas en esta área. La mayoría de estas especies son aves andinas de amplia distribución y su presencia en el área de estudio era esperada. Varias otras especies, tales como *Otus alboguralis* (lechuza de cuello blanco) y *Basileuterus luteoviridis* se encuentran típicamente en elevaciones mucho mayores en otras partes del Perú. Se encontró *Grallaria erythroleuca* con relativa frecuencia en el Campamento Dos. Esta especie tiene una distribución muy restringida, y sólo se la conoce en las cadenas montañosas de Vilcabamba y Vilcanota. Aunque no se han hecho comparaciones detalladas, la población de la Cordillera de Vilcabamba podría representar un taxon que aún no ha sido descrito.

Los estudios de las poblaciones de mamíferos estuvieron concentrados en la variedad de hábitats con bosque, y solamente se colocaron unas cuantas trampas y una línea de trampas de caída libre en el cenegal de *Sphagnum*. La misma especie de roedor, *Akodon torques* que fue dominante en el Campamento Uno también resultó dominante en este campamento. Se capturaron cinco marsupiales a 2050 m.s.n.m., incluyendo dos especies de zarigüeyas de cola corta (*Monodelphis* spp.). Para una de las especies, los especimenes representan el tercer y cuarto individuos muestreados en el Perú. Se capturaron varias zarigüeyas de cola corta en el borde del cenegal de *Sphagnum*. Los densos matorrales de bambú estaban habitadas por el cono-cono peruano (*Dactylomys peruanus*); este lugar bien podría ser una nueva extensión de la distribución de esta especie que ha sido colectada en raras ocasiones. Se observaron por lo menos dos especies de primates, monos araña y monos nocturnos, así como posiblemente monos capuchinos. Como los primates han sido reducidos severamente por la caza en las tierras bajas habitadas por humanos, las elevaciones más altas de la Cordillera de Vilcabamba son un refugio local importante.

La herpetofauna muestreada en el Campamento Dos estuvo compuesta por 13 especies (11 especies de rana y dos de serpientes). La fauna de ranas en el sitio fue básicamente similar a la que se encontró en elevaciones similares dentro del Parque Nacional Manú, pero fue raro no haber encontrado especímenes de *Hyla* o *Phrynops*, que normalmente están presentes en altitudes similares. Es posible que la ausencia de *Hyla* se deba a la falta de hábitat adecuado. Hasta dos terceras partes de las especies de ranas encontradas pueden ser nuevas para la ciencia.

Debido a lo limitado de los esfuerzos de muestreo, las mariposas recolectadas en el Campamento Dos (19 especies) representan solamente una pequeña muestra de la fauna que debe estar presente en el área. La mayoría de las especies recolectadas están ampliamente distribuidas en los bosques montañosos de los Andes, aunque se recolectó un nuevo registro para *Pedaliodes*, conocida solamente en los valles de Vilcanota y Santa María en Cusco. Las mariposas de esta área de estudio están relacionadas biogeográficamente a la fauna de Chanchamayo.

Estudio SI/MAB, Llactahuaman, 1717 m.s.n.m. (1998)

Las parcelas de estudio de vegetación en Llactahuaman tienen una mayor proporción de árboles grandes que en Wayrapata (a 2445 m.s.n.m.), pero un número mucho menor que el registrado en la locación del pozo de Pagoreni, en la zona baja del Urubamba, que fue estudiado durante la fase IV del proyecto SI/MAB.

Se observó un total de 111 especies de aves pertenecientes a 28 familias. La familia más representada fue Emberezidae con 27 especies (23.9% del total), seguido por Tyrannidae con 17 especies (15.0%), Formicaridae con 12 especies (10.6%), y Trochilidae con 9 especies (8.0%).

Se halló en Llactahuaman 19 especies de mamíferos pequeños. La mayoría de especies de mamíferos pequeños fue típica de los bosques montañosos aunque algunas especies son de bosques pre-montanos o tienen un amplio rango de distribución altitudinal, que incluye varios tipos de bosques de montaña. No se encontró en esta zona el género *Thomasomys*, muy diverso en especies, que generalmente se encuentra en bosques de montaña de las laderas orientales de Perú. Se registraron 12 especies de mamíferos grandes. El método más efectivo para registrar estos animales fue la estación de olor donde se registraron cinco especies.

También se registró visualmente tres especies de mamíferos grandes.

Se registraron ocho especies de anfibios (todos anuros) y nueve especies de reptiles (cuatro lagartijas y cinco serpientes). Como en Wayrapata, se encontró la mayor diversidad en la familia de ranas de Leptodactylidae, registrándose cinco especies.

Dos especies de peces, *Trichomycterus* sp. y *Astroblepus* sp., fueron registradas en tres estaciones de muestreo en Llactahuaman, ambas pertenecientes al suborden de Suliformes. En la quebrada Bagre se encontró una densidad relativamente baja de macroinvertebrados acuáticos: 96 individuos/m^2.

Se muestrearon 60 especies de arañas distribuidas en 16 familias y 22 especies de grillos en cuatro familias. El equipo de trabajo también registró una alta diversidad de escarabajos (166 especies de 21 familias), cantidad mucho mayor que la encontrada en Wayrapata, quizá debido a la mayor diversidad de hábitats especializados presentes en el área de estudio. Así mismo, se encontró un total de 102 especies de abejas y avispas (Hymenoptera-Aculeata, sin incluir hormigas) pertenecientes a 10 familias.

Estudio CI-RAP, Campamento Tres, 1000 m.s.n.m. (1998)

El Campamento Tres incluye una amplia variedad de tipos de hábitats, desde cañones húmedos, sombreados y profundos con una alta diversidad de plantas herbáceas, y afiladas crestas de piedra caliza, donde las palmeras son dominantes, hasta bosques de neblina en laderas con gran diversidad de especies arbóreas. Aunque la mayoría de los especímenes vegetales colectados todavía deben ser identificados a nivel de especies, se registraron aproximadamente 220 morfoespecies de árboles con dap mayor a 10 cm, de un total de 500 árboles estudiados. Además, se identificaron en esta área de estudio casi 90 familias de plantas con flores. No se conoce todavía el número de familias de pteriodofitas (helechos), aunque es muy probable que éste sea significativo.

La información generada en los transectos y observaciones generales indican que ocurre muy poca sobreposición en las especies vegetales encontradas en cada elevación muestreada en este campamento. Probablemente, estas diferencias se deben al tipo y profundidad del suelo y a los factores climatológicos. Las familias herbáceas con el mayor número de especies en el área son: Acanthaceae, Aracaceae, Bromeliaceae, Cyclanthaceae, Gesneriaceae, Maranthaceae, Orchidaceae y Piperaceae.

El área en los alrededores del Campamento Tres representa el límite más alto de la avifauna de la Amazonía y el límite más bajo de las especies de aves andinas. En el valle del alto Río Apurimac, o en las laderas occidentales de la Cordillera de Vilcabamba, se encontró por lo menos 36 especies de aves que no habían sido reportadas por Terborgh y Weske. La mayoría de estas especies amazónicas son de distribución amplia, y su aparente ausencia en el alto Apurímac puede deberse al efecto "filtro" que ejerce el bosque deciduo más seco del Río Ene, ubicado río abajo del Apurímac, sobre las especies amazónicas en el valle del Río Apurímac.

Se observó en este campamento varias especies de aves "del trópico alto", incluyendo *Phaethornis koepcheae*, *Heliodora branickii*, *Hemitriccus rufigularis*, *Phylloscartes parkeri*, *Pipreola chlorolepidota*, *Ampelidoes tschudii*, *Lipaugus subularis*, y *Oxyrunus cristatus*. La mayoría de estas especies se aprecian sólo en algunas localidades del Perú y varias de éstas, están aparentemente restringidas a estribaciones de baja altitud, al pie de los Andes. Consecuentemente, sus distribuciones son excepcionalmente restringidas y naturalmente fragmentadas, determinando que estas especies sean de particular interés para la conservación. La protección y conservación de las crestas que mantienen estas raras especies locales debe de ser una prioridad.

Por otra parte, se registraron 58 especies de mamíferos en el Campamento Tres. Como se esperaba, la mastofauna está compuesta principalmente por géneros y especies de tierras bajas. La rata espinosa, *Proechimys simonsi,* domina en términos numéricos la fauna pequeña del área, especialmente en los bosques de bambú, donde también se encontraron densas poblaciones del cono-cono boliviano, *Dactylomys boliviensis*. Sin embargo, especies que son comunes en las tierras bajas como *Oryzomys megacephalus* fueron escasas en esta área de estudio. Por otra parte, *Oryzomys macconnelli*, que es una especie rara en las tierras bajas, fue sumamente abundante. Dentro del rango de elevación estudiado, las regiones altas presentan una composición de fauna más singular. De esta manera, la mastofauna de la Cordillera de Vilcabamba se hace más distintiva y divergente conforme aumenta la elevación.

Se registraron relativamente pocas especies de anfibios y reptiles en este sitio (cinco especies de sapos, 24 de ranas, cinco de lagartijas y cuatro especies de serpientes). Más aún, dos de estos registros son de gran interés. Una rana marsupial, *Gastrotheca* sp., es quizá una especie que aún no se ha descrito y puede representar una de las localidades más bajas para los miembros de este género en Perú. La otra es una especie de *Colostethus* sp., que tampoco ha sido descrita y que pertenece a la familia de ranas venenosas (Dendrobatidae). El género *Colostethus* tiene unas pocas especies grandes en la base de los Andes, pero su taxonomía es poco conocida y se necesita hacer más estudios sistemáticos.

Estudio CI-RAP, Tributarios del río Picha, 500 m.s.n.m. (1997)

Se tomaron muestras de peces en varios tributarios del Río Picha, cerca de las comunidades nativas de Camaná, Mayapo y Puerto Huallana. Durante el estudio, se registraron 86 especies de peces de 19 familias. La composición de las especies de peces estuvo dominada por characidos, pimelodidos y loricaridos. La ictiofauna aquí es muy similar a la que se encontró en el Río Camisea, un tributario del Río Urubamba, y tiende a compartir varias especies con la cuenca del Río Manú (Departamento de Madre de Dios). Es posible que varias especies de peces, incluyendo dos de los géneros *Scopaeocharax* y *Tyttocharax*, sean nuevas para la ciencia. La primera parece estar relacionada a *S. atopodus*, una especie conocida en el Río Huallaga, Tingo María. La especie sin describir del género *Tyttocharax* es cercana a *T. tambopatensis*, para Madre de Dios. Muchas de las especies encontradas, especialmente aquéllas de la familia Characidae, viven en una cuenca similar a la del Río Urubamba y han sido descritas previamente a partir de investigaciones que se llevaron a cabo durante la primera mitad del siglo pasado.

Además de los estudios biológicos, los antropólogos del equipo también hicieron estudios rápidos de campo sobre la etnoecología y el uso de recursos en las poblaciones Matsigenka de las tierras bajas del Río Picha, adyacentes a la Reserva Comunal propuesta. Los antropólogos de CI-RAP estuvieron de cuatro a seis días en cada una de las comunidades Matsigenka de Camaná, Mayapo y Puerto Huallana durante el mes de mayo de 1997. También visitaron las comunidades de Nuevo Mundo, Nueva Luz y la Misión Católica de Kirigueti (todas a lo largo del Río Urubamba). Las tres principales comunidades de estudio tienen poblaciones de más de 300 personas cada una, bastante numerosas para el promedio de cualquier asentamiento tradicional Matsigenka (entre 30 y 80 personas). La población de comunidades nativas del Río Picha y sus tributarios es de 1500 habitantes aproximadamente. Para aprovechar los recursos diversos y dispersos de los bosques tropicales, los Matsigenka de las comunidades del Río Picha mantienen una fuerte tradición de excursiones estacionales y migraciones frecuentes.

La agricultura de los Matsigenka, al igual que la de otros grupos en la Amazonía (Posey y Balée 1989, Boster 1984), se caracteriza por un área de bosque disturbado relativamente pequeña, policultivos, gran diversidad genética de cultivos y un rápido proceso de regeneración del bosque. La mayoría de los requerimientos proteicos de los Matsigenka no se satisfacen a través de la agricultura, sino que provienen de recursos silvestres como peces, animales de caza, frutas, nueces y corazones de palma. Se construyó un mapa para cada comunidad con la ayuda de pobladores indígenas, quienes detallaron los nombres locales de los ríos, montañas, asentamientos humanos actuales y pasados, huertos, trochas, recursos de plantas y animales y otros rasgos del terreno que rodean a cada comunidad. Se realizó un estudio preliminar sobre la clasificación Matsigenka de los bosques. Los Matsigenka distinguen una gran diversidad de tipos de hábitats, incluyendo bosques de neblina de altura, pajonales y vegetación altoandina.

LITERATURA CITADA

Alonso, A. and F. Dallmeier (Eds.). 1998. Biodiversity assessment and monitoring of the Lower Urubamba Region, Perú: Cashiriari-3 well site and the Camisea and Urubamba Rivers. SI/MAB Series #2. Washington, DC. Smithsonian Institution/MAB Biodiversity Program.

Alonso, A. and F. Dallmeier (Eds.). 1999. Biodiversity assessment and monitoring of the Lower Urubamba Region, Perú: Pagoreni well site assessment and training. SI/MAB Series #3. Washington, DC. Smithsonian Institution/MAB Biodiversity Program.

Baekeland, G. B. 1964. By parachute into Peru's lost world. National Geographic 126: 268-296.

Boster, J. S. 1984. Classification, cultivation, and selection of Aguaruna cultivars of *Manihot esculenta* (Euphorbiaceae). Advances in Economic Botany 1: 34-47.

Dallmeier, F. and A. Alonso (Eds.). 1997. Biodiversity assessment of the Lower Urubamba Region, Perú: San Martin-3 and Cashiriari-2 well sites. SI/MAB Series #1. Washington, DC. Smithsonian Institution/MAB Biodiversity Program.

Emmons, L. H. 1999. A new genus and species of a brocomid rodent from Peru (Rodentia: Abrocomidae). American Museum Novitates, Number 3279.

Foster, R. B., J. L. Carr, and A. B. Forsyth (eds.) 1994. The Tambopata-Candamo Reserved Zone of southeastern Perú: a biological assessment. RAP Working Papers Number 6. Washington, DC. Conservation International.

Myers, N., R. A. Mittermeier, C. G. Mittermeier, G. A. B. da Fonseca, and J. Kent. 2000. Biodiversity hotspots for conservation priorities. Nature. 403: 853-858.

Posey, D. and W. Balée (eds.). 1989. Resource Management in Amazonia: Indigenous and Folk Strategies. Advances in Economic Botany, Vol. 7. New York Botanical Gardens. New York.

Rodríguez, L. O. 1996. Diversidad biológica del Perú: zonas prioritarias para su conservación. Lima: Proyecto Fanpe (GTZ and INRENA): Deutsche Gesellschaft für Technische Zusammenarbeit (GTZ) and Ministerio de Agricultura, Instituto Natural de Recursos Naturales (INRENA).

Schulenberg, T. and K. Awbrey (Eds.). 1997. The Cordillera del Condor region of Ecuador and Peru: A biological assessment. RAP Working Papers Number 7. Washington, DC. Conservation International.

Stotz, D. F., J. W. Fitzpatrick, T. A. Parker III, and D. K. Moskovits. 1996. Neotropical birds: ecology and conservation. Chicago: University of Chicago Press.

Terborgh, J. 1985. The role of ecotones in the distribution of Andean birds. Ecology 66: 1237-1246.

Terborgh, J., and J. S. Weske. 1969. Colonization of secondary habitats by Peruvian birds. Ecology 50: 765-782.

Terborgh, J., and J. S. Weske. 1972. Rediscovery of the Imperial Snipe in Peru. Auk 89: 497-505.

Terborgh, J., and J. S. Weske. 1975. The role of competition in the distribution of Andean birds. Ecology 56: 562-576.

PROPOSED PROTECTED AREAS IN THE CORDILLERA DE VILCABAMBA, PERU

Conservation International –Perú (CI-Perú), Asociación para la Conservación del Patrimonio del Cutivireni (ACPC), and Centro para el Desarrollo del Indígena Amazónico (CEDIA)

OVERVIEW OF EFFORTS TO PROTECT THE CORDILLERA DE VILCABAMBA

Interest in protecting the pristine and unique forests of the Cordillera de Vilcabamba dates back over 30 years. This long-term interest has developed into a collaborative process involving many different groups working for biodiversity protection and representing the indigenous peoples of the region.

In October 1963, the Peruvian federal government passed Supreme Resolution 442-AG-63, which created the Bosque Nacional Apurímac (Apurimac National Forest). This national forest contained two sectors, the Ene sector on the western margin of the Río Ene, and the Apurimac sector, containing the Cordillera de Vilcabamba between the ríos Apurímac, Ene, Tambo, and Bajo Urubamba. All together, the Bosque Nacional Apurímac encompassed 2,071,700 ha and was the largest of three national forests created by this resolution.

In 1975, the Ley Forestal y de Fauna Silvestre (Decreto Ley 21147) defined Bosque Nacional as natural forests declared suitable for the permanent production of wood, other forest products, and wildlife, whose utilization can only be conducted directly and exclusively by the State. In addition, scientific, educational, tourist and recreational activities were permitted, but strictly regulated, within the forest. The native populations within the area maintained their right to their traditional uses of the forest, as long as they were compatible with the overall objectives of the area.

On April 28, 1988, the Peruvian government issued Supreme Resolution 186-88-AG/DGFF, which superceded the previous resolution. The Ene sector of the Bosque Nacional Apurímac was removed from protection since it contained operating logging concessions owned and operated by private companies and not by the government. The Apurimac sector, comprising 1,669,300 ha, retained protected status, with the new designation as the Zona Reservada del Apurímac. A Zona Reservada (Reserved Zone) is an area of transitory status. This designation provides protection of the biological diversity of the area until a formal declaration of status is declared. Under this status, the zone is to be studied and evaluated to determine a permanent category and a management plan. Meanwhile, no new development activities are allowed; only subsistence hunting and agriculture by native peoples are permitted. Any other activities and uses of the forest while under this status must be compatible with the conservation objectives of the area.

In 1988, with the creation of the Zona Reservada Apurímac, a collaborative process was initiated among many diverse groups to plan and regulate activities within the region. These groups included the following non-governmental organizations (NGOs): Asociación para la Conservación del Patrimonio de Cutivireni (ACPC), Centro Amazónico de Antropología y Aplicación Práctica (CAAAP), Centro para el Desarrollo del Indígena Amazónico (CEDIA), Centro de Investigación y Promoción Amazónica (CIPA), Fundación Peruana para la Conservación de la Naturaleza (FPCN), Instituto Lingüístico de Verano (ILV), as well as the following indigenous groups: Central Asháninka del Río Tambo (CART), Confederación de Nacionalidades Amazónicas del Perú (CONAP), Consejo de Comunidades Nativas Nomatsiguengas y Asháninkas de Pangoa (CONOAP), Federación de Comunidades Nativas Yanesha (FECONAYA), Organización Aguaruna del Río Mayo (OAAM), and the Organización Central de Comunidades Aguaruna del Alto Marañón (OCCAAM). The following governmental organizations also participated: the Programa Nacional Parques Nacionales-Perú of the Dirección General de Forestal y Fauna (DGFF), Dirección de Manejo Forestal of the DGFF, ONERN and the

Universidad Nacional Agraria-La Molina. The Dominican missionaries of the Río Urubamba area also participated.

In addition to official governmental actions, many groups and individuals have made significant efforts to draw attention to the importance of protecting the Vilcabamba region. These efforts include:

1964 Ing. Alfonso Rizo Patrón presented a plan to the Peruvian government, in which he analyzed aerial photographs of the Cordillera de Vilcabamba taken by the Hunting Co. in 1961, and in which they determined the existence of the natural bridge "Pavirontsi".

1965 The Organization of American States (OAS) proposed the creation of the Parque Nacional de Cutivireni.

1974 The Franciscan Mission through the vicary of San Ramón, proposed to the military government the creation of areas of protection.

1982 The Centro para el Desarrollo del Indígena Amazónico (CEDIA) initiated a program of granting land title to the native communities in the Urubamba Valley, between the Río Mishagua and the Pongo del Mainique.

1987 The Asociación para la Conservación del Patrimonio de Cutivireni (ACPC) completed travels across the Cordillera de Vilcabamba, reaching the "Pavirontsi" and confirming that it is the largest natural bridge in the world. Upon returning, ACPC held a press conference where they stressed the importance of conserving this unique region and submitted a proposal to the government.

1988 In response to ACPC's proposal, the Dirección General de Forestal y Fauna (DGFF) conducted an overflight of the region by helicopter.

1988 (June) The Centro de Desarrollo Rural de Satipo, pertaining to the Unidad Agraria Departamental XVI de Junín del Ministerio de Agricultura, produced a report about the lands governed by the Dirección General Forestal y de Fauna.

1988 (October) A workshop entitled "Bosque Nacional del Apurímac: Reality and Perspective" was organized and participants included all those listed above. The results were submitted to the Ministerio de Agricultura. In the Zona Reservada del Apurímac, the following natural areas were proposed: El Santuario Nacional de Cutivireni proposed by ACPC, the Reserva Comunal de Cutivireni proposed by INRENA and the Ashaninka communities, the Reserva Comunal de Vilcabamba proposed by CEDIA, Parque Nacional de Vilcabamba proposed by Conservation International-Perú, and a recommendation to create the Reserva Nacional del Apurímac (CEDDRE, IDEFE, Asociación Amazonia).

1989 DGFF formed a convenio with ACPC to conduct ecological studies of the flora and fauna of the Cutivireni zone.

1990 The Fundación Peruana para la Conservación de la Naturaleza (FPCN) conducted a field study to analyze the situation of the Zona Reservada del Apurímac. Dr. Dirk McDermott, a consultant with World Wildlife Fund participated in this study and produced the report "Technical Concerns and Recommendations for the Camisea Gas Field, Peru." The extraction of natural gas from the neighboring Camisea gas fields will involve a pipeline that could cross the Cordillera de Vilcabamba.

1991 CEDIA published the book "El area de influencia del Proyecto Gas de Camisea- Territorio Indígena."

1991 FPCN published "Document of Conservation #5: El Gas de Camisea, Reflexiones sobre el impacto ambiental de su exploitación y cómo reducirlo a mínimos aceptables."

1992 FPCN signed a convenio with ACPC to put forward proposals for protected areas within the Vilcabamba region in a forum with other institutions and with the DGFF.

1993 After the forum of 1992, ACPC, CEDIA, FPCN and SPDA presented to the DGFF a "Propuesta para el establecimiento de Areas Naturales Protegidas en la Zona Reservada Apurímac." Their proposal included the creation of a strict protected Sanctuary, a Reserva Comunal along the Río Urubamba, and expansion of the territories of the indigenous communities along the Río Tambo.

1995 CEDIA presented a proposal for the creation of the Reserva Comunal del Urubamba.

1996 ACPC and Pronaturaleza (FPCN) developed a map of the proposed protected areas of the Cordillera de Vilcabamba, incorporating natural resource use and indigenous communities of the region.

1997 Conservación Internacional-Perú (CI-Perú) and ACPC conducted a rapid biodiversity assessment (RAP) of the Cordillera de Vilcabamba to collect data on the biodiversity of the region.

1998 The Asháninka communities of the Río Tambo formed CART (Central Asháninka del Río Tambo) and met in an extraordinary congress in the community of Otica. The community leaders pledged their support for the proposed protection of the Cordillera de Vilcabamba, particularly the Reserva Comunal Asháninka Tambo Ene.

1998 CI-Perú and ACPC carry out a second phase of rapid biodiversity assessment (RAP) in the Cordillera de Vilcabamba.

1998 ACPC and CARE organized a meeting of representatives of many indigenous groups of the area, including communities of the Río Ene, Río Apurimac, and Río Tambo, CART, OARA, FPCN, CI-Perú, ARPI – AIDESEP. The group discussed the proposal for protected areas, which includes two Communal Reserves and a National Park for the Cordillera de Vilcabamba. The group endorsed the proposal.

1998 SI/MAB conducted biodiversity assessments of the Southern Cordillera de Vilcabamba with support from Shell Prospecting and Development-(Perú) B.V. (SPDP) to understand and minimize disturbance to the area from potential natural gas pipeline construction.

1999 CI-Perú, ACPC, and CEDIA organized a meeting along the Río Urubamba with representatives of the communities and organizations of the Urubamba valley to discuss the 1998 proposal. After some modifications, the group also endorsed the proposal.

1999 CI-Perú, ACPC, and CEDIA started facilitating a consultative process to collect socio-economic information on the region, to carry out consultations with local communities and NGOs in the area, and to prepare a draft rezoning of the Zona Reservada Apurímac. Contact CI-Perú for more information about this on-going process.

PROPOSAL FOR TWO COMMUNAL RESERVES AND A NATIONAL PARK

In light of the history outlined above and the weak transitory protected status of the Zona Reservada Apurímac within the Cordillera de Vilcabamba, CI-Perú, ACPC, and CEDIA are working jointly with the native communities and Federations of the three rivers that border the Zona Reservada Apurímac (ZRA) and with relevant government agencies (INRENA) in the process of establishing three protected areas, two indigenous Communal Reserves (Reservas Comunales) and a National Park (Parque Nacional, an area of strict protection) in the Cordillera de Vilcabamba (CI-Perú et al. 1999). This process includes a consultation process to rezone the ZRA, to develop management plans for the protected areas, to form a management committee, and to promote healthy native communities. The project also supports sustainable economic and subsistence activities such as recovering and revitalizing traditional crafts and ethno-botanical knowledge.

The consultative process is strongly supported by many NGOs, including ProNaturaleza and the Sociedad Peruana de Derecho Ambiental (SPDA) as well as the local indigenous communities. The overall objective of the protection of this region is to change the status of the Zona Reservada del Apurímac to three areas of more permanent protected area status in order to assure the protection and sustainable development of the natural resources contained therein. Specific objectives include developing a legal means of protecting montane and lowland forest that would assure the maintenance of freshwater and natural resources for use by the local indigenous communities.

The three protected areas would be located in the departments of Cusco, Ucayali, and Junín and would consist of land irrigated by the Río Bajo Urubamba, Río Tambo, Río Ene y Río Apurímac and their respective tributaries. The design of the proposed protected areas includes a core area of strict protection bordered by multiple use zones, which corresponds to the Biosphere Reserve model recommended by UNESCO and the International Union for the Conservation of Nature (IUCN) .

The three proposed protected areas, all located in the northern section of the Cordillera de Vilcabamba, are:

1. **A Communal Reserve on the Eastern Side** of the Cordillera de Vilcabamba (approximately 205,218 ha).
2. **A Communal Reserve on the Western Side** of the Cordillera de Vilcabamba (approximately 230,980 ha).
3. **A National Park**, located along the summit of the northern section of the Cordillera de Vilcabamba, delimited to the east and west by the two Communal Reserves (approximately 272,500 ha.).

Ninety-seven percent (97%) of the Cordillera de Vilcabamba is forested land that protects the headwaters of the rivers that form the Río Apurimac, Río Ene, Río Tambo and Río Bajo Urubamba. The area proposed as National Park is located in the central part of the Cordillera de Vilcabamba, along the backbone of the mountain range. This area is mostly steep slopes and is virtually uninhabitable. The local indigenous groups use this region as a travel route from one side of the Cordillera to the other but do not live or practice agriculture in the area.

The establishment of Communal Reserves would allow the local indigenous communities to continue to harvest forest products to sustain their traditional ways of living. The Communal Reserve to the east of the national park would include land occupied by the Ashanínka, Matsigenka, and Yine groups along the western banks of the Río Bajo Urubamba. Fifteen native communities would benefit directly from the establishment of this reserve, which would aid in preventing large scale colonization by people from the high Andean areas and would provide regulations governing petroleum exploration and exploitation in the Camisea area to the east.

The Communal Reserve on the western side of the Cordillera de Vilcabamba would benefit over 20 Ashanínka communities. The chiefs of each of these communities have pledged their support for the creation of this communal reserve.

LITERATURE CITED

ACPC, INRENA, CEDIA, CI-Peru, CEDORE, IDEFE, and Asociación Amazonia. 1988. Bosque Nacional del Apurímac: Reality and Perspective. Document submitted to the Ministerio de Agricultura.

ACPC, CEDIA, FPCN, and SPDA. 1993. Informe Técnico: Propuesta para el establecimiento de Areas Naturales Protegidas en la Zona Reservada Apurímac. Document presented to INRENA.

CEDIA. 1991. El área de influencia del Proyecto Gas de Camisea – Territorio Indígena.

CEDIA. 1995. Proyecto: Reserva Comunal de Vilcabamba "Pavlik Niktine".

CI-Perú, ACPC, and CEDIA. 1999. Términos de referencia para el establecimiento de dos reservas comunales y un parque nacional en la Zona Reservada de Apurimac. Document presented to INRENA.

FPCN. 1991. El Gas de Camisea, Reflexiones sobre el impacto ambiental de su exploitación y cómo reducirlo a mínimos aceptables. Document of Conservation #5.

Pan American Union. 1965. Resource conservation and the establishment of national reserves in Latin America, the Cutivireni National Park: a pilot project in the selva of Peru.

INTRODUCTION TO THE CORDILLERA DE VILCABAMBA, PERU

Based on CI-Perú, ACPC, and CEDIA (1999)

LOCATION AND CURRENT PROTECTED STATUS

The Cordillera de Vilcabamba, comprising more than 3 million hectares (approximately 2.4% of Peru's land area) is situated in the Andes Mountain chain in south-eastern Peru (Maps 1 and 2). 1,699,300 hectares of the Cordillera de Vilcabamba was set aside by the Peruvian government in 1988 as the Zona Reservada del Apurímac (ZRA). The ZRA is located in the departments of Cusco (Región Inka), Ucayali (Región Ucayali), and Junin (Región Andres Avelino Caceres) and contains the watersheds of the Río Bajo Urubamba, Río Tambo, Río Ene and Río Apurímac and their respective tributaries. A Zona Reservada (Reserved Zone), one of several categories of natural protected areas currently recognized by the Peruvian government, is a transitory category that temporarily preserves an area while research is conducted to determine the status and boundaries of permanent protected areas within the zone.

CONSERVATION IMPORTANCE

Peru contains over 40% of the remaining Andean tropical forests (Young and Valencia 1992). The Tropical Andes region has been designated as one of Conservation International's 25 global biodiversity "hotspots," areas of exceptional concentrations of biological diversity and endemic species that are experiencing high rates of habitat loss (Myers et al. 2000). Of the 25 global hotspots, the Tropical Andes contains the highest percentage of the world's endemic plant (7%) and vertebrate (6%) species. The fact that the Tropical Andes contains only 25% of its original forest cover and has only 25% of its current forests under protection makes it one of the most important regions for biodiversity conservation in the world (Myers et al. 2000).

Within the Tropical Andes region, the forests of the Cordillera de Vilcabamba are critical to the healthy functioning of both highland and lowland regions of not only Peru, but a large part of South America. The waters that originate from the high peaks of the Cordillera de Vilcabamba form four important rivers of south-eastern Peru, the Río Apurimac, Río Ene, Río Tambo and Río Urubamba (Map 2). These rivers flow northward, eventually joining the Río Amazonas and flowing through Brazil to the Atlantic Ocean. The Cordillera de Vilcabamba protects the headwaters of these rivers, which maintain a clean supply of freshwater and are the source of fertile soils in the lowlands below. The continued health and survival of all of these rivers and the human communities that utilize them, depends on the future condition of the forests in the mountains of the Cordillera de Vilcabamba. These forests filter the water, control floods, and provide drinking, cooking and cleaning water for local populations. The health of the rivers will also be determined by the wise use of the aquatic resources by local human populations and by the petroleum companies that currently are exploring the lowlands to the east of the Cordillera. The forests, the water, and the local people are closely linked.

Tropical forests perform the extremely important global function of absorbing carbon dioxide and producing oxygen for the atmosphere. The forests of the Cordillera de Vilcabamba contain some of the highest vegetation biomass in the world (51-124 tons/ha, Houghton and Hackler 1996). The presence and preservation of these forests is therefore not only of regional and national importance but also global importance for climate regulation. Only five Peruvian departments contain montane or lower montane forest (1500-3500 m) and only 5% of these forests are protected in national parks (Manu, Yanachaga, and Abiseo National Parks; Young and Valencia 1992).

The Cordillera de Vilcabamba is unique in terms of its biological diversity and endemism due to three main factors (CI-Perú et al. 1999):

1. Isolation from the main Andean Range as it penetrates the Amazon Basin, converting it into a land archipelago that extends into the lowland tropical rain forest. Access to the region has been limited due to dangerous stretches of whitewater ("pongos").
2. Wide variations in altitude and rainfall, from the highest peaks above 4,300 m of altitude, to the lower river basins at around 400 to 500 m of altitude. Rainfall ranges from 1,200 mm/year (47.24 inches/year) in the areas classified as SubTropical Humid Forest to a high of 5,600 mm/year (220.47 inches/year) in areas of Pluvial Montane Tropical Forests.
3. Calcareous soils that act as water traps and provide unique nutrients and edafic conditions for plant development.

The range of elevations and resulting variety of habitat types of the Cordillera de Vilcabamba are important to the movement and survival of a great diversity of migratory and resident animals (Young 1992). The Cordillera de Vilcabamba may contain over 50% of all bird species known from Peru. The area is important for birds that migrate from lower to higher elevations and from northern to southern regions of Peru. The steep slopes and isolation of the Cordillera de Vilcabamba have created the setting for the evolution of a high number of plant, amphibian, insect, and mammal species, many of which are found no where else on earth and are therefore endemic to the cordillera. Scientists consistently discover new species as they explore the montane forests of this area. The results of the RAP and SI/MAB surveys presented in the chapters here attest to the extraordinary biodiversity of the Cordillera de Vilcabamba.

The Cordillera de Vilcabamba contains seven life and two transition vegetation zones (Holdridge 1967). A total of only eight life zones and four transition zones are represented in the Protected Areas System of Peru. Designation of greater protective status to the Zona Reservada Apurimac as proposed by CI-Peru, ACPC and CEDIA (CI-Perú et al. 1999, see Proposed Protected Areas chapter) would provide protection to several additional life zones. The Cordillera de Vilcabamba is primarily covered by Pre-montane Moist Tropical Forest and Moist Tropical Forest (Young and Valencia 1992). In other parts of Peru, these forest types have disappeared or are highly threatened by human activities. Several are found only in this region of Peru; it is therefore of high importance to conserve these fragile and unique life zones within the Cordillera de Vilcabamba.

The exceptional natural beauty of the Cordillera de Vilcabamba has surprised and delighted explorers since the 1500s. National Geographic Magazine twice has featured this spectacular region as an example of one of the few remaining pristine tropical forest areas on earth (Baekland 1964, Morell 1999). The presence of over 55 waterfalls that range from 80 to 267 meters tall, the world's longest natural bridge, and innumerable unexplored lagoons, canyons, subterranean rivers and forests make this region one of the world's most biologically fascinating and important for conservation. Due to the outstanding natural features and beauty, the region has great potential for developing an eco-tourism industry.

Because of its high levels of biological and cultural diversity and significance, the Cordillera de Vilcabamba region demands a conservation effort. Protecting the Cordillera de Vilcabamba is necessary and urgent. Its isolation, due to both geographical barriers and the presence of terrorism until recently, has sheltered this region from significant impact by the logging, ranching and oil firms so prevalent in the rest of Amazonia. Thus, large areas of pristine tropical forest still exist. Furthermore, the Cordillera de Vilcabamba is recognized nationally and internationally as an area of high conservation importance due to its great indigenous diversity. The traditional local inhabitants are native Amazonian groups, Asháninka, Matsigenka (or Matsiguenga), Nomatsiguenga, and Yine, whose diversified economies and disperse settlement pattern have had little impact on biodiversity.

CHALLENGES AND THREATS IN THE CORDILLERA DE VILCABAMBA

Unfortunately, the relative economic and cultural isolation of the Cordillera de Vilcabamba is changing rapidly. The native indigenous groups, particularly the Matsigenka and the Asháninka, have been confronted with several challenges over the past century. During the rubber boom between 1870 and 1912, hundreds of Matsigenka were forced to work for local rubber patrons (Pennano 1988, Dourojeani 1988). This destroyed many communities and drove hundreds more into the headwaters of the Río Urubamba's tributaries. More recently, hydrocarbon exploration, colonization, political conflict, and the drug trade have had an impact on many local groups. For example, the Asháninka lost part of their territory to the armed guerrilla group Sendero Luminoso (Shining Path) and are now trying to reclaim their territory and prevent the immigration of outsiders to keep the drug cartels out of the area (Benavides 1991, Rojas Zelezzi 1994).

Oil Extraction

The largest known deposit of natural gas in South America is located in Camisea, on the eastern fringes of the Zona Reservada Apurímac (ZRA), along the Río Urubamba (CEDIA 1991, FPCN 1991, Rivera 1991). In 1995 and 1996, the Peruvian government signed two hydrocarbon development contracts with Chevron (in the Picha region) and Shell/Mobil (in the Camisea region). After initial work by the Shell/Mobil consortium, the Peruvian government recently has selected new petroleum companies to extract the gas and will select a new company to construct the gas pipeline, a section of which threatens to cut across the southern region of the Cordillera de Vilcabamba. The commitment of these new companies to cultural and biological conservation will play a major role in determining the future health and viability of the Cordillera de Vilcabamba.

Oil firms operating both along the Río Ene (Elf) and the Río Urubamba (Shell and Chevron) have contributed to rapid urban growth of towns like Satipo and Sepahua. The area is attracting Andean migrants from the southern highland provinces, where pervasive rural poverty and land fragmentation prevail. Government presence is also growing through investments in education, health and poverty alleviation programs coupled to resettlement plans in areas that were abandoned during the years of terrorist activity, especially along the Río Tambo (Alvarez Lobo 1998, Zarzar 1989).

Timber Extraction

Selective harvesting of timber has been an important economic activity in the Río Bajo Urubamba region for over 25 years, with the village of Sepahua the commercial center for logging. Early timber harvesting was conducted by individuals; more recently large logging companies lease forest concessions from native communities in exchange for a percentage of the extracted timber. Some native communities, such as Miaría, have given up their lands for commercial exploitation. The timber industry has been an extremely important economic activity in the area in terms of establishing a permanent presence (roads, villages, etc.) and employing local people and colonists. Tree species most often exploited for timber extraction in the Cordillera de Vilcabamba are listed in Table 1.

Table 1. Tree species most often exploited for timber extraction in the Cordillera de Vilcabamba.

Tree Species	Spanish Common Name
Ocotea sp.	moena
Cedrelinga cataneiformis	tornillo
Juglans neotropica	nogal
Manilikara bidentata	quinilla
Inga sp.	shimbillo
Cedrella sp.	Cedro de altura
Cordia sp.	cumula
Ochroma sp.	topa

Between 1974-1986 there was a reduction in the number of hectares requested for logging concessions in the region, which resulted in higher intensity logging in the existing forestry concessions. During this time, logging became more concentrated in the region between the Río Urubamba and Río Sepahua and between the Río Inuya and Río Mishagua. These changes were also related to the migration of many Andean peoples toward the Rio Inuya. In 1984, contact between loggers from Sepahua and the native Nahua people resulted in high sickness and mortality among the Nahua.

Currently to the northeast, a logging company is constructing a road from Puerto Ocopa, a community only three hours by road from a major commercial town (Satipo), to the Río Tambo that flanks the Zona Reservad Apurímac to the North. This road advances almost 150 meters a day.

Cattle Ranching

Cattle ranching was introduced to the Cordillera de Vilcabamba region, primarily along the Río Bajo Urubamba, by early colonists and missionaries. Most cattle ranching is conducted on a small to medium scale along the Río Bajo Urubamba (95%), Río Tambo, and Río Ene. The data on the extent of cattle ranching in the region is poor and unreliable (Zarzar 1989). The actual extent of lands recently or historically cleared for cattle ranching in the Cordillera de Vilcabamba is unknown.

Most cattle ranches are located along rivers, on alluvial soils, which are the best agricultural soils in the region. Cattle ranching on these soils may constitute a poor use of land in the region. Deforestation along river banks for cattle causes erosion and flood control problems. Based on information for 22 cattle ranches in the Bajo Urubamba region, 66% of the cattle are housed on only 3 ranches. The remaining 34% of cattle are distributed among the 19 other ranches. There are another 18 ranches in the region for a total of 40 ranches, with a total of 1,722 head of cattle. Assuming a ratio of 2 ha per cattle, approximately 3,400 ha of land has been taken out of agricultural production for use by cattle.

Economically, cattle ranching plays a secondary role in the region, behind agriculture and timber extraction. However, its ecological impact is primary, in terms of enhancing soil erosion along rivers, eliminating habitat for native flora, fauna and people, and taking land out of agricultural production.

CULTURAL OVERVIEW

History of Human Activities in the Region

Incas in the Vilcabamba region

Historical accounts by early Peruvian writers, including Cieza de León and the Inca Garcilaso de la Vega report that the Inca State, known as Tahuantinsuyo (1430-1532), was centered in Cusco, Peru and from there departed routes to the four corners of the State (Ferrero 1996). The route that headed toward the east passed through the region of Antisuyo, which included the area and populations of the Cordillera de Vilcabamba.

Most historians agree that the natives of the Vilcabamba region were able to keep themselves on the outskirts of the Inca State (Camino 1977). Therefore, the inhabitants of this area were known by the generic name of "Andes" or "Anticunas" (people of the eastern Andes).

In the mid 1400s, the Incan ruler, Pachacutec, apparently penetrated the Cordillera de Vilcabamba through the Valley of Tambo, retreating later through this same valley (Camino 1977). There was much commercial interchange between the Incas and the lowland native groups, principally in the Pongo de Mainique, at the limit of the Río Bajo and Río Alto Urubamba (Savoy 1970; Camino 1977; Lyon 1981). This narrow canyon through the mountains was, and still is, an area held sacred by the local peoples on both sides. The largest settlement outside the Cordillera de Vilcabamba was at Antisuyo, whose inhabitants traded with the Incas products such as coca, gold, honey, bushmeat, and feathers.

The arrival of the Spanish in the 1530s basically ended the Inca State. In 1536, the Incan ruler, Manco Inca, took refuge in the forests of the Cordillera de Vilcabamba. Some believe that he occupied the small city of Vilcabamba, located to the north-east of Machu Picchu, developing stronger contacts and interchange with the inhabitants of the Amazon lowlands (Guillén 1994). His retreat to this area, although possibly not his first choice, bestowed on this area a great importance. Some people say that his refuge in the zone was due to his search for the magical-religious roots of the Incan people in order to rescue the Incan identity, which had been affected and partially broken by the Spanish Conquest. The symbolism and religious prestige of the Cordillera de Vilcabamba appears to be derived from the fact that the region, called "Anti," is considered the native land of the sun. Therefore, in accordance with the cyclical character of the Andean timeframe, the sun returns there for its immersion into the other side of the word, from which the splendid Inca State was destined to return.

Manco Inca resisted the Spanish from the Cordillera de Vilcabamba from 1537 through 1544. The Incan resistance continued until 1572 with the capture and death of the Incan ruler, Tupac Amaru I, by order of the Spaniard Virrey Toledo. For more than 40 years, the Spanish tried to conquer the Cordillera de Vilcabamba but eventually abandoned their efforts due to the difficult terrain of the area, as well as the warlike nature, low population densities and dispersion of native communities (Camino 1977).

Religious presence in the region

Friars and other clergymen of various Christian denominations explored the Cordillera de Vilcabamba with the aim of converting the natives to Christianity, of verifying the myths and legends of the region, and of increasing the zone of influence of their respective congregations. Some of the earliest records are of the Augustines of Cuzco in the Urubamba Valley in the 1570s. In 1594, five indigenous leaders from the Cordillera de Vilcabamba area traveled to Lima and asked Virrey Marques de Canete to send religious representatives to teach them about Christianity. The brother of the Virrey sent two Jesuit fathers, Juan Font and Nicolas Mastrillo on this mission. In 1595 they entered the region through the Jauja valley, with the intent of making contact with the indigenous Asháninka.

At first, the beauty and wildness of the region inspired the Jesuits in their evangelism. However, the Jesuits decided not to create a mission in the area because the indigenous population was very small, with dispersed settlements of less than 20 persons, and the indigenous people did not show any indication that they were interested in converting to Christianity. In addition, the area was very inaccessible such that it would be difficult to establish towns near the indigenous settlements. The reported barbarity of the indigenous groups also halted their efforts, particularly a rebellion in the Vilcabamba area led by the native leader Francisco Chichima. In 1602, Father Font attempted a second entry into the region, this time from Huamanga (Ayacucho). Once again the sparse populations of native peoples and the harsh natural conditions caused the crusade to fail.

After 1635, the Franciscans tried to enter the area from Huanuca passing through Huancabamba and el Cerro de la Sal (Salt mountain; Biedma 1981). They chose this route because the experiences of the Jesuits had shown them that it was almost impossible to access the area from Jauja, and also because of the domination of the areas of Tarama and Quirimi by the Dominicans. In 1673, the Franciscan missionaries, Friar Manuel and Friar Juan de Ojeda, made contact with the first indigenous settlements, headed by the cacique Tonte. They toured the region together, funded by the mission of Santa Cruz del Espíritu Santo de Sonomoro.

Early missionaries reported "this nation is very extensive and its language runs more than 400 leagues from one end to the other. In the north are the comonomas, Callisecas and Cunibos...in the south confined by the Andes are the Huanta and Huamanga who speak the same language and reach as far south as Cuzco" (Biedma 1981). The first Franciscan missions were established and attracted various communities of the Asháninkas, including the Pangoas, Menearos, Anapatis and Pilcusunis, who came from the south. The northern communities sent the Satipos, Capiris and the Tomirisatis. In 1709 the Franciscans attempted another crusade and established 38 missions in the region.

According to Biedma (1981), the Franciscan Mission was visited by 18 or 20 indigenous groups, among them representatives from the famous settlements of Picha and Masadobeni of King Enim, who came from east of the Río Ene.

In 1742, Juan Santos Atahualpa unified the Asháninkas and the Matsigenkas by presenting himself as a descendant of the Incas and led a revolt, expelling the missionaries and

Spanish colonists (Renard-Casewitz, 1977). For 15 years, Atahualpa dominated an extensive area of the lowlands, with the support of the Asháninkas and Matsigenkas as well as other ethnic groups. Atahualpa succeeded in reconstructing ancient, ancestral alliances between these diverse groups through his affiliation with the Inca State.

By the middle of the 1800s, after the death of Atahulpa, the missionaries and colonists had returned and resumed their activities. For example, in 1806, Father Ramón Bausquet traveled the Urubamba River representing the Cocabambilla Mission (Renard-Casewitz, 1977). Currently, the presence of missionaries in the Cordillera de Vilcabamba is very strong, mostly along the principal rivers of the region. The Christian doctrine is clearly being consolidated into the native communities.

Recent History

The past two decades have been a period of extreme stress for the indigenous inhabitants of the Cordillera de Vilcabamba. In the early 1980s, the terrorist group, the Sendero Luminoso (Shining Path), penetrated the Asháninka territory in the western part of the Cordillera (Benavides 1991, Dávalos 1996). The area under control by the Sendero Luminoso extended south from the Alto Río Tambo to the community of Poyeni, to the Río Perené and the settlements of Puerto Ocopa and Puerto Prado, to both sides of the Río Satipo, and also along the main road to the settlements of Satipo, Mazamari and San Martín de Pangoa in the east. They also invaded the western and southern parts of the Cordillera de Vilcabamba, including the Bajo Río Apurimac Valley and the communities of Luisiana, San Francisco, Quimbiri, and Pichari. The Sendero Luminoso divided the indigenous people into two commands, the first composed principally of highland peoples and a few Asháninka who had the privileges of food and health care, and the second consisting of native peoples reduced to conditions of slavery. The defeat of the Sendero Luminoso in the region in the 1990s was possible due to the formation of Committees of Self-defense (Rondas Campesinas) of people from several Asháninka communities (Benavides 1991). These committees were implemented and directed by the Peruvian Army.

Another threat to the security of the indigenous cultures of the area has been the cultivation of coca, with the aim of producing and selling the base for cocaine, principally in the Apurímac Valley. Coca production has brought violence and corruption to the region, changing the traditional way of living and economy of the native people. There is a general fear that the narcotics trafficking will spread to the zone of the Río Tambo and Río Perené, which are already used as transport routes for coca to Lima.

Most recently, oil, logging and ranching have been changing the nature of the Cordillera de Vilcabamba (see above).

Scientific Exploration of the Vilcabamba region

While the remoteness and relative inaccessibility of the Cordillera de Vilcabamba have precluded much scientific exploration, naturalists, scientists and adventure seekers have been attempting to penetrate its vast forests in search of biological and cultural treasures for over a hundred years. The earliest records are of missionaries and sailors, such as Captain Chávez and Domingo Estrella, who in 1846 departed from Mainique and traveled the length of the Río Bajo Urubamba. In 1857, Fustino Maldonado navigated the Río Urubamba to the Pongo de Mainique, disembarking in Illapani and then travelling over land to Cusco. In 1868, a team from the Hidrographic Commission of the Amazon navigated parts of the Río Urubamba and Río Tambo. Carlos Fermin Fitzcarrald traveled through the Urubamba region in 1890 toward the headwaters of the Río Camisea. He reached a tributary of the Río Manu and, at the river's mouth at the Río Madre de Dios, discovered an isthmus that now holds his name.

In 1963, two explorers sponsored by the National Geographic Society and the New York Zoological Society attempted to cross the Cordillera de Vilcabamba but encountered more difficult terrain and harsher conditions than they expected (Baekland 1964). They parachuted into the highlands and had planned to be picked up by airplane 15 days later. However, the wet and steep ground conditions prevented an aircraft from landing. They hiked through the forests for 61 days until they encountered native Matsigenka communities at lower elevations. It took them 18 more days via land and river to reach the Dominican Mission at the confluence of the Río Sepahua and Río Urubamba. These explorers learned first-hand why most of the Cordillera is uninhabited - the region above 2000 m is cold and perpetually wet, contains little game, and the soil is largely unfit for agriculture. Despite their tough journey, articles published in National Geographic magazine (Baekeland 1964) described the beauty of the Cordillera de Vilcabamba and sparked international interest in conserving the area as a national park.

Other expeditions included exploration by Dr. Wolfram Drewes and Ing. José Lizárraga in 1961, who investigated areas of potential colonization and discovered two large waterfalls on the eastern flank of the Cordillera de Vilcabamba. In 1964, members of a team from the InterAmerican Development Bank (IDB) observed 12 large waterfalls in the area with the potential for hydroelectric energy production. All 12 waterfalls were on a tributary of the Río Cutivireni. The French explorer Jacques Cousteau visited the area in 1984 during his travels to the headwaters of the Río Amazonas.

Despite these explorations, very little biological data have been collected from the Cordillera de Vilcabamba. Following his discovery of the ruins of Machu Picchu in 1911, Hiram Bingham of Yale University organized two large multidisciplinary scientific expeditions to the Upper Urubamba basin around Machu Picchu in 1912 and 1915 (Eaton 1916; Thomas 1917, 1920; Chapman 1921). In the late 1960s and early 1970s, John Terborgh and John Weske surveyed birds and mammals, mainly around the community of Luisiana on the Río Apurimac, between 600 and 3520 m (Terborgh 1971, 1973, 1977; Weske 1972). The purpose of their surveys was to investigate the ecological aspects of bird community structure along an elevational gradient.

NATIVE INDIGENOUS GROUPS

The Cordillera de Vilcabamba has traditionally been occupied by a diverse array of indigenous groups, including the Asháninka, Matsigenka, Yine, and Kugapakori of Arawark origin, and the Amahuaca, Yaminahua and Nahua of Pano origin. The Kakinte, a group related to the Asháninka, live in the upper Río Poyeni river area. The Asháninka are the most numerous indigenous group of the Amazon region and inhabit the western side of the Cordillera, along the ríos Ene, Apurimac, Tambo, Picha, Alto Perene, Ucayali and Gran Pajonal. The Matsigenka (or Matsiguenga) and Yine have traditionally occupied the eastern side of the Cordillera de Vilcabamba, mainly along the Río Bajo Urubamba and the Río Manu. The Matsigenka and Asháninka are quite similar in many ways; they not only have a similar language but through ancient and frequent contact with one another have developed strong ties (Chirif and Mora 1977).

The native population of the Zona Reservada Apurímac as of 2000 is presented in Table 2. Both the Asháninka and Matsigenka have a high proportion of young people, indicating that populations will likely increase. Tables 3 and 4 present information on the population size, total area inhabited, educational resources, and health services of the indigenous communities of the Cordillera de Vilcabamba.

Patterns of settlement by Native Groups

Both the Ashaninka and Matsiguenga are traditionally semi-nomadic, typically remaining in one area to practice agriculture for a period of three to five years, and hunting and fishing in the surrounding areas (Snell 1964). Traditionally, groups lived in scattered single family groups or small extended family hamlets (Posey and Balée 1989, Boster 1984). Today, they still practice a combination of subsistence horticulture, hunting, fishing and gathering.

Social organization of the Asháninka has changed due to both the needs to interact with the larger society as a group and to defend themselves from terrorism. Every independent community has a leader called "pinkathari" and several other authorities who operate as a council for common affairs but who have little power over private matters (Rojas Zelezzi 1994). Some communities also have a traditional healer or shaman called "seripigari" who acts as medicine man and as a link between the natural spirits and humans. In the last decade, the need for protection has promoted the formation of defense groups called "ronderos" or local policemen who have had the most active role in the fight against terrorists (Rojas Zelezzi 1994). The Asháninka communities along the Río Ene and Río Tambo are members of second degree organizations: CARE (Central Asháninka del Río Ene and CART (Central Asháninka del Río Tambo).

Social and spatial organization has also been changing among the Matsigenka since the 1940s when Dominican Catholic missions were established in Timpía and Kirigueti along the Río Urubamba. The features of these missions that have been attractive to the Matsigenka include education and health services, which today are run under agreements with government agencies. The missions are a focus of cultural change through radio and solar powered TV. They also serve as centers of limited economic innovation through introduction of new economic activities (cattle raising) and as a market for a series of products (timber, fish, fruits, etc.), although on a very small scale (Rosengren 1987, Baksh 1995).

Table 2. Native Indigenous Populations in the Zona Reservada Apurímac (ZRA) of the Cordillera de Vilcabamba.

	ASHANINKA	MATSIGENKA	YINE	KAKINTE (Campa Caquinte)
In Peru (% of total)	52,461 (21.9%)	8,679 (3.6%)		
Number of communities	306	32		
% under 15 years old	48%	49.6%		
In the ZRA (% of Peru)	53.4%	42.7%	3.9%	
# Communities				
Adjacent to the CRW	21	0	0	0
Along the Río Tambo	8	0	0	0
Along the Río Ene	9	0	0	0
Along the Río Apurímac	3	0	0	0
Adjacent to the CRE	3	9	1	3
Along the Río Urubamba	1	12	1	2

CRW = Communal Reserve proposed on the Western Side of the Cordillera
CRE = Communal Reserve proposed on the Eastern Side of the Cordillera
Source: Mora et al. 1997; updated by CI-Perú (1998) and ACPC (2000)

Table 3. Indigenous Communities Adjacent to the Proposed Communal Reserve on the Eastern Side of the Cordillera de Vilcabamba

COMMUNITY	ETHNIC GROUP	LOCATION	ESTIMATED POPULATION	TOTAL AREA (ha)	EDUCATION (# Students)
Puerto Rico	Asháninka	Río Bajo Urubamba	No info.	16,655.00[3]	No info.
Miaría	Yine	Río Bajo Urubamba	No info.	10,380.31[1]	No info.
Porotobango Porotobango-ampliación	Matsigenka Campa Caquinte	Río Uitiricaya	75[4]	2,400.00[1] 16,836.91[1]	Nursery[2]
Kitepampani	Campa Caquinte	Confluence of Río Ayeni with Río Mipaya	85 (census 1993)	12,808.60[1]	Primary (20 students)[2]
Taini	Campa Caquinte	Río Ayeni	79[2]	6,371.03[3]	Primary (20 students)[2]
Tangoshiari	Asháninka	Río Pagoreni	305[2]	21,220.76[3]	Primary (98 students)[2]
Kochiri/ Campo Verde Kochiri- ampliación	Matsigenka Asháninka	Río Pagoreni	91 (census 1993) 120[4]	13,414.90[1] 14,387.56[1]	Kochiri: Nursery, Primary (32 students)[2] Campo Verde: Nursery, Primary (30 students)[2]
Mayapo	Matsigenka	Río Picha	243 (census 1993) 252[4]	21,200.65[1]	Nursery (21 students) Primary (68 students) Secondary (56 students)[2]
Camaná	Matsigenka	Río Picha	263(census 1993)	26,042.42[1]	Nursery (28 students) Primary (74 students)[2]
Timpía Timpía –ampliación	Matsigenka	Río Bajo Urubamba	379 (census 1993) 516 (local census 1998)[5]	32,893.21[1] 1,124.16[1]	Nursery (42 students) Primary (131 students)[2]
Poyentimari Poyentimari - ampliacion	Matsigenka	Río Alto Urubamba	280[3]	8,245.78[1] 7,619.92[3]	40[3]
Chakopishiato	Matsigenka	Río Alto Urubamba	46[3]	148.53[1]	06[3]
Tipeshiari	Matsigenka	Río Alto Urubamba	51[3]	65,775.60[3]	24[3]
Monte Carmelo Monte Carmelo - ampliación	Matsigenka	Río Alto Urubamba	459[3]	8,527.00[1] 7,683.80[3]	119[3]

1 Mora et al. 1997. Amazonía Peruana, comunidades indígenas, conocimientos y tierras tituladas. Atlas y base de datos.
2 Data provided by CHEVRON, obtained in the field in 1996.
3 Data provided by CEDIA, obtained in the field in 2000, as part of the *Participatory Conservation and Sustainable Development Program with Indigenous Communities in Vilcabamba*, funded by GEF/WB and co-executed by CI, CEDIA, and ACPC.
4 Data provided by CI, obtained in the field in 1998.

Table 4. Indigenous Communities Adjacent to the Proposed Communal Reserve on the Western Side of the Cordillera de Vilcabamba

COMMUNITY	ETHNIC GROUP	LOCATION	ESTIMATED POPULATION	TOTAL AREA (ha)	EDUCATION (# Students)
Tsoroja	Asháninka	Río Tambo	No info.	25,957.61[1]	No info.
Poyeni	Asháninka	Río Tambo	858 (Census 1993) 888[4]	10,712.17[1]	481[4]
Cheni	Asháninka	Río Tambo	256 (Census 1993) 272[4]	12,010.98[1]	128[4]
Anapati	Asháninka	Río Tambo	95 (19 families)[2] 164[4]	8,988.98[1]	91[4]
Oviri	Asháninka	Río Tambo	281[4]	7,298.09[1]	160[4]
Otica	Asháninka	Río Tambo	40 (Census 1993) 238[4]	18,630.57[1]	94[4]
Coriteni - Tarso	Asháninka	Río Tambo	82[4]	8,079.33[1]	27[4]
Shimavenzo	Asháninka	Río Tambo	No info.	22,342.50[1]	No info.
Samaniato	Asháninka	Río Ene	No info.	5,128.21[1]	No info.
Caperucía	Asháninka	Río Ene	No info.	9,305.28[4]	No info
Meteni	Asháninka	Río Ene	No info.	14,924.90[1]	No info.
Kiteni	Asháninka	Río Ene	100 (20 families)[4]	15,219.44[1]	Primary[3]
Cutivireni	Asháninka	Río Ene	197 (Census 1993) 1,350 (Local census)	33,340.26[1]	Nursery (16 students) Primary (119 students) Secondary (34 students)[3]
Camantavishi	Asháninka	Río Ene	138 (Census 1993) 220[5]	11,269.33[2]	Primary (82 students)[3]
Kempiri	Asháninka	Río Ene	83 (Census 1993)	19,202.63[4]	No info.
Kimaropitari	Asháninka	Río Ene	No info.	2,853.50[1]	No info.
Shirotiari	Asháninka	Río Ene	No info.	No info	No info.
Shinongari	Asháninka	Río Apurimac	No info.	17,819.72[4]	No info.
Otari	Asháninka	Río Apurimac	No info.	364.86[4]	No info.

1 Mora et al. 1997. Amazonía Peruana, comunidades indígenas, conocimientos y tierras tituladas. Atlas y Base de Datos.
2 GIS (Geographic Information System).
3 Data provided by ACPC, obtained in the field, in 1998.
4 Data provided by ACPC, obtained in the field, in 2000, as part of the *Participatory Conservation and Sustainable Development Program with Indigenous Communities in Vilcabamba*, funded by GEF/WB and co-executed by CI, CEDIA, and ACPC.
5 Data provided by CI, obtained in the field, in 1998.

The Matsiguengas in the Urubamba region have strong local democratic organizations headed by a leader, "itingami," usually a young male who is more cognizant of western ways. Their communities are also part of native federations such as CECONAMA, founded 23 years ago, which includes as members the majority of communities in Lower Urubamba, and COMARU, a recent federation (7 years old) which originated in the Upper Urubamba and is supported by the NGO, CEDIA. A certain degree of antagonism exists between these two federations, but the Matsigenka recognize both as legitimate.

Recently, oil and gas exploration to the east of the Cordillera de Vilcabamba, in the Camisea region, has created a limited labor market for a few natives as non-skilled workers. However, this has created high expectations more than real opportunities among migrants due to the "offshore" policies of the oil companies (Mobil , Shell, and Chevron). More significant is perhaps the effects on social and infrastructure investment made possible by the compensation models that oil firms must negotiate with native communities affected by their activities (Chevron 1996, Shell 1996). Most of the compensation has been focused on sanitation, school and health posts, repairs, and support in the form of medicines and some health equipment (e.g. solar powered refrigerators).

In the past, the migration of the Asháninka and the Matsigenka was fairly infrequent and was due more to cultural tradition than to the reduction of resources. Native groups had no more than 20 or 30 members. Local oral history recounts that the Matsigenka used to inhabit a greater area within the Cordillera de Vilcabamba including the headwater regions and make frequent travels to the mountains as far as the Río Tambo and Río Mantaro (Rosengren 1987).

Immigration into the Region

Even though the montane forests of the Cordillera de Vilcabamba have not been under as strong pressure from expanding human populations as have the coastal or highland regions of Peru, human populations of the mountain regions are increasing. Colonization is variable between regions, for example, five colonist settlements are located along the eastern side of the Río Bajo Urubamba. Over the last few years, the population of these settlements has only slowly increased due to the remoteness of the settlements and lack of infrastructure. In contrast, there has been a great influx of colonists along the Río Tambo, near the confluence with the Río Urubamba, as a result of a colonization program known as "Earthly Paradise" sponsored by the Agrarian Coffee Satipo Cooperative Ltd. Similarly, heavy colonization is occurring along the Río Bajo Urubamba by individual colonists and through the sponsorship of the Alto Urubamba Cooperative.

The area along the Río Apurímac has perhaps had the highest colonization rate by Andean campesinos from the high Andes to the west, who have invaded the territories of the native groups in order to cultivate coca. Since 1970, the Río Ene region has experienced increased pressure from colonists along the Río Perené, Río Satipo and Río Mazamari from the west and the Río Apurimac from the south. The areas that have been most affected by human settlement in Peru are the lower parts of the Río Huallaga and Río Urubamba, the high regions of Chachapoyas province, and the higher streches along the Río Tambo. In terms of demographics, compared to the overall population of Peru, which increases at an annual rate of 1.7%, the human populations in montane forests are growing quite slowly. However, three of six mountainous provinces (the eastern portion of Huallaga, Tambo and Urubamba) have exponential human population growth rates (Young 1992). This increase in human populations may be a result of development programs implemented by the Peruvian government during the 1960s in these three regions.

Resource Use by Native Groups

Traditional Matsigenka agriculture appears to have a minimal impact on the forest: small plots of land are cleared and planted with several crops (Renard-Casewitz, 1972). After the primary production of the garden ends, the forest is allowed to regenerate and fruit trees are planted and left to mature. Eventually, these trees become part of the forest vegetation that still produce for many years after the decline in primary crops. Staple foods include manioc, maize, plantains, and bananas. This sustainable form of agriculture has been successful in providing an adequate source of nutrition for the Matsigenka for several reasons, including the gardens' relatively small area and the use and rotation of several crops.

Equally important have been the frequent migrations of the family groups in search of appropriate plots of land to establish new gardens and in search of the diverse and scattered resources of the forest. When a human population is small, the land and forest resources are generally utilized in a sustainable manner. But when the population is greater than the carrying capacity of the environment, resources are used more intensely. Therefore, after a while the land, terrestrial flora and fauna and aquatic resources diminish and human populations have to travel farther to obtain their resources. The native populations therefore need to migrate to new lands and extend the territory of their communities. While the establishment of permanent settlements of native groups has brought education and health services to them, it has also brought challenges associated with obtaining sufficient products from agriculture, fishing, and hunting (Baksh 1995, Denevan 1972, Rosengren 1987).

Agricultural practices of the native groups of the Cordillera de Vilcabamba are generally a mix of crops grown in 1/4 to 1/3 hectare plots. The crops have a high genetic diversity and a rapid regeneration time (Posey and Balée 1989). Fruit trees are also grown to provide food and attract animals that can be hunted and eaten. A wide variety of crops are grown, including manioc, plantains, pineapple, papaya, sugar cane and medicinal plants (Johnson 1983, Rojas Zelezzi 1994). Manioc is the principal crop, with more than 30 varieties grown. Commercial crops such as achiote, coffee and cocoa are now also cultivated. Native groups traditionally do not use artificial pesticides or fertilizers on their crops.

This type of agriculture has had minimal impacts on the environment, especially compared to habitat destruction caused by large scale monoculture agriculture and cattle ranching practiced by the colonists (Rosengren 1987). However, with the new more stationary lifestyle of the native groups, communities are reducing their fishing, hunting and non-timber forest product collection, which had been their predominant activities, and have turned more toward agriculture. The Asháninka and Matsigenka are facing a crisis in their agricultural practices due to the shortage of suitable lands, thereby putting at risk their native migratory lifestyle (Johnson 1983; Henrich 1997).

Traditional hunting has long been practiced by native groups primarily for personal consumption. A variety of mammals and birds of diverse sizes are hunted (Shepard 1997). The patterns of species distributions within the Cordillera de Vilcabamba must be considered when designing conservation plans for the fauna. Hunting is conducted using a variety of bows and shotguns. The native communities have difficulty obtaining guns and bullets but they are often shared among hunters.

Like hunting, fishing is an essential activity for native groups. Of 81 species of fishes known from the Río Picha and its tributaries, at least 35 species are regularly consumed (see Chapter 12). Fishing is especially important during the dry season, when water levels are low and the native fishing techniques are most effective. However, fishing is conducted all year long, which has an impact on fish populations over the long term (Shepard 1998).

Native groups also collect a wide variety of fruits, nuts, palms, mushrooms, construction materials, medicinal plants, dyes, and adornments from the forest. Palms are used frequently by native groups for construction; of the 30 palm species of the region of the Río Bajo Urubamba, 20 are heavily utilized, either for their fruits or for construction materials (see Chapter 16).

Land Use by Colonists

The colonist communities located along the Río Apurimac and Río Ene and some of the colonists along the Río Tambo and the Río Bajo Urubamba have been cultivating coca and products for self-consumption for the past ten years. Cultivation of coca leaves has become the principal agricultural product for the colonists, which introduces a significant change in land use patterns. Coca may be the most environmentally damaging crop grown in Peru, due to the cultivation practices used and habitat types selected for its cultivation (Dourojeanni 1988).

GEOLOGICAL/ ECOLOGICAL OVERVIEW

Geology

The Cordillera de Vilcabamba is a 250-km mountain range with an area of about 3 million ha in south-eastern Peru (Maps 1 and 2). This lone massif is separated from the eastern and western foothills of the Andes mountains by steep canyons and rivers. The Río Tambo-Ene system on the northwestern side and the Río Apurímac on the west have carved steep canyons. The Río Urubamba defines the northeast limits of the range. The rivers of the region are notorious for their dangerous stretches of white water rapids.

The Cordillera de Vilcabamba spans an altitudinal range from 400 to over 4,300 meters above sea level. Much of the Cordillera de Vilcabamba is composed of granite and amphiboles, as well as metamorphic and calcareous rock dating from the Paleozoic (570 million years ago). Soils are poorly developed, and tend to be shallow and acidic. These characteristics, combined with a rough topography and high rates of erosion make the soil poor for agriculture.
The Cordillera de Vilcabamba contains unique tropical karstic formations that developed during the Pleistocene. Karst topography results from calcareous, limestone bedrock. For this reason, calcium, an element that is very rare in the Amazon, is found here in natural form. The only other place in the tropics where this is found is on the Island of Borneo. The alkaline nature of these karstic formations not only create a unique environment for plants and animals but also confer an alkaline quality to many of the region's rivers. These sources of calcium balance the soil's acidity and increase the pH of waters that eventually flow into the Río Ucuyali and the Río Amazonas.

The eastern side of the Cordillera is steeper and more heterogeneous than the western side, which slopes gradually toward the Río Ene, with some large canyons carved out. Altitudinal transects on the eastern side of the mountain reveal bands of rock of Cretaceous, Devonian, Carboniferous, and Permian origin while on the western side, only bands of Permian are revealed except in the deep canyons or along the walls of the Río Ene.

The Cordillera de Vilcabamba contains some of the world's most spectacular natural formations. In the headwaters regions of the Río Cutivireni alone there are at least 12 spectacular waterfalls, several of which have brought international attention to the area (Baekland 1964). The world's largest natural bridge, the Pavirontsi, is located over the Río Cutivireni. The Río Cutivireni has formed a tunnel more than 67 meters high, with a mouth 63 meters wide and a length of more than 220 meters. Two narrow canyons within the region, the Pongo de Pakitza and the

Pongo de Mainique, are not only geologically unique but also culturally important since they are considered sacred by native indigenous groups.

The Cordillera de Vilcabamba contains three general landscape zones: alluvial, hilly, and montane. Table 5 shows the area covered by each type of landscape within the Zona Reservada Apurímac of the Cordillera de Vilcabamba.

Table 5. Landscape Types in the Zona Reservada Apurímac (ZRA) of the Cordillera de Vilcabamba (ACPC, CEDIA, FPCN, and SPDA 1993).

Landscape Type	Area Ha	covered%
Agriculture	18,195	1.09
Alluvial	174,108	10.0
Hilly	1,202,897	72.0
Montane	274,099	16.0
Total	1,699,300	100%

Alluvial (Paisaje Aluvial): Formed by ancient sediments and deposits from rivers and streams that cross the area. Slope ranges from 5-10%.

Hilly (Paisaje Colinoso): undulating relief and uneven microtopography, with maximum elevations that range from 500-800 meters above the Río Urubamba. These are formed from clay, sand, and decomposed arsenic, which are all heavily eroded by rainfall. This landscape is the most common, and is characterized by small chains that run parallel to the highest mountains of the Cordillera and descend gradually toward the east in a series of steppes.

Montane (Paisaje Montañoso): from strong undulating hills to steep slopes over 70%. In some areas sheer inaccesible rockfaces line canyons and deep rivers, narrow valleys and rocky outcrops with slopes greater than 90%. The highest point in the Cordillera de Vilcabamba is over 4300 meters above sea level and the lowest point around 400 m.

Hydrology

The Zona Reservada Apurímac (ZRA)of the Cordillera de Vilcabamba is delimited by the sub-basins of the ríos Urubamba, Apurimac, Ene, and Tambo (Map 2), which all pertain to the watershed of the Río Ucuyali, whose waters eventually flow north to join the Río Amazonas.

The Río Urubamba originates from the Río Vilcanota in the Department of Cusco and, with a length of 862 km, is the sixth longest river in Peru. As it flows north past the Pongo de Mainique, the Río Urubamba becomes known as the Río Bajo Urubamba (Map 2). The Río Alto Urubamba is fast and wide, with a steep incline and many rocky outcrops that render it un-navigable. The Río Bajo Urubamba is navigable for its entire extension. The Río Bajo Urubamba has at least eleven major tributaries to the west, which originate in the Cordillera de Vilcabamba. The principle tributaries include the ríos Saringabeni, Picha, Pagoreni, Mipaya, Huitiricaya, Sensa, and Miaría (Map 2). Characteristics of the most important tributaries are given in Table 6.

Table 6. Characteristics of the principal tributaries of the Río Bajo Urubamba (western side) in the Zona Reservada Apurímac (ZRA) of the Cordillera de Vilcabamba.

River (length in km)	Water flow (m^3/second)	Fish and other aquatic species used by local communities	Characteristics
Río Saringabeni (12 km)	Low 2-3 High 15	Limited due to the low water flow and steep incline. High fishing pressure. Fish species include *carachamas, mojarras, boquichico,* and *sábalo.*	High water velocity, Many rocks and pebbles
Río Picha (122 km)	Low 65 High 100	High abundance of fishes, molluscs, and crustaceans. Presence of *mijanos.*	High water flow and velocity all year, Navigable by canoe, Utilized by local peoples from July to September. 60% of the rivers and streams from the western slopes of the Cordillera flow into this river.
Río Mipaya and Río el Tzoyeni (56 km each)	Low 18 High 45	Fishing for *boquichico, sábalo, sababanti, carachama,* shrimp and molluscs. Kakinte peoples migrate in the northern parts of the watershed.	Medium size, navigable in small boat, except between July and September. These are sacred rivers and are the origin of the god "Tasorinchi"
Río Huirinticaya (30 km)	Low 3 High 20	Rich in small fish species such as *sábalo, boquichico, sababamti, carachama,* as well as larvae, shrimp and snails of various sizes.	
Río Sensa (54 km) and Río Miaría (45 km)	Low 15 High 35-40	Low fish diversity; mainly small fish species.	

Source : CEDIA, 1995.

The Río Tambo originates from a union of the Río Ene and the Río Perené (Map 2). Between July and November, the waters are clear. The river is navigable to its endpoint at the Río Ucayali and is joined by at least 17 tributaries. The Río Apurímac joins with seven tributaries from the mountains of the Cordillera de Vilcabamba, including the ríos Churitiari, Quisto, Samaniato, Quimbiri, and Pichari (Map 2).

The mountain peaks of the Cordillera de Vilcabamba are the source of water for approximately 50 rivers, large and small, that travel toward the west, east and north. These rivers maintain the hydrological system of the region and are the source of water for more than 40 communities in the southern part of the Cordillera. Furthermore, the rivers have a cultural importance in that many, such as the Río Tzoyeni, are considered sacred by local indigenous groups.

Climate and Life Zones

Rainfall is an important component of the Cordillera de Vilcabamba's biophysical characteristics, as it is extremely variable. In areas classified as Sub Tropical Humid Forest the average yearly rainfall is about 1200 mm, whereas in Pluvial Pre Montane Tropical Forests the average is 5600 mm/yr. Even in the dry season, May through August, the rain and mist are almost constant. Median annual temperatures vary from 6^0 to 22^0 C.

The Zona Reservada Apurímac (ZRA) of the Cordillera de Vilcabamba contains seven life zones as described by Holdridge (1967) and two transition zones. The life zones pertain to three biogeographical provinces: Tropical Yungas, Sub-tropical Yungas, and Subtropical Amazon. Tropical and Sub-tropical Yungas are recognized as biogeographical provinces with a great diversity of life zones (17 and 11 respectively; CDC 1991). However, only eight life zones and four transition zones are represented in the Protected Areas System of Peru.

Most of the Zona Reservada Apurimac is made up of Pre-montane Moist Tropical Forest (36%) and Moist Tropical Forest (21%) (Dávalos 1996). The region largely contains humid and rainy areas; only a small area (0.16%) is Dry Forest. More than 76% of the zone corresponds to pre-montane, montane, or lower montane forest owing to the steep mountainous terrain of the region.

Habitat Diversity

The Cordillera de Vilcabamba has largely been separated from the central Andes mountains, adjoining only by a small portion at its southern end. This isolation has created unique biogeographic characteristics of the Cordillera resulting in a unique and highly endemic flora and fauna.

On the western and eastern slopes of the Cordillera de Vilcabamba there are five principal types of vegetation: dwarf high-elevation forest, cloud forest, montane forest, lowland tropical forest, and lowland grasslands. The montane forests on the eastern slopes have a high density of bamboo patches.

The montane forest vegetation on the western slopes of the Cordillera Vilcabama to the Río Apurimac was described by Terborgh (1971). Terborgh affirmed the presence of a very tall montane forest, with some trees exceeding 35 meters in height and a meter in diameter. He also noted an abundance and high diversity of arboreal ferns. It is interesting that while the highest parts of the Cordillera de Vilcabamba are not inhabited by local indigenous groups, the Matsiguenga from the eastern slopes recognize several high-altitude habitat types including *pajonal* and Sphagnum grasslands (see Chapter 16). The Matsigenkas distinguish a great number of habitat types (Appendix 32).

The lowland forests contain trees up to 60 meters in height with an high abundance of epiphytic plants (e.g. bromeliads, ferns). Forests on the eastern side of the Río Urubamba (east) contain *Guadua*, a native bamboo species that exists in large extensions known as *pacales*. The lowland forests of the Bajo Urubamba can be divided into three principal types: *pacal*, semi-*pacal*, and pure forest, depending on the density of bamboo. The pure *pacal* can be seen in satellite photos as yellow patches amidst the green pure forest.

Due to the difficulty of separating *pacal* and semi-*pacal*, the two were lumped (Houghton and Hackler 1996) during a study of the biomass of the forests of the Río Bajo Urubamba.

Table 7. Biomass of the forests in and around the Bajo Urubamba region (Houghton and Hackler 1996).

	Bajo Urubamba Region			Nearby Forest	
	Biomass (tons/ha)	**Total Area (%)**	**Total Biomass in millons of metric tons (%)**	**Total Area ha (%)**	**Total Biomass in millons of metric tons (%)**
Pure Forest	124	869,000 (65%)	108 (85%)	240,000 (56%)	30 (81%)
Pacales (forest dominated by bamboo)	51	353,000 (27%)	18 (15%)	138,000 (32%)	7 (9%)
Open areas or areas not classified		111,000 (8%)		49,000 (11%)	
TOTAL		**1,333,000**	**126**	**427,000**	**37**

The forests dominated by *pacales* contain a lower biomass per hectare (51 tons/ha) compared to pure forests (124 tons/ha). The total tree biomass of pure forest is five times greater than that of *pacales* (108 vs. 18 million metric tons; Table 7).

LITERATURE CITED

ACPC, CEDIA, FPCN, and SPDA. 1993. Informe Técnico: Propuesta para el establecimiento de Areas Naturales Protegidas en la Zona Reservada Apurímac. Document presented to INRENA.

Alvarez Lobo, R.. 1998. Sepahua (Vols. I – V). Lima: Misioneros Dominicos.

Baekeland, G. B. 1964. By parachute into Peru's lost world. National Geographic 126: 268-296.

Baksh, M. S. 1995 Change in resource use patterns in a Machiguenga community. *In:* Sponsel, L. (Ed.) Indigenous Peoples and the Future of Amazonia: An Ecological Anthropology of an Endangered World. Tucson: University of Arizona Press.

Benavides, M.. 1991. Autodefensa Ashaninka en la Selva Central. Amazonía Indígena No. 11, Vols. 17 and 18.

Biedma, M. 1981. La Conquista Franciscana del Alto Ucayali. Lima: Milla Batres.

Boster, J. S. 1984. Classification, cultivation, and selection of Aguaruna cultivars of *Manihot esculenta* (Euphorbiaceae). Advances in Economic Botany 1: 34-47.

Camino, A. 1977. Trueque, correrías e Intercambio entre los Quechas Andinos y los Piros y Machiguenga de la montaña peruana. Amazonia Peruana 1(2): 123-140.

CDC. 1991. Informe del Plan Director – Sistema Nacional de Unidades de Conservación. Lima: Universidad Nacional Agraria La Molina.

CEDIA. 1991. El área de influencia del Proyecto Gas de Camisea – Territorio Indígena.

Chapman, F. M. 1921. The distribution of bird life in the Urubamba valley of Peru. United States National Museum Bulletin 117.

Chevron. 1996. Estudio de Impacto Ambiental para la Exploración Sísmica del Lote 52. Unpublished Report. Lima.

Chirif, A. and C. Mora. 1977. Atlas de Comunidades Nativas. Lima: SINAMOS.

CI-Perú, ACPC, and CEDIA. 1999. Términos de referencia para el establecimiento de dos reservas comunales y un parque nacional en la Zona Reservada de Apurimac. Document presented to INRENA.

Dávalos T. L. 1996. Diagnóstico Socioeconómico de l a Cordillera de Vilcabamba, Bajo Urubamba. Unpublished Report. Lima: CI-Peru.

Denevan, W. M. 1972. Campa Subsistence in the Gran Pajonal, Eastern Peru. Actas y Memorias del XXXIX Congreso Internacional de Americanistas. Lima: IEP.

Dourojeani, M. 1988. Si el Arbol de la Quinua Hablara. Lima: FPCN/GEF/PNUD/UNOPS.

Eaton, G. F. 1916. The collection of osteological material from Machu Picchu. Memoirs of the Connecticut Academy of Arts and Sciences 5: 1-96 (plus 39 plates).

Ferrero, A.. 1996. Los Machiguenga. Tribu Selvática del Suroriente Peruano. Lima: Instituto de Estudios Tropicales "Río Aza."

FPCN. 1991. El Gas de Camisea, Reflexiones sobre el Impacto Ambiental de su exploitación y cómo reducirlo a mínimos aceptables. Document of Conservation #5.

Guillén, E.. 1994. La Gerra de Reconquista Incka. Vilcabamba: Epílogo Trágico del Tawantinsuyo. Lima.

Henrich, F. 1997. Market incorporation, agricultural change, and sustainability among the Machinguenga Indians of the Peruvian Amazon. Human Ecology. 25(2): 319-351.

Holdridge, L. R. 1967. Life Zone Ecology. Rev. ed. San José, Tropical Science Center.

Johnson, A. 1983. Machiguenga gardens. *In* Hames, R., and W. Vickers (eds) Adaptive responses of Native Amazonians, pp. 29-63. New York: Academic Press.

Lyon, P. J. 1981. An imaginary frontier: Prehistoric highland-lowland interchange in the southern Peruvian Andes. *In*: Francis, P.D., F.J. Kense and P.G. Duke (Eds.) Networks of the Past: Regional Interaction in Archeology. Calgary: University of Calgary Archeological Association.

Mora, C., A. Zarzar, G. Huamaní, Anto. Brack, and E. Caparo. 1997. Amazonía Peruana, comunidades indígenas, conocimientos y tierras tituladas. Atlas y base de datos. Lima: GEF-PNUD-UNOPS.

Morrell, V. 1991. Wilderness Headcount. National Geographic (February). 195 (2): 32-41.

Myers, N., R. A. Mittermeier, C. G. Mittermeier, G. A. B. da Fonseca, and J. Kent. 2000. Biodiversity hotspots for conservation priorities. Nature. 403: 853-858.

Pennano, G. 1988. La Economía del Caucho. Iquitos: CETA

Posey, D. and W. Balée (Eds). 1989. Resource Management in Amazonia: Indigenous and Folk Strategies. Advances in Economic Botany, Vol. 7. New York Botanical Gardens. New York.

Renard-Casewitz, F.M. 1972. Los Machiguengas. Boletín de la Sociedad Geográfica de Lima.

Renard-Casevitz, F.M. 1977. Fronteras de la Conquista Española en la Montaña Alta

Rivera, L. 1991. El área de influencia del Proyecto Gas de Camisea – Territorio Indígena. Cusco: CEDIA

Rosengren, D. 1987. Concepciones de Trabajo y Relaciones Sociales en el Uso de la Tierra entre los Machiguenga del Alto Urubamba. Amazonía Peruana 8(14), pp. 39-59.

Rojas Zelezzi, E.. 1994. Los Ashaninka, Un pueblo tras el bosque. Fondo Editorial de la Pontificia Universidad Católica del Perú. Lima, Perú.

Savoy, G. 1970. Antisuyo. New York: Simon and Schuster.

Shell. 1996. Campaña de Perforación Exploratoria Camisea. EIA. Lima.

Shepard, G.H, Jr. 1997. History and ecology of the eastern slopes of the Vilcabamba Cordillera. Unpublished report to Conservation International (Peru). To be published elsewhere.

Shepard, G. H. Jr. 1998. Psychoactive plants and ethnopsychiatric medicines of the Matsigenka. Journal of Psychoactive Drugs 30(4): 321-332.

Snell, W. W. 1964. Kinship relations in Machiguenga. M.A. The Hartford Seminary Foundation.

Terborgh, J. 1971. Distribution on environmental gradients: theory and a preliminary interpretation of distributional patterns in the avifauna of the Cordillera Vilcabamba, Peru. Ecology 52: 23-40.

Terborgh, J. 1973. Biological Exploration of the Northern Cordillera Vilcabamba, Peru. National Geographic Society Research Reports. 1966 Projects: 255-264.

Terborgh, J. 1977. Bird species diversity on an Andean elevational gradient. Ecology 58: 1007-1019.

Thomas, O. 1917. Preliminary diagnoses of new mammals obtained by the Yale-National Geographic Society Peruvian expedition. Smithsonian Miscellaneous Collections 68(4): 1-3.

Thomas, O. 1920. Report on the Mammalia collected by Mr. Edmund Heller during the Peruvian Expedition of 1915 under the auspices of Yale University and the National Geographic Society. Proceedings of the United States National Museum 58: 217-249.

Weske, J. S. 1972. The distribution of the avifauna in the Apurimac Valley of Peru with respect to environmental gradients, habitats, and related species. Unpublished Ph. D. thesis. Norman, Oklahoma: University of Oklahoma.

Young, K. R. 1992. Biogeography of the montane forest zone of the Eastern slopes of Peru. Pages 119-140 in: Biogeografia, Ecología y Conservación del Bosque Montano en el Perú. Memorias del Museo de Historia Natural 21. (K. R. Young and N. Valencia, eds.). Universidad Nacional Mayor de San Marcos, Lima.

Young, K. R., and N. Valencia. 1992. Introduccíon: los bosques montanos del Perú. Pages 5-9 in: Biogeografia, Ecología y Conservación del Bosque Montano en el Perú. Memorias del Museo de Historia Natural 21. (K. R. Young and N. Valencia, eds.) Universidad Nacional Mayor de San Marcos, Lima.

Zarzar, A. 1989. Identidad y Etnicidad en Sepahua: un Pueblo de Frontera en la Amazonía Peruana. Revista Peruana de Ciencias Sociales 1(2): 83-152. Lima.

MAP 1. Location of the Cordillera de Vilcabamba in south-eastern Peru, South America.

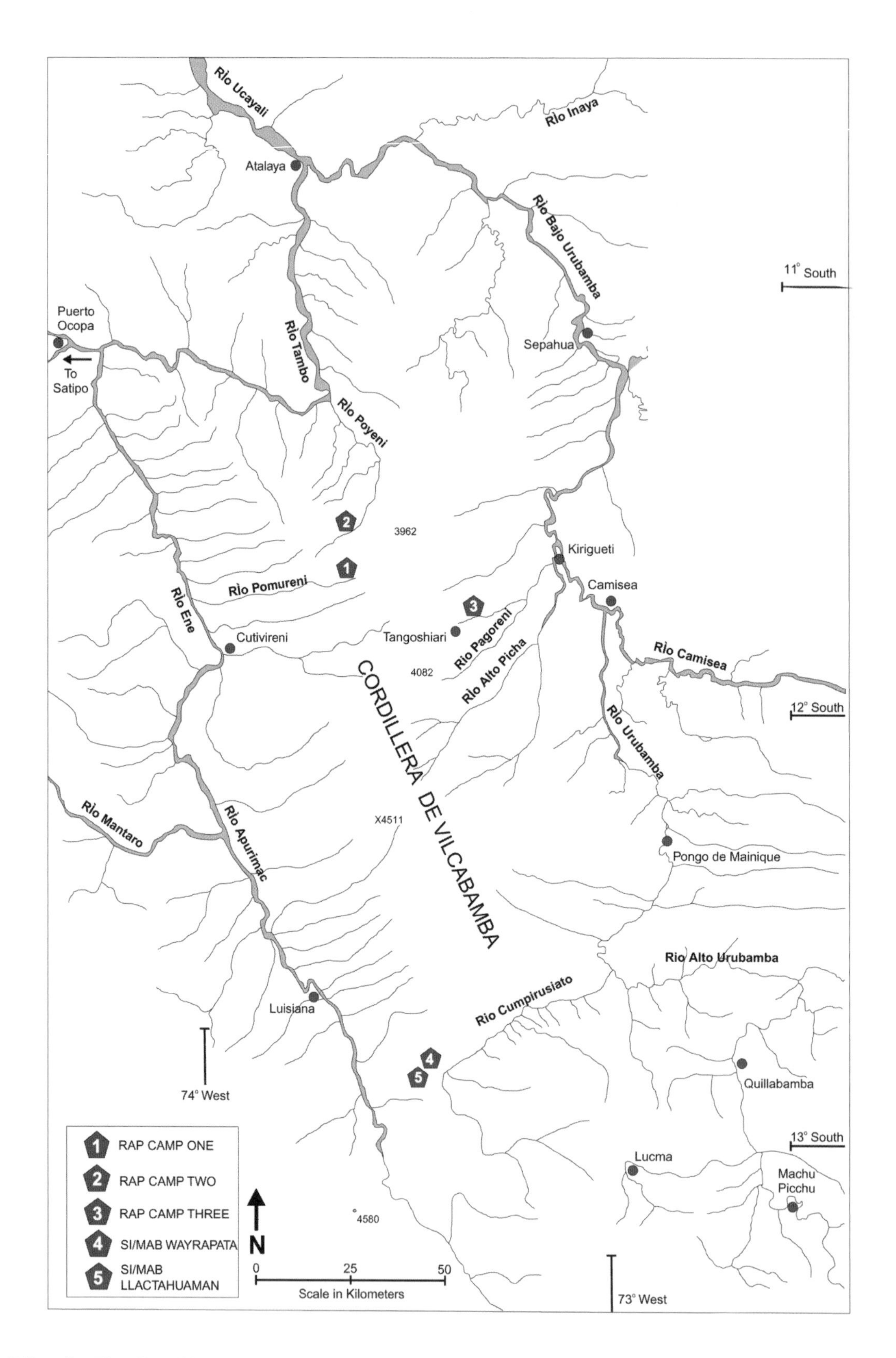

Map 2. The Cordillera de Vilcabamba, Peru showing the approximate locations of the field sites visited during the 1997 and 1998 CI-RAP expeditions and the 1998 SI/MAB expedition.

T. Schulenberg

Pajonal - *Polylepis* forest mosaic at RAP Camp One (3350 m).

T. Schulenberg

Tall ridge to the west of RAP Camp Two; helicopters landed in the bog in the foreground (2050 m).

B. Holst

Stream forest at RAP Camp Three (1000 m).

B. Holst

Bamboo forest (pacal) of *Guadua* sp. (Poaceae) near RAP Camp Three (1000 m).

T. Schulenberg

Spectacular rocky outcrops of the Northern Cordillera de Vilcabamba.

L. Emmons

Cuscomys ashaninka, a new genus and species of rodent discovered at RAP Camp One (3350 m).

A. Chicchón

Women from the Matsigenka community of Mayapo.

SI/MAB Expedition

All photos taken by F. Dallmeier

F. Dallmeier

An aerial view of the region of research in the Llactahuaman Forest – the clearing for the helipad is evident in the landscape.

F. Dallmeier

The helipad cleared in the Llactahuaman Forest serves as a more sustainable alternative to constructing roads through the forest.

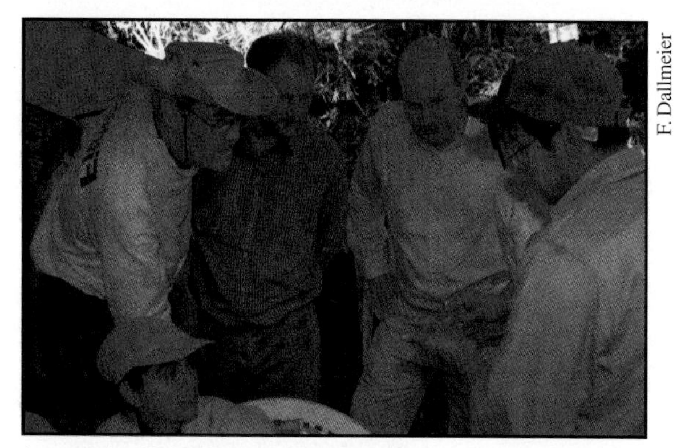

F. Dallmeier

An example of scientific and industry collaboration: Jose Santisteban, an entomologist, shows members of the Smithsonian and SPDP team a newly identified species from the forest.

F. Dallmeier

Biodiversity teams use a path cut by hand during the pipeline proposed designation. In the foreground a yellow pan trap has been placed by the arthropod research team to sample flying insects.

H. Plenge

Ornithologist Tatiana Pequeño weighs a bird caught in the mist net taken out by other members of the team.

F. Dallmeier

Tatiana Pequeño and Edwin Salazar display a beautiful Band Winged Nightjar - an insectivorous species in the Llactahuaman Rainforest – before the animal is set free.

F. Dallmeier

A malaise trap - set up to trap flying insects – serves as a primary sampling tool for insect assessments.

CHAPTER 1

NOTES FROM AN OVERFLIGHT OF THE NORTHERN CORDILLERA DE VILCABAMBA, PERU

Robin B. Foster

The overflight took place on 2 May 1997, in the early afternoon. The flying time was approximately 2.5 hours. I was accompanied by Thomas Schulenberg and Mónica Romo, Conservation International, and the pilot was Enrique Tante of Alas de Esperanza.

Route

From the Sepahua airstrip on the Río Urubamba, in a south-southwestern direction crossing the Río Sepahua and Río Miaría to just beyond the Río Censa, then west, recrossing the alto Censa and following the alto Río Poyeni into the mountains. Above 2000 m, visibility was obstructed by clouds but with intermittent openings and views of the row of highest ridges below the crest of the mountain range (Reference 1: approximately 11°29'69"S, 73°39'52"W; 2500 m?). After circling, then south along the flanks of these ridges, circling over a broad secondary ridge with suitable landing site (Reference 2: 11°33'25"S, 73°38'9"W; 2000 m?). Then southwest to the crestline, with good visibility of the entire western flank of the Cordillera de Vilcabamba. Then circling northwest over an isolated sloping *pajonal* just below the crest on the western slope (Reference 3: 11°39'54"S, 73°39'38"W; 3200 m). Then southeast following the western edge of the crestline to a large flat area of *pajonal* (Reference 4: approx. 12°04'40"S, 73°24'74"W; 3000 m?). Then south around a high clouded ridge with steep pajonal into the valley of the alto Río Picha. Then southwest and west up to the highest jagged peaks of the Vilcabamba and headwaters of the Río Picha over to the drainage into the Río Ene, circling over a suitable landing site (Reference 5: 12°19"55"S, 73°35'45"W; 4000 m?). Then turning back to follow the length of the Picha valley northeast to the mouth of the Río Picha, then over the low plateau to the Río Parotori. Then north across the low ridges on the eastern side of the Río Parotori. Then northwest across the union of the Río Parotori and the Río Picha to the valley of the Río Pangoreni and over the settlement of Tangoshiari. Then northeast to Kiringueti for refueling and north-northeast back to Sepahua.

The Mountains

The Cordillera de Vilcabamba appears to be mainly a big block lifted up on its eastern side. The central west side is for the most part relatively flat and gradually sloping from the crest down to near the Río Ene, except for the deep, steep-walled canyons reminiscent of the Grand Canyon in North America but with a sloping surface plain. The angle of the western slope is much steeper in the northern and southern ends of the Cordillera. The eastern side is mostly much steeper and more heterogeneous than the western side. Based on a limited geological map of the area, one can see that rising from the low Tertiary hills and modern floodplains of the Urubamba Valley, an altitudinal transect passes through bands of Cretaceous, Devonian, Carboniferous, and Permian rock. On the western side, the Permian band slopes from the crest almost down to the Río Ene, only exposing the other bands on the walls of the steep canyons. One of these lower rock bands is significantly more resistant than the others (probably because of sandstone), and is apparently the cause of the distinctive and spectacular waterfalls over the lip of a curving wall which we saw both in the upper levels of the Poyeni and Picha valleys.

The Mountain Vegetation

The central part of the Cordillera de Vilcabamba that we visited seems highly likely to have its greatest vegetation (and species) diversity on the eastern slope because of the different geological exposures. This heterogeneity was visually apparent to us on the eastern side in comparison to the seeming uniformity of vegetation on much of the western side. The region of the highest peaks in the southern end of the range, however, seems highly heterogeneous, although the altitude is such to restrict development of much woody vegetation. Although we were mostly too high to distinguish tree species, especially on the lower slopes, the conspicuous white-leaved *Cecropia* trees were plainly visible on both slopes at mid- to high-elevation, and were clearly restricted to certain geological bands.

The vegetation along the flat crestline of the central Vilcabamba and on many secondary ridges is mostly elfin cloud forest (mostly 2-7 m tall) or shrubland (*Polylepis*) with relatively few woody species. A few "islands" of dwarf forest within this complex sometimes appeared much more diverse (based on canopy form and color) than the rest, perhaps because of an outcrop of different rock. "*Pajonales*," herbaceous meadows, are common along the crest, and occasional on the secondary ridges. These probably are caused and maintained by fire started by lightning during unusually dry years, but the role of humans cannot be dismissed, especially given the relative ease of walking around the flattish central crestline area, more so once it is burned. There also is an existing trail over the cordillera in this region and this may have been an important communication route between the two large river systems (Río Urubamba and ríos Ene-Apurimac), although the lowest divide is further to the southeast.

The vegetation of the higher southern region of the Cordillera de Vilcabamba is noticeably different in appearance from that to the north. The elfin forest seems taller and more diverse, and the herbaceous vegetation a different color, presumably reflecting different dominant species. The ridge slopes seem more prone to landslide, and species of small bamboo are more in evidence, sometimes dominating the steepest slopes, sometimes dominating the understory (different species). The origin and maintenance of these extensive bamboo (*Guadua*) stands is still a mystery. The non-bamboo areas are mostly on the highest, steepest hills such as the bands on either side of the Río Censa, but there are plenty of exceptions such as the low hills east of the Río Parotori, and there are numerous high ridges with bamboo. *Guadua* seems to prefer sandier, well-drained sediments and needs significant amounts of light to get established. Major flood erosion/deposits and human settlement on the floodplains would explain the dense riverside stands. My own hypotheses for the rest are that the area is largely made of sandy Tertiary deposits, and either rare severe drought caused significant tree die-off on these areas to allow the bamboo to penetrate, or major earthquakes caused significant microlandslips and treefalls in the areas geologically most prone to earthquake vibration. I doubt that the soils were ever so fertile on these hills and human populations so large as to cause such a large uniform disturbance. Probably no single phenomenon will explain the dominance of bamboo for the region. Large patches of bamboo appear from the air to be in a state of dieback, possibly having just finished a first wave of flowering and resprouting before the second lower, stature wave of reproduction that was observed in the bamboo stands on the Río Tambopata.

CHAPTER 2

VEGETATION OF TWO SITES IN THE NORTHERN CORDILLERA DE VILCABAMBA, PERU

Brad Boyle

INTRODUCTION

The Cordillera de Vilcabamba is a large isolated mountain range extending northeastward from the main chain of the Peruvian Andes into the Amazon basin. Its steep slopes still are clothed in the species-rich montane forests that once ran the length of the eastern Andes from northern Venezuela south to Bolivia. Few botanical collections have been made in the area, in part due to its remoteness and rugged terrain. In June and July 1997, as part of the RAP expedition, we sampled vegetation from two locations at the northern end of the range. Camp One was at 3320-3400 m at the Upper Montane forest-paramo boundary, just below the summit ridge; Camp Two was at 1850-2090 m in the Lower Montane Life Zone along the Amazonian slope.

Using a combination of general collecting and rapid transect inventories (Foster et al. in prep.), we documented the presence of 53 families, 125 genera and approximately 247 species of vascular plants at Camp One (Upper Montane-paramo). An additional 5 families and 35 species of non-vascular plants also were collected. Among dicotyledon families, species richness was highest for Asteraceae (23 species), followed by Ericaceae (14 species), Melastomataceae (7 species) and Rosaceae (7 species). Among monocotyledons, the richest families were Orchidaceae (10 species), followed by Poaceae (7 species), Cyperaceae (7 species), and Bromeliaceae (5 species). Pteridophyte species totalled 22. By growth form, by far the largest number of species of vascular plants were terrestrial herbs (70 species). Species richness in the remaining habit categories were as follows: trees, 38 species; shrubs, 34 species; epiphytes (woody and herbaceous), 22 species; lianas, 12 species; hemiepiphytes, 3 species.

At Camp Two (Lower Montane), we found 75 families, 202 genera and about 428 species of vascular plants. Non-vascular plants totalled 16 species in four families. The richest dicotyledon family was Melastomataceae (with 20 species), followed by Rubiaceae (18 species), Asteraceae and Lauraceae (both with 11 species), Piperaceae (9 species) and Ericaceae (8 species). By far the richest monocotyledon family was the Orchidaceae (38 species), well ahead of the Poaceae and Bromeliaceae, which had 6 and 5 species, respectively. Ferns were conspicuous and diverse , with a total of 57 species. In terms of species, trees were the predominant growth form at this elevation, with a total of 99 species. The next richest habit category was epiphytes (65 species), followed by terrestrial herbs (50 species), shrubs (37 species), lianas (31 species) and hemiepiphytes (22 species).

At both sites, overall structure, taxonomic composition and alpha diversity (species richness within individual samples) were typical of Amazonian-slope wet montane forests within their respective life zones. However, the total number of species collected was high given the short period of time spent at each site (two weeks) and the limited area sampled (mostly within a 2.5 km radius surrounding each camp). For example, the species total for the high elevation site compared favorably with the 94 families and 450 species of flowering plants known from all of Manu National Park above 2650 m (Cano et al. 1995). This diversity at the landscape scale is in part the result of a high degree of habitat heterogeneity, and ultimately may reflect the complex geology of the underlying substrates. Although final determinations of many of the 898 collections still are pending, wetness, geographic isolation from the central Andean chain, and the scarcity of earlier botanical collections from the Cordillera de Vilcabamba indicate that the number of new records and undescribed species resulting from this expedition will likely also be high.

METHODS

Vegetation was sampled according to a modified version of the rapid transect methodology of Foster et al. (in prep.). Each transect consisted of separate subsamples designed to apportion sampling effort equally across a continuum of vascular plant size classes and growth habits (the normal growth form of an adult plant). Six subsample categories were based on size, as follows: (1) < 0.5 m high, (2) ^{3} 0.5 m high and < 1 cm dbh (diameter at breast height), (3) ^{3} 1 to < 2.5 cm dbh, (4) ^{3} 2.5 to < 10 cm dbh, (5) ^{3} 10 to < 30 cm dbh, and (6) ^{3} 30 cm dbh. A seventh category included all non-self supporting plants-lianas, vines and hemiepiphytes, as well as true epiphytes not potentially included under the other six categories (e.g., epiphytes growing above breast height). For simplicity, we refer to this category as the "Epiphytes" subsample, category 1 as the "Herbs" subsample, category 2 as the "Shrubs" subsample; however, no subsample was restricted to a particular habit. Thus, a seedling of a large tree was counted in the Herbs subsample if less than 0.5 m high. Likewise, a hemiepiphyte with aerial roots 3 cm diameter at breast height would be censused in category 4.

We recorded growth habit of each individual censused. As many species exhibited multiple growth habits, each species was assigned a category corresponding to its most common growth habit. This was the value used to calculate diversity within habit categories (Appendix 5). Habit categories were as follows: herb (terrestrial only), shrub (multi-stemmed terrestrial woody plant), tree (single-stemmed or single-stem-dominant terrestrial woody plant), lianas and vines (climbers, initiating growth at ground-level; lianas woody, vines mostly herbaceous), hemiepiphytes (climbers and other non-self-supporting plants, initiating growth on host tree and connected to ground via adventitious roots), epiphytes (woody or herbaceous, no connection to ground).

In most cases, sampling effort was equalized by counting a constant number of individuals (50) within each subsample. For the smallest size classes, individuals were difficult to distinguish (e.g., Herbs and Epiphytes subsamples, sometimes also Shrubs) and we therefore recorded species presence within subsections of the transect (Herbs subsample) or on several host trees (Epiphytes subsample). Each species occurrence in a given subsection or host tree was counted as an "individual". Within the Herbs subsample, we frequently exceeded 50 "individuals" (sometimes continuing up to 100 or more) in order to obtain a more complete representation of the taxa present. For standardized comparisons of species richness, we report species totals at 50 and 25 individuals (Appendix 5). In many subsamples, we censused less than the standard number of individuals. This was done mostly for habitats where individuals in the largest size classes were at such low density as to require disproportionate time and effort to sample (e.g., shrubs or trees in grassland).

All subsamples for a given transect were collected along a single center line. Width was allowed to vary between subsamples such that all of them included roughly the same number of individuals while traversing the same linear distance. Sampling widths for larger size classes were as follows: size category 3: 2 m wide; category 4: 2 m wide; category 5: 4 m wide; category 6: 8 m wide. Depending on the density of individuals within category 3 (1-2.5 cm dbh), this subsample was normally broken into subunits spaced several meters apart. Smaller size class subsamples (Herbs and Shrubs categories) were also broken into subunits (0.5 x 2 m, and 1 m x 2 m, respectively). Epiphyte subsamples were not taken within a fixed area; rather, we measured species frequency by recording presence (once per species per tree) on each of several host trees haphazardly selected from along the center line. Dimensions of all transects and subsamples are given in Appendices 2 and 3.

We surveyed 13 transects at Camp One and six transects at Camp Two. The overall total of 19 rapid transects consisted of 66 separate subsamples. With two exceptions, we censused the full spectrum of size classes present in each habitat. Exceptions were Transect 9 (mixed *Polylepis*-elfin forest), which did not include an Epiphytes subsample, and Transect 13 (elfin forest), where only plants > 1 cm dbh were censused (i.e., no Herbs, Shrubs or Epiphytes subsamples). At both camps, we attempted to do at least one transect in each visually distinctive vegetation type. For some vegetation categories (e.g., *pajonales* subtypes at Camp 1), we did replicate transects at different locations within the same habitat.

Calculation of species overlap

As a measure of species overlap between samples, we used the Sorensen Index, I_s (Magurran 1988). $I_s = 2S_c / (S_1 + S_2)$, where S_1 = number of species in sample 1, S_2 = number of species in sample 2, and S_c = number of species common to both samples. Note that I_s varies between 0 (no overlap) and 1 (total overlap). Throughout this report, species overlap is expressed as a percent, equal to I_s x 100.

Collections

We collected herbarium specimens of all species censused. We also did general collecting to voucher additional species, as well as to collect better material of species represented from transects only by sterile or incomplete material. Collection numbers totaled 898: 349 from Camp One and 549 from Camp Two. At least three duplicates were collected for most species. The first set was deposited at the Herbarium of the Universidad de San Marcos in Lima (USM) and the second set at the Field Museum of Natural History in Chicago (F), with additional sets to be distributed by F. A list of species collected will be placed on Conservation International's web site (www.conservation.org/RAP).

A note on vegetation classification

Throughout this report, I use the term "Upper Montane forest" in the specific sense equivalent to "Montane forest" of the Holdridge Life Zone classification system (Holdridge 1967). This is to avoid confusion with the general term "montane forest" (spelled without capitals), which refers to all non-lowland forests, regardless of Life Zone.

DESCRIPTION OF THE VEGETATION: CAMP ONE

Camp One was located at 3300 m on the western, or Río Ene, side of the Cordillera de Vilcabamba. Within the vicinity of the camp, which straddled a subsidiary, westward-trending ridge just below the main divide, a series of sandstone ridges and inclined flats ascended gradually to the main crest of the Cordillera. The slopes of most of these ridges were gentle to moderately steep, although many featured terraced cliff bands along their southern and southeastern sides. Cutting across this sandstone were occasional bands of limestone. These either were fractured into coarse angular blocks or dissected by deep trenches or fissures. Some of these fissures were wide enough to constitute small vertical-walled canyons; many became river courses at lower elevations. Other fissures, while less than a meter wide, still were several meters deep. Frequently, what appeared from a distance to be an easy stroll through grassy meadows turned out to be more like negotiating crevasses on a glacier. Shallow pools, some ephemeral, were common on relatively flat areas over sandstone. On the broad ridge immediately above camp (to the northeast), was a series of small but deep lagoons. These appeared to be water-filled limestone sink holes.

From the air, this mosaic of substrates was reflected in the patchy distribution of the three principal vegetation categories: *Pajonales* (herbaceous meadows with scattered low shrubs), *Polylepis* forest (pale green forest dominated by one to two species of *Polylepis*), and mixed species Upper Montane forest (dark green, stunted to moderately tall mixed-species). Our collections revealed considerable variation in taxonomic composition within each of these broad categories. For example, two tall forest types, almost indistinguishable from the air, shared only a small minority of species. It is likely that this variation reflects not only differences in soil nutrients but also patterns of slope and drainage characteristic of the different substrates. Below, I describe the various vegetation types within these categories.

Pajonales

Several types of mostly herbaceous meadows occurred as a mosaic with forested vegetation. These meadows, which we collectively termed "*pajonales*", tended to occur on flatter areas with poor drainage, on shallow rocky soils, and occasionally on steep slopes. Although the different types of *pajonal* were easily recognized and appeared somewhat discrete from one another, they overlapped broadly in species composition. Much of their apparent discreteness is attributable to variation in abundance of a relatively small set of visually-distinctive shrubs and tall grasses. Although distributional boundaries of these species were abrupt in places, presumably reflecting sharp differences in soil saturation or underlying substrates, at the community level species composition varied more or less continuously among *pajonales*.

A core set of species was common to almost all *pajonales*. These included the grass *Calamagrostis* sp. 1 and several low to prostrate subshrubs in the Ericaceae (*Gaultheria vacciniodes*, *Gaultheria glomerata, Pernettya prostrata*) and Gentianaceae (*Gentiana* sp. 1, *Gentianella* sp. 1). *Diplostephium* sp. 2 (Asteraceae) was a 0.5 to 1.5 m high shrub that grew as isolated individuals throughout *pajonal*. Common low herbs included *Bartzia pedicularioides* (Scrophulariaceae), *Hieracium* sp. 1 (Asteraceae) and a probably undescribed species of valerian (*Valeriana* sp. 2). These usually were set in a matrix of tiny mat-forming plants such as *Xyris subulata* var. *breviscapa*, *Xyris* sp. 2 (Xyridaceae), *Paepalanthus* aff. *pilosus* (Eriocaulaceae; probably undescribed) and an unidentified species of Apiaceae (Indet. sp. 1). Other species distributed more sparsely throughout *pajonal* included *Chaptalia* sp. 1 (Asteraceae), *Rhychospora* sp. 2 (Cyperaceae), *Geranium* sp. 1 and *Grammitis moniliformis* (Pteridophyta).

Dwarf *pajonal*

"Dwarf *pajonal*" was the name we assigned to the low mat-forming vegetation commonly occurring near the bottoms of drainages or at the margins of lagoons, as well as in flat areas of exposed sandstone (Transects 1, 2, 4, 7 and 10; Appendix 2). The tallest components of this vegetation were 10 to 30 cm high plants of the bunch-forming grass *Calamagrostis* sp. 1 and scattered subshrubs of the above-mentioned species of Ericaceae and Gentianaceae. Two tiny shrubs largely restricted to dwarf *pajonal* were *Gentianella* sp. 1 (Gentianaceae) and *Myrteola nummularia* (Myrtaceae). The herbaceous composite *Oritrophium* sp. 1 was found mostly in this association. Although not nearly as common as the ubiquitous *Diplostephium* sp. 2, *Blechnum auratum*, a low (< 1.5 m high) stout tree fern, was a conspicuous element in dwarf *pajonal*. One peculiar feature of these *pajonales* was the presence of a layer of algae (*Nostoc* sp.?) just below the surface. This algae would ooze to the surface through footprints, building up into gelatinous masses over frequently traveled areas.

Areas with saturated soils and standing water, such as at the margins of ponds (e.g., Transects 2, 10, and to a lesser extent Transect 1; see Appendix 2), had fewer subshrubs and greater representation of Cyperaceae (three species of *Carex*) and Juncaceae (*Luzula* sp. 1). We referred to this vegetation as "boggy dwarf *pajonal*". Isolated plants of *Senecio hygrophyllus* (Asteraceae) also were characteristic of wetter sites in dwarf *pajonal.*

Dwarf *pajonales* on better-drained and/or rocky areas frequently were dominated by a small (to 10 cm high) terrestrial bromeliad, *Puya* sp. 1 (Transects 4 and 7; Appendix 2D and 2G). This vegetation was typical of cliff terraces and certain low-angled ridge crests, where areas of thin soils alternated with flat bare rock and shallow ephemeral pools. Such vegetation also contained a greater representation of shrubs, as well as slightly higher overall species diversity than the other types of dwarf *pajonal.* The low shrub *Hypericum struthianum* (Hypericaceae) largely was restricted to this association.

Tall *pajonal*

Waist-high grasslands, which we termed tall *pajonales*, were found on steeper slopes as well as on flatter areas that in many cases appeared to be underlain by irregularly-fractured limestone (Transects 3, 5 and 6; see Appendix 2C, 2E-F). This vegetation was considerably taller than the dwarf *pajonales* (1 to 1.5 m high) and was dominated either by a single species of bunch-forming grass (*Calamagrostis* sp. 2) or a bambusoid grass (*Chusquea* sp. 1). These tall grasses were set in a deep, spongy matrix of moss (*Sphagnum magellanicum*). Although the *Sphagnum* appeared to exclude many small herbs, most species could be found at low density in the moss layer. Overall species richness in this vegetation thus differed little from dwarf *pajonal.*

Rosette plant association

A distinctive and highly local plant community occasionally was found on saturated soils at the bottoms of drainages. This "rosette plant formation" consisted of a tightly-interlocking mat of *Plantago tubulosa*, *Viola* sp. 1, and two Asteraceae, *Oritrophium peruvianum* and *Hypochaeris* sp. (4240). All four species have remarkably similar stiff-textured leaves arranged in a low spiral. Although isolated individuals of *Bartzia pedicularoides*, *Castilleja nubigena* and *Halenia* cf. *weddelliana* sometimes could be found growing through this mat, other species were generally excluded from this vegetation. Furthermore, the four rosette-forming species almost always occurred together (their presence in Transect 6 resulted from the transect line passing briefly from tall *pajonal* into a patch of rosette-plant formation). Of particular interest was *Viola* sp. 1. This species differed from *Viola pygmaea* (found locally throughout dwarf *pajonal*) in that its flowers were "tipped back" into the non-resupinte position: what normally would be the lip was displayed vertically, with the two remaining petals held together horizontally as (presumably) a landing platform for pollinators.

Shrubby *pajonal*

"Shrubby *pajonal*" (Transect 8; Appendix 2H) was dominated by the tall (2-3 m high) composite shrub *Loricaria lucida.* The brownish-green foliage of this shrub, and its distinctive branching pattern, all in a single plane, made it conspicuous even from the air. These dark-colored patches of *Loricaria*-dominated grasslands were what observers in the initial aerial reconnaissance originally mistook for "burned-over areas". Several other species of shrubs also common in this association included *Vaccinium floribundum* (Ericaceae), *Tibouchina* sp. 1 (Melastomataceae), *Hesperomeles weberbaueri* (Rosaceae) and occasional juvenile individuals of *Myrteola phylicoides* var. *glabrata* (Myrtaceae) and *Weinmannia fagaroides/ microphylla* (Cunoniaceae). Also of interest were several species of low shrubs and herbs rarely found in neighboring habitats. These included *Hesperomeles cuneata* (Rosaceae; a prostrate shrub), *Lycopodium clavatum* (Pteridophyta) and *Senecio argutidentatus* (Asteraceae).

Cliff vegetation in limestone gullies

A small set of species was associated with the vertical walls of limestone gullies to the west of camp. These included four species of trailing herbs: *Hydrocotile urbaniana* (Apiaceae), *Muehlenbeckia volcanica* (Polygonaceae), *Ourisia chamaedrifolia* (Scrophulariaceae) and *Galium* sp. (4448; Rubiaceae). More robust plants of two species of Asteraceae, *Senecio* sp. (4444) and *Hypochaeris* sp. (4446), grew in soil-filled depressions in the rock face. Four ferns, *Asplenium extensum*, *A. triphyllum*, *Polystichum nudicaule* and *Cyathea* sp. (4447), were collected only in this habitat. One of these (*Asplenium triphyllum*) grew in semi-darkness within the mouths of small, shallow caves at the foot of the cliff.

Elfin forest islands

Small patches of forested vegetation were scattered throughout the *pajonales.* Most of the tree species in these habitat islands of forest, for example, *Weinmannia fagaroides/microphylla* (Cunoniaceae), the tiny-leaved *Symplocos nana* (Symplocaceae) and a still-undetermined *Oreopanax* with leathery, orange-pubescent leaves (*Oreopanax* sp. 1; Araliaceae), also were found in the nearby elfin forest (Transect 13). Others, such as *Blechnum auratum*, *Diplostephium* sp. 2, and *Vaccinium floribundum*, were more characteristic of *pajonal.* The few species observed only in elfin forest islands, e.g., *Bomarea coccinea*, *B. cornuta* (Alstroemeriaceae) and *Ribes albifolium* (Grossulariaceae), were quite rare; it is possible they may have turned up in other forest habitats with further collecting. Several orchid species (*Epidendrum* sp.1, *Epidendrum*. sp. 2, *Pachyphyllum* sp. 1, *Stelis* sp. 2 and one to two species of *Lepanthes*), although also occurring in closed forest, were particularly abundant on the exposed moss-carpeted branches of trees in forest islands.

Polylepis forest

Polylepis forest almost invariably occurred over highly fractured, blocky limestone. The most extensive stands were dominated by a single species of *Polylepis* (*Polylepis* sp. 1; see Transect 11, Appendix 2K-L) with rather small, silvery-tomentose leaflets. This species forms low (4 - 5 m high) open woodlands with an understory of scattered low shrubs (mainly Ericaceae), small herbs, and the same species of tall grasses characteristic of tall *pajonal* (*Calamagrostis* sp. 2 and *Chusquea* sp. 1). Some of the commoner shrubs and herbs shared between *Polylepis* forest and *pajonal* include *Bartzia pedicularoides*, *Geranium* sp. 1, *Gaultheria vaccinioides* and *Diplostephium* sp. 2. Relatively few species were shared with the tall hilltop forest described below; *Sphyrospermum cordifolium* (Ericaceae) and the trailing herb *Hydrocotile* sp. 1 (Apiaceae) are among the only examples. In general, the commonest species of herbs and shrubs were unique to *Polylepis* forest. Examples of *Polylepis* forest herbs include *Valeriana jasminoides*, *Nertera granadensis* (Rubiaceae), *Arenaria lanuginosa* (Caryophyllaceae), *Oxalis phaeotricha* (Oxalidaceae) and *Luzula* sp. 1 (Cyperaceae) and the ferns *Elaphoglossum* sp. 3 and *Jamesonia alstonii*. Shrubs collected exclusively in *Polylepis* forest include *Senecio* sp. 1 and *Baccharis* sp. 1 (both Asteraceae), *Miconia* sp. 4 (Melastomataceae), *Arcytophyllum* sp. 1 (Rubiaceae) and *Ribes incarnatum* (Grossulariaceae). Although *Berberis saxicola* (Berberidaceae) was censused only in the *Polylepis* forest transects (9 and 11), it also was observed occasionally in tall *pajonal*.

As with elfin forest islands, all branches in *Polylepis* forest were covered thickly with mosses and lichens. However, compared with all other forest types, *Polylepis* forest was strikingly poor in both species richness and abundance of vascular epiphytes. More than half of the species, and by far the greatest number of individuals, of vascular plants found growing on *Polylepis* were ferns. Two of these appeared to be specific to this forest type: *Elaphoglossum* sp. 1 and *Grammitis variabilis*. With the possible exception of two individuals of Ericaceae (Indet. sp. 1 and *Sphyrospermum cordifolium*), all of the flowering plants observed growing epiphytically were more commonly observed as terrestrials in the same forest, or in neighboring tall *pajonal*.

A second species of *Polylepis*, with fewer, darker green and nearly glabrous leaflets (*Polylepis sericea*) also occurred patchily within stands of *Polylepis* sp. 1. Although *P. sericea* also grew primarily in monospecific stands on blocky limestone, these were not as extensive as those of *Polylepis* sp. 1. In addition, unlike the latter species, *P. sericea* occasionally was found on other substrates, sometimes growing within the edges of mixed-species forest. Transect 9 (Appendix 2I) sampled such a mixed *Polylepis* forest. Results of this census are somewhat misleading, however, as it traversed from a relatively pure stand of *Polylepis sericea* into a neighboring mixed species elfin forest (which included individuals of *P. sericea*). *Polylepis sericea* stands were the common *Polylepis* forest type on the side of the ridge occupied by camp. The more extensive *Polylepis* forests on the opposite (North) side of the ridge mostly were dominated by *Polylepis* sp. 1.

In contrast to the two species just mentioned, *Polylepis* cf. *pauta* was a common tall tree (to 25 m high) in the tall hilltop forest to the east of camp (Transect 12; see below). There, it was second in abundance only to *Weinmannia fagaroides/microphylla*. However, this species also was found nearby in rather small monospecific stands. In general it had a sporadic distribution, with isolated individuals occurring in a variety of locations (for example, in elfin forest islands).

We suspect that the three *Polylepis* species encountered may constitute some sort of successional sequence, or at least that their optima are spaced along an environmental continuum that features *pajonal* at one end and closed canopy forest at the other. This continuum, which might also represent spatial rather than temporal variation in some environmental factor, was vividly illustrated along the steep slope below the ridge where we sampled Transect 12. Here, the valley bottom was occupied by pure tall *pajonal*, punctuated with a few small trees of *Polylepis* sp. 1. At the base of the slope, *pajonal* gave way to a band, perhaps 20 m wide, of pure *Polylepis* sp. 1. This graded into a similar band of *Polylepis sericea*, which in turn yielded to nearly pure *Polylepis* cf. *pauta*. The slightly different colors of the foliage of each species of *Polylepis* made this gradation obvious even from a distance. About halfway up the slope, *Polylepis* cf. *pauta* intergraded with the tall mixed-species forest that occupied the upper portion of the hill. Although mature individuals of *Polylepis* cf. *pauta* also grew on the hilltop, where they were among the commonest large trees, the absence of juvenile individuals suggests that this species does not persist within mixed-species forest.

Elfin forest

Transect 13 (Appendix 2P) stretched from a narrow ridge-crest and hilltop just to the WNW (288°) of camp. The forest along this transect, which we referred to as "elfin forest", had the typical near-treeline forest physiognomy of inclined multiple trunks and twisted branches heavily burdened with epiphytic mosses and vascular epiphytes. Although individuals as high as 15 m were encountered

(mostly *Weinmannia fagaroides/microphylla*, *Clethra cuneata*, and *Oreopanax* sp. 1), trees rarely exceeded five to eight meters in height. This was the predominant forest vegetation on slopes and narrow ridge crests.
Higher taxonomic composition was typical of the uppermost elevational fringe of most Neotropical montane forests (Gentry 1992, 1995) with *Clethra*, *Weinmannia*, *Ilex*, *Oreopanax*, *Symplocos*, Asteraceae (in this case, *Gynoxys*) and *Mrysine* predominating. Species of Asteraceae (*Diplostephium* sp. 1, *Senecio* sp. 2) and Ericaceae (*Gaultheria buxifolia*, *Siphonandra elliptica*) were the major components of the shrub layer, along with the smaller species of *Ilex ovalis* (Aquifoliaceae). Woody epiphytes and hemiepiphytes consisted of numerous individuals of only two species of Ericaceae (*Gaultheria erecta* and Indet. sp. 1).

At the family level, species richness in elfin forest was highest for Asteraceae (seven species), with Aquifoliaceae and Ericaceae second (four species each), followed by Melastomataceae (three species). Of note was the high species diversity within some genera: four species of *Ilex*, three species of *Gynoxys* and four highly-distinctive species of *Symplocos* all were found within the same small area of forest.

Tall hilltop forest

Transect 12 (Appendix 2M-P) sampled a tall (to 25 m high) forest atop a flat ridge crest approximately 2 km to the east of camp. By far the commonest large tree species in this forest was *Weinmannia fagaroides/microphylla* (Cunoniaceae). The remainder of the canopy and subcanopy trees consisted mostly of *Polylepis* cf. *pauta* (Rosaceae), *Miconia* sp. 2 (Melastomataceae), *Gynoxys* sp. 2 and sp. 3 (Asteraceae) and *Myrsine dependens* (Myrsinaceae). Common understory trees and shrubs were *Hedyosmum* sp. 1 (Chloranthaceae), *Symplocos reflexa* (Symplocaceae), *Miconia* sp. 1 (Melastomataceae), *Centropogon peruvianus* (Campanulaceae) and a single species of rubiaceous shrub (*Psychotria/Palicourea* sp. 1). Vines and woody lianas were more abundant and diverse than in other habitats observed near Camp One. These included *Munnozia* sp. 1 (Asteraceae), *Manettia* sp. 1 (Rubiaceae), *Rubus* sp. 1 (Rosaceae), *Bomarea coccinea* (Alstroemeriaceae) and *Valeriana clematitis* (Valerianaceae). Understory ferns (e.g., *Asplenium delicatulum*, *Camplyloneurum*, *Trchomanes diaphanum*) also were distinctly more prevalent than in other sites at this elevation. Common low herbs included *Pilea diversifolia*, *Pilea* sp. 3 and sp. 2 (Urticaceae), *Hydrocotile* sp. 1 (Apiaceae) and *Peperomia* sp. 2 (Piperaceae).

Structurally, Transect 12 also differed considerably from the "typical" elfin forest near camp. In contrast to Transect 13, the forest sampled in Transect 12 had numerous tall (to 25 m) trees with relatively straight large-diameter trunks (mostly *Weinmannia* and *Polylepis* cf. *pauta*). Several individuals were encountered with dbh > 1 m. Despite the greater abundance of large diameter trees (dbh > 30 cm; 2.2 individuals per 100 m^2 in Transect 12 as opposed to only 0.8 in Transect 13) medium sized trees (10-30 cm dbh) were rare (3.6 per 100 m^2 in Transect 12 versus 11.2 per 100 m^2 in Transect 13). Another unusual feature of the forest sampled in Transect 12 was the presence of large gaps filled with *Chusquea* sp. 3.

Overall woody plant diversity was higher in elfin forest (Transect 13) than in the tall forest censused in Transect 12 (Herbs, Shrubs and Epiphytes subsamples were not censused in Transect 13, therefore comparisons between Transects 12 and 13 are limited to size classes > 1 cm dbh). Whereas Transect 12 contained a total of 23 species among the 189 individuals censused among size classes ³ 1 cm dbh, 31 species were recorded within these same subsamples in Transect 13, this despite a lower number of individuals (cf. Appendices 2N and 2P). Much of the "extra" diversity of Transect 13, however, resided in the small tree component (< 10 cm dbh). Comparing only within a single size class, species richness of trees 10 - 30 cm dbh was slightly higher in Transect 12 than in Transect 13 (11 species per 25 individuals in Transect 12 versus 9 species per 25 individuals in Transect 13; Appendix 4).

Differences in structure and diversity between the two forests were mirrored by differences in species composition: only 26% of the species in their combined woody floras (³ 1 cm dbh) were shared between the two sites (Appendix 7). Contrast this with the 53% overlap in individuals ³ 1 cm dbh between Transect 13 and the mixed *Polylepis*-elfin forest sampled in Transect 9. One major difference was the presence of *Polylepis* cf. *pauta* as one of the commonest canopy trees in Transect 12. *Polylepis* cf. *pauta* was absent in Transect 13, although *Polylepis* sp. 2 occurred there as "strays" from an adjacent monospecific stand. Also absent from Transect 12 were the three species of *Symplocos* and four species of *Ilex* present in Transect 13. The only representative of these normally ubiquitous Upper Montane forest genera in Transect 12 was a different species of *Symplocos* (*S. reflexa*). Indeed, over half (12 out of 23) of the species encountered in the > 1 cm dbh size classes were found only in this transect. Examples of other common species unique to Transect 12 include *Hedyosmum* sp. 1 (Chloranthaceae), *Psychotria/Palicourea* sp. 1 (Rubiaceae), *Freziera revoluta* (Theaceae), *Vallea stipularis* (Elaeocarpaceae), *Gynoxys* sp. 3 (Asteraceae) and *Viburnum ayavasense* (Caprifoliaceae).

Within the Herbs and Shrubs size classes, common species unique to Transect 12 included three species of fern (*Asplenium delicatulum*, *Camplyoneurum amphostenon*, and *Blechnum* sp. 1), *Bomarea coccinea* (Alstroemeriaceae), *Manettia* sp. 1 (Rubiaceae), *Pilea* sp. 3 (Urticaceae), *Centropogon peruvianus* (Campanulaceae), *Rubus* sp. 1 (Rosaceae) and a large terrestrial bromeliad, *Greigia* sp. 1.

Although we did not collect quantitative data on epiphyte abundance for Transect 13, based on general observations it appears that many species of epiphytes, in particular orchids, are restricted to one or the other of these forests. A new species of bromeliad was found here and recently described as *Greigia raporum* (Luther 1998).

In general, structure and higher taxonomic composition of the tall forest sampled in Transect 12 more closely resembled "typical" Upper Montane forest (e.g., in the middle of its elevational range) than did the elfin forest sampled in Transect 13. Examples of Upper Montane (as opposed to elfin) forest characteristics include the higher abundance of Rubiaceae and Melastomataceae, the decreased prevalence of Aquifoliaceae and Symplocaceae, the greater abundance and species richness of lianas and understory ferns, and the apparent higher diversity of epiphytic flowering plants (Gentry 1995, Boyle 1996). Nonetheless, other characteristics of Transect 12, (such as the scarcity of trees in the middle size classes, the lack of juveniles of the common large tree *Polylepis* cf. *pauta*, the lack of adults of other tree species common in the understory (e.g., *Freziera revoluta*) and the numerous *Chusquea*-filled gaps) suggest that the taxonomic composition of this forest may be in a state of flux. If this forest is undergoing a process of active succession, the nature of the disturbance initiating such succession remains uncertain. One possibility might be locally higher frequency and/or intensity of wind-throw. In Ecuador, I have observed forests with the same peculiarities (taller-than-average stature combined with large *Chusquea*-filled gaps) growing along similarly exposed ridge crests. According to local informants, these forests are subject to extensive blow downs during rare but violent wind storms.

General conclusions: Camp One

Although the three vegetation types visible from the air (*pajonales*, *Polylepis* forest and mixed-species forest) had lead us to expect considerable taxonomic heterogeneity, the great variety among, and in some cases within, these habitat classes was surprising. Some of this variation may result from heterogeneity in the underlying substrate, as suggested by the close association between blocky limestone and *Polylepis* forest. Distributions of other vegetation types, such as dwarf *pajonales*, may be more strongly determined by differences in soil saturation. Other distributional patterns are harder to explain. For example, elfin forest and tallgrass *pajonales* occurred as a mosaic on well-drained slopes and ridge crests. Although this pattern may reflect unobserved properties of the underlying substrates, the role of fire in maintaining the grassland vegetation should also be considered.

DESCRIPTION OF THE VEGETATION: CAMP 2

In contrast to the relatively gently sloping Río Ene side of the Vilcabamba, the Amazonian slope drops off precipitously from its summit in a series of spectacular waterfalls and cliffs totaling nearly 1000 m. Below these walls, the Río Poyeni begins its descent via a deep, narrow canyon. A broad plateau or bench, perched at the edge of this canyon, features an unusual series of openings in the otherwise continuous cover of forest. These openings consist of low boggy meadows, surrounded by a narrow zone of stunted forest and much more extensive bamboo thickets. Camp Two, at 2050 m, was situated in patchy forest near the edge of one of these meadows.

Geology of the area surrounding Camp Two remains mostly unknown. Highly eroded, exposed sandstone was encountered near cliff edges along the eastern perimeter of the plateau. Although soil nutrients may also be important, drainage patterns appear to govern the distribution of the principal vegetation types in this area. The plateau forms a catchment basin for the steep-sided ridge that towers above it to the west. Other low rises act as berms along the plateau's eastern and northern edges. Precipitation thus accumulates in the slightly concave center of the plateau, draining via a low point at its southern end. Soils in the middle of the plateau are waterlogged and shallow but get deeper and progressively better drained toward its outer edges, where tall forested vegetation occurs.

Pajonal (*Chusquea-Sphagnum* bog)

The open vegetation in the center of the plateau (sampled in Transects 16 and 17; see Appendices 3M-N) was strikingly similar to the *pajonales* sampled over 1000 m higher at Camp One. Most of the genera, and at least four species (the dominant grass, *Chusquea* sp. 1, *Blechnum auratum*, *Gaultheria* sp. 1 and the tiny, mat-forming monocot *Xyris subulata* var. *breviscapa*) were identical between the two elevations. This between-elevation overlap in species composition (from 5 to 18% for *pajonal* transects; see Appendix 6) was much higher than for other habitats sampled. Structurally, the main difference was the taller stature of the lower-elevation *pajonal*, where *Chusquea* sp. 1 frequently exceeded 2 m in height. Interestingly, species diversity was lower at 2000 m: 9 and 13 species per 50 individuals, respectively, for Herbs subsamples in Transects 16 and 17, versus an average of 18.7 species per 50 individuals in all *pajonal* transects at 3300 m (Appendix 4).

Common shrubs in *pajonal* included *Befaria aestuans* (Ericaceae), unusual for its large pink gamopetalous flowers, the tree fern *Blechnum auratum*, *Weinmannia crassifolia* (Cunoniaceae), *Hesperomeles heterophylla* (Rosaceae; also collected at 3300 m in *Polylepis* forest), and *Cybianthus peruvianus* (Myrsinaceae). The herb layer was composed mostly of the ferns *Elaphoglossum* sp. 6 and *Trichomanes diaphanum*, set in a matrix of *Sphagnum* moss (*S. erythrocalyx* rather than *S. magellanicum*) and *Xyris subulata* var. *breviscapa* and *Xyris* sp. 2 (Xyridaceae). A few trees, mostly *Clusia weberbaueri* (Clusiaceae), could be found growing as isolated and somewhat stunted individuals in *pajonal*.

The lower diversity of the *pajonal* vegetation at 2000 m is in striking contrast to the usual pattern of increasing species richness with decreasing elevation, and may reflect the relative rarity of this habitat at lower elevations. Whereas *pajonales* are extensive at high elevations in the Cordillera de Vilcabamba, most vegetation at 2000 m is tall cloud forest on moderate to steep slopes. Flat waterlogged terrain, along with the low, mostly herbaceous vegetation it appears to favor, is probably rare below the summit crest of the Cordillera. Thus, *pajonales* within the Lower Montane belt appear to be small, isolated habitat islands. Although small populations may in some circumstances promote speciation, the limited total area of this habitat suggests that a high rate of local extinction may be the over-riding effect of this isolation.

Vegetation of the forest-*pajonal* ecotone

Surrounding these open boggy *pajonales* was a mosaic, somewhat intergraded, of three vegetation types of medium stature: 4 to 8 m high "dwarf *Clusia* forest", dominated by *Clusia* and *Podocarpus;* dense 4 to 5 m high "*Chusquea* thickets" consisting mostly of the large bambusoid grass *Chusquea* sp. 6; and lower stature (to 4 m high) "herbaceous thickets" of *Chusquea* mixed with a rich variety of ferns and robust herbaceous flowering plants, especially orchids. Dwarf *Clusia* forest and herbaceous thickets appeared to be restricted, along with many of their constituent species, to the edges of *pajonales*. Bamboo thickets were widespread throughout the area.

Herbaceous thickets

Three species of bambusoid grass grew in herbaceous thickets. The most common was *Chusquea* sp. 4, followed by the *pajonal* dominant *Chusquea* sp. 1 (the latter predominating at the interface with *pajonal*). *Chusquea* sp. 6 (which comprised almost monospecific stands in the adjacent bamboo thickets) also occurred locally in patches. In addition to isolated trees of *Clusia weberbaueri*, several shrubs formed a sparse "emergent" layer above the head-high tangle of herbaceous plants. These included *Psychotria steinbachii* (Rubiaceae), *Desfontainea spinosa* (Loganiaceae), *Hedyosmum racemosum* (Chloranthaceae), and two species of Melastomataceae (*Miconia* sp. 16 and 18). A terrestrial Cyclanthaceae, *Sphaeradenia* sp. 1, was common throughout this habitat, as well as in the neighboring dwarf *Clusia* forest. Among the numerous vines were *Dioscorea acanthogene* (Dioscoreaceae), *Munnozia* sp. (Asteraceae; 4905), *Muehlenbeckia* sp. (Polygonaceae; 4964), *Manettia* sp. (Rubiaceae; 4985) *Passiflora pascoensis* (Passifloraceae) and *Rubus roseus* (Rosaceae). *Anthurium nigrescens* (Araceae) commonly was found trailing throughout this vegetation. Characteristic ferns of this habitat were *Blechnum binervatum*, *B. auratum* (also found in *pajonal*), *Elaphoglossum paleaceum*, *E. cuspidatum* and *Grammitis moniliformis*. The large tree fern *Sphaeropteris elongata* occasionally was found in this habitat as well as in dwarf *Clusia* forest and in *Chusquea* thicket.

Orchids collected in herbaceous thicket included *Epidendrum elongatum*, *Maxillaria aurea*, *M. augustae-victoriae*, *Odontoglossum ligulatum* and *O. wyattianum*, and several still unidentified Pleurothallidinae. Although herbaceous thickets structurally were very distinctive, many of its species were shared with dwarf *Clusia* forest (see below).

Dwarf *Clusia* forest

Around its perimeter, *pajonal* gave way to a low, tangled forest that gradually increased in stature, up to about 8 m high. This forest, which we referred to as "dwarf *Clusia* forest" (Transect 18, Appendices 3O-R) was dominated by *Clusia weberbaueri*, intermixed with *Podocarpus oleifolius* (Podocarpaceae), a single species of Lauraceae (*Nectandra* sp. 1) and the tree ferns *Sphaeropteris elongata* and *Cyathea* sp. 1. *Chusquea* was intermixed throughout this forest, progressing from *Chusquea* sp. 1 at the interface with *pajonal*, to mostly *Chusquea* sp. 4 in the forest interior, to *Chusquea* sp. 6 at the interface with *Chusquea* thicket vegetation. The small tree and shrub layer was comprised of *Prunus pleiantha* (Rosaceae), *Lycianthes acutifolia* (Solanaceae), a few still undetermined Melastomataceae (*Miconia* sp. 18, *Clidemia* sp. 1, Indet. sp. 1) and juvenile trees in the families Myrtaceae (Indet. sp. 1), Sapotaceae (Indet. sp. 1) and Euphorbiaceae (*Alchornea* sp. 2). Several Ericaceae normally occurring as woody hemiepiphytes in tall forest (e.g., *Cavendishia bracteata*, *Diogenesia octandra* and *Psammisia ulbrichiana*) grew terrestrially as gnarled shrubs intertwined with the other vegetation.

This forest had a dark interior, with relatively few non-climbing terrestrial herbs, mostly *Pilea* sp. 1 (Urticaceae), *Oxalis* sp. 1 (Oxalidaceae) and the ferns *Blechnum binervatum* and *Hymenophyllum plumieri*. By far the majority of herbaceous plants in this habitat were either vines, such as *Cissus* sp. 1 (Vitaceae), *Dioscorea acanthogene* (Dioscoreaceae) and *Manettia* sp. 2 (Rubiaceae), or epiphytes. Common herbaceous epiphytes included *Maxillaria augustae-victoriae* and *Pleurothallis* sp. 3 (Orchidaceae), a single species of bromeliad (*Tillandsia* sp. 3) and several ferns (*Elaphoglossum cuspidatum*, *Elaphoglossum* sp. 7, *Grammitis* sp. 1, *Hymenophyllum* sp. 2).

Species overlap with the taller forest transects (Transects 14 and 15, see below) was between 18 % and 24 %. Overlap with both of the *pajonal* transects was 8 % (see Appendix 6). Although many species collected only in dwarf *Clusia* forest likely would have turned up elsewhere with more extensive collecting (e.g., most orchids), other common taxa, especially among the trees and shrubs, appeared to be genuinely restricted to this vegetation type, e.g., *Nectandra* sp. 1 (Lauraceae), *Lycianthes acutifolia* (Solanaceae), and *Vaccinium* sp. 1 (Ericaceae). In addition, several tree species collected in other habitats were more abundant in dwarf *Clusia* forest (for example, *Clusia weberbaueri*, *Podocarpus oleifolius* and *Prunus pleiantha*). Based on general observation, species overlap with the adjacent thicket vegetation appears to be quite high; indeed, this latter vegetation was in some respects a subset of the non-tree component of dwarf *Clusia* forest.

Dwarf *Clusia* forest appears to be restricted to poorly drained and/or acidic soils at the edges of *Chusquea-Sphagnum* bogs. Therefore, it probably is somewhat rare, at least in the area of Camp Two. Unlike the *pajonales* themselves, diversity in dwarf *Clusia* forest was relatively high. Most of this diversity resides in the herb/liana/small shrub component: the 26 species per 50 individuals recorded in the > 0.5 m high size class was comparable to values obtained in adjacent tall forest (20 and 28 species per 50 individuals, respectively, in Transects 14 and 15; Appendix 4).

Tall humid forest

Transect 15 ("tall humid forest"; Appendices 3G-L) was located directly at the base of a steep north-south trending ridge, approximately two kilometers to the west of camp. This forest had a typical Lower Montane forest structure and composition (Gentry 1988, Boyle 1996) with such families as Rubiaceae, Meliaceae, Annonaceae, Lauraceae, Icacinaceae, Euphorbiaceae, Moraceae and Cunoniaceae predominating among the large tree taxa. Canopy height was 25 to 28 m or more, with individuals commonly exceeding 30 cm dbh, and a few greater than 1 m. Trunks were relatively straight, and bore numerous hemiepiphytic Araceae. Large tree ferns formed a conspicuous structural component of this forest, and small terrestrial and epiphytic ferns were particularly diverse. Understory trees included numerous fruit-bearing Rubiaceae, Solanaceae, Melastomataceae and Piperaceae. A single species of *Evodianthus* (Cyclanthaceae) was abundant both terrestrially and as an epiphyte on trunks.

Common large trees included *Elaeagia utilis* (Rubiaceae), *Guarea kunthiana* (Meliaceae), *Weinmannia sorbifolia* (Cunoniaceae) and *Nectandra lineatifolia* (Lauraceae). In addition to juveniles of large trees, the "midstory" consisted mostly of *Faramea multiflora* (Rubiaceae), along with seven species of tree ferns, e.g., *Alsophila engelii*, *Cyathea pallescens* and *Dicksonia sellowiana*. Common understory trees and shrubs were *Cestrum megalophyllum* (Solanaceae), *Saurauia biserrata* (Actinidiaceae), two undetermined species of *Palicourea* (sp. 1 and sp. 2; Rubiaceae) and one Melastomataceae (*Miconia* sp. 13). Among understory herbs, the genus *Pilea* (Urticaceae) was particularly conspicuous, with five fairly common species: *P. diversifolia*, *P. henkii* (actually a small shrub) and three undetermined taxa, *Pilea* sp. 2, sp. 4 (mostly trunk-epiphytic) and sp. 5. Ferns were exceptionally rich in species, with a total of 33 species recorded among all habit and size classes.

Ridge crest forest

Structure and taxonomic composition of Transect 14 ("ridge crest forest"; Appendices 3A-F) was surprisingly different from those of the tall humid forest censused in Transect 15. This sample was taken along the crest of a low ridge to the east of the *pajonal*, immediately above the edge of Río Poyeni canyon. Overall stature of the forest, up to 20 m high, was much lower than in Transect 15. Few individuals exceeded 30 cm dbh. The understory featured numerous small trees and shrubs, but very few large-leaved herbs or trunk epiphytes. Canopy epiphytes and woody hemiepiphytes, in particular Ericaceae, were abundant and diverse. Principal large tree genera were *Miconia*, *Alchornea*, *Gordonia*, *Clusia* and *Weinmannia* (three species). Conspicuously absent were Lauraceae, Annonaceae and Meliaceae, as well as understory Rubiaceae and hemiepiphytic Araceae. Such structure and taxonomic composition, especially the presence of such families as Clethraceae, Podacarpaceae, Theaceae and Symplocaceae and Styracaceae, are typical of Upper Montane forests, rather than of the Lower Montane Life Zone in which this forest occurs (Gentry 1988).

Common trees in ridge crest forest included *Miconia* sp. 10 (Melastomataceae), *Alchornea* sp. 2 (Euphorbiaceae), *Gordonia fruticosa* (Theaceae), *Clethra revoluta* (Clethraceae), *Clusia weberbaueri* (Clusiaceae) and *Weinmannia sorbifolia* (Cunoniaceae). Trees ferns were relatively uncommon, and represented by only two species: *Cyathea pallescens* and *Spahaeropteris elongata*. The understory tree and shrub component was dominated mostly by juvenile individuals of the canopy trees, with lesser numbers of *Faramea multiflora* (Rubiaceae), *Meliosma* sp. 1 (Sabiaceae) and several species of Melastomataceae. Climbers (vines and lianas) were diverse and relatively common, and included a *Celastrus* (sp. 1; Celatraceae), *Moutabea* cf. *aculeata* (Polygalaceae), *Siphocampylus angustiflorus* (Campanulaceae) and several Asteraceae (mostly *Pentacalia* and *Munnozia*). Principal hemiepiphytic shrubs included a *Clusia* (sp. 1; Clusiaceae), a Margraviaceae (*Norantea* sp. 1) and several Ericaceae, especially *Cavendishia bracteata* and *Diogenesia octandra.*

A total of 35 species of epiphytes were collected in this forest (Appendix 5), which was higher than the number of species collected along Transect 15. The majority of these were orchids, many of which were collected only in Transect 14. Although many were so rare that they might have turned up in other habitats with additional collecting, some were quite abundant in Transect 14, suggesting that they may be specific to this habitat. Examples include *Maxillaria* sp. 4, *M. acuminata* (Orchidaceae), *Polypodium cacerisii*, *Peltapteris moorei*, and *Grammitis* sp. 2 (Pteridophyta). A new species of bromeliad was found in this habitat type, recently described as *Greigia vilcabambae* (Luther 1998).

Comparison of ridge crest and tall humid forest

In general, diversity was similar between tall humid and ridge crest forest. However, certain growth forms, in particular medium-sized trees (10-30 cm dbh), were less diverse (15 species per 50 individuals in tall humid forest, versus 27 in ridge forest; see Appendix 4). Species overlap between the two forests was very low: only 7% of the species in their combined floras were shared (Appendices 6 and 7). By contrast, ridge forest and the nearby dwarf *Clusia* forest (Transect 18) shared 12% of their species. Only 2% of species were shared between the elfin swamp forest and tall humid forest.

As with Camp One, one of the biggest surprises was the lack of overlap in species composition between the two transects in tall forest (Transects 14 and 15). One possible explanation for these differences may be small-scale variation in microclimate and its effects on soil saturation and evapotranspiration rates. The forest sampled in Transect 15 lay at the very base of the 300 m high ridge that towered over the west end of the plateau. Precipitation from clouds rising up this ridge may result in slightly higher local precipitation and humidity. In addition, the low ridge sampled in Transect 14 at the upper edge of Río Poyeni canyon probably is more exposed to the evaporative effects of wind. Subjectively, the latter forest "felt" less humid; this impression agrees with the scarcity of large leaved, moisture-loving taxa such as hemiepiphytic Araceae and ferns. In addition to microclimate, substrate also may play a role in determining the differences between these two forests. The only extensive outcrops of limestone observed in the vicinity of Camp Two were located along the ridge crest sampled in Transect 14.

Whatever the reason for such differences in taxonomic composition, they probably have important consequences for animals utilizing these two forests. Both the abundance and diversity of fruit bearing trees, shrubs and epiphytes was higher in the tall humid forest than in the ridge crest forest. Examples include small trees in the genera *Palicourea* and *Psychotria* (Rubiaceae), *Miconia* (Melastomataceae) and several species of hemiepiphytic *Anthurium* and *Philodendron* (Araceae). Fleshy fruit-bearing canopy trees common in Transect 15 but absent or rare in Transect 14 include several species of Lauraceae as well as the annonaceous genus *Guatteria*.

HABITAT HETEROGENEITY: CONSERVATION IMPLICATIONS

At both Camps One and Two, the high levels of habitat heterogeneity (beta diversity) were surprising. This is particularly true for the two taller forest samples from Camp Two (Transects 14 and 15). Despite appearing similar from a distance and separated by no more than 3 km, they shared only 21% of their species. Comparing only among size classes > 1 cm dbh, the 15% overlap between Transects 14 and 15 is even less than the 26% similarity observed the two mixed-species forests sampled at Camp One (Transects 12 and 13; see Appendix 6). In addition, alpha (local, or within-sample) diversity was higher at 2050 m than at 3500 m: with the exception of *pajonal* habitats, samples from comparable vegetation types were approximately twice as rich in species at the lower elevation. At the landscape scale, the high alpha diversity is further magnified by the already high beta diversity observed at this elevation.

Relative to the total number of species collected in both camps, species overlap between the two elevations was almost nil. This result agrees with other studies of altitudinal distributions in Neotropical montane forests: typically, species overlap drops to near zero with about 1000 m of elevational separation (Boyle 1996, Lieberman et al. 1996). Even in what appears to be homogeneous forest, a change of only 200-300 m in elevation brings dramatic differences in species composition. This was illustrated clearly at Camp Two, where a single collecting trip down to 1700 m produced numerous species not encountered in nearly two weeks of sampling at 2050 m.

The discovery of two new species of bromeliads from these two camps (Luther 1998) highlights the need for continued inventory work and conservation activities. Although nearly 700 plant species were collected over the course of this expedition, consideration of all three components of diversity (alpha diversity, beta diversity, and elevational species turnover), suggest that many more species await to be discovered within an area as large, as poorly known, and as topographically and geologically diverse as the Cordillera de Vilcabamba.

LITERATURE CITED

Boyle, B. L. 1996. Changes on altitudinal and latitudinal gradients in Neotropical montane forests. Ph. D. thesis. St. Louis, Missouri: Washington University.

Cano, A., K. R. Young, B. León and R. B. Foster. 1995. Composition and diversity of flowering plants in the upper montane forest of Manu National Park, southern Peru. *In* Churchill, S. P., H. Balslev, E. Forero, and J. Luteyn (editors). Biodiversity and conservation of Neotropical montane forests. New York: New York Botanical Garden.

Gentry, A. H. 1988. Changes in plant community diversity and floristic composition on geographical and environmental gradients. Annals of the Missouri Botanical Garden 75: 1-34.

Gentry, A. H. 1992. Diversity and floristic composition of Andean cloud forests of Peru and adjacent countries: implications for their conservation. Memorias del Museo de Historia Natural, Universidad Nacional Mayor de San Marcos (Lima) 21: 11-29.

Gentry, A. H. 1995. Patterns of diversity and floristic composition in neotropical montane forests. *In* Churchill, S. P., H. Balslev, E. Forero and J. Luteyn (editors). Biodiversity and conservation of Neotropical montane forests. New York: New York Botanical Garden.

Holdridge, L. R. 1967. Life zone ecology. San José, Costa Rica: Tropical Science Center.

Lieberman, D., M. Lieberman, R. Peralta, and G. S. Hartshorn. 1996. Tropical forest structure and composition on a large-scale altitudinal gradient in Costa Rica. Journal of Ecology 84: 137-152.

Luther, H. E. 1998. Miscellaneous new taxa of Bromeliaceae (XIII). Selbyana. 19(20): 218-226.

Magurran, A. E. 1988. Ecological diversity and its measurement. Princeton, New Jersey: Princeton University Press.

CHAPTER 3

VEGETATION OF AN OUTER LIMESTONE HILL IN THE CENTRAL-EASTERN CORDILLERA DE VILCABAMBA, PERU

Bruce K. Holst

INTRODUCTION

The eastern slope of the Cordillera de Vilcabamba arises from a series of rolling hills, the last of which before the Cordillera form a low chain, cresting at 1400-1500 m. Inside of this chain the topography is varied from small forested plains to deep, rock-walled gorges. These steep, outer hills are composed at least in part of limestone, and are thinly connected to the main Vilcabamba massif by narrow saddles, after which the massif steeply rises to its often cloud-covered heights.

Major rivers and their tributaries that flow off of the eastern slope are, from south to north, the Paratori, Picha, Pagoreni, and Hupaya rivers. These, in turn flow in to the Río Urubamba. According to local sources, water is found year-round, at least at the point where the water courses empty out of the limestone hill series, but is sporadic in the upper canyons, as we witnessed in the limestone creek bed next to our camp.

We conducted the work for the 1998 RAP expedition to the Cordillera de Vilcabamba along one of these outer limestone hills, in the region of the upper Pagoreni above the Ashaninka community of Tangoshiari. Three main vegetation types on and around the limestone hill could be discerned from the air: tall slope forest, limestone ridge forest, and bamboo forest (locally termed *pacales*). However, dramatic and abrupt changes in soil type and depth were responsible for a much greater forest and vegetation diversity when viewed at the ground level. In general, forests in this area are 25 to 30 m tall, evergreen to semi-evergreen, and with an irregular, though mostly closed canopy. Emergents are up to 45 m tall and with diameters of up to 1.5 m at breast height. The bamboo forests are usually monospecific (*Guadua* sp., Poaceae) and present formidable barriers to traverse due to the stout, recurved spines on their main stems and smaller branches. The same species of bamboo also appears to be invading the contiguous broadleaf forests, with its impressive culms reaching the top of the canopy.

Other than the number of flowering plant families (88) and approximate number of trees species > 10 cm dbh collected during the expedition (ca. 220 species), numbers of taxa collected for various taxonomic groups could not be calculated since the majority of the specimens have not yet been studied. However, a few general observations on the herbaceous flora and the transect data on the arboreal flora are summarized here. The herbaceous flora was very diverse as one would expect from this elevation in the Andes, though the pteridophyte flora surprised us in its scope. Hymenophyllaceae and Aspleniaceae were perhaps the most diverse fern families, and various tree ferns were present in nearly all of the vegetation types. Numerous rock outcrops and deep, shaded, moist canyons provided abundant habitat for pteridophytes. Other speciose or abundant herbaceous vascular plant families included Acanthaceae, Araceae, Begoniaceae, Bromeliaceae, Campanulaceae, Commelinaceae, Costaceae, Cucurbitaceae, Cyclanthaceae, Cyperaceae, Gesneriaceae, Heliconiaceae, Marantaceae, Orchidaceae, Piperaceae, Pteridophyta, and Solanaceae. Common or speciose arborescent plant families were Arecaceae, Bombacaceae, Caryocaraceae, Cecropiaceae, Chrysobalanaceae, Clusiaceae, Elaeocarpaceae, Euphorbiaceae, Fabaceae, Flacourtiaceae, Lauraceae, Lecythidaceae, Melastomataceae, Meliaceae, Moraceae, Myristicaceae, Myrtaceae, Nyctaginaceae, Poaceae, Rubiaceae, Sapindaceae, and Sapotaceae.

Rainfall amounts for the area are not known, but the rich vegetation of terrestrial and epiphytic herbs attest to a frequently humid environment. The presence of some gray *Tillandsia* species (Bromeliaceae) and occasional deciduous trees (e.g. *Enterolobium* sp., Fabaceae-Mim.), however, provide some evidence that the dry season is marked. Palms (Arecaceae) also were speciose and abundant at all levels of the forest. Understory, acaulescent or short-caulescent palms (*Geonoma* spp.) often formed sizeable colonies on deeper soils. Other caulescent, solitary or colonial *Geonoma* species were common small trees in the area. *Iriartea deltoidea*, and to a lesser extent, *Wettinia augusta*., were common canopy trees, especially on limestone ridges. Epiphyte diversity in the area appears to be high, though it was inadequately sampled by us. This inadequacy was made clear when we took advantage of investigating some canopy trees, recently felled for agricultural purposes, in the vicinity of Tangoshiari. Careful exploration of the branches of dozens of these trees revealed a high number of small branch and twig epiphytes, particularly Orchidaceae that we had not previously collected. Several of these species are new to science (J. T. Atwood, pers. com.), and most have not yet been studied.

Night-blooming, bat-pollinated Bromeliaceae (*Guzmania* spp., *Werauhia* spp.) were the dominant epiphytes in the mid- to low-canopy on trunks and branches. Gesneriaceae also were conspicuous in nearly all of the available habitats and provided the most abundant and varied color, as well as important food sources for birds and bats. Hemi-epiphytic and epiphytic Cyclanthaceae were found to be colonizing virtually all tree trunks in certain areas. Terrestrial and lithophytic pteridophytes were astonishingly abundant and diverse, and bryophytes were abundant in canyons and on rocky, moist slopes.

Above Tangoshiari, little human-caused disturbance was seen. Community residents commonly hunt in the limestone hill area and recently have been selectively logging nearby forests for wood for dwellings. Regenerating *Euterpe* palms are found throughout the area, and we suspect that they were a much more common forest element prior to the founding of the Asháninka community of Tangoshiari only several years ago. Young *Euterpe* palms are harvested for the edible and desirable hearts of their stems. Only one mature individual was seen in the most remote area that we visited. Other palms, particularly *Geonoma* spp., are used for roofing materials.

Closer to Tangoshiari, forests are strongly fragmented due to clearing for agricultural purposes. Yuca (*Manihot esculenta*) is the main crop, with bananas, citrus, anatto, papaya, topiro (*Solanum sessiliflorum*) and sweet potato (*Ipomoea batatas*) commonly cultivated. Some attempt is being made to grow coffee for commercialization, but the expense of transporting the coffee beans to market in Satipo has been prohibitive.

METHODS

B. Holst, M. Arakaki, and H. Beltrán conducted 10 rapid transects of 50 individuals each (>10 cm dbh) at elevations between 600 and 1400 m. General collections also were made of fertile plants outside of the transect areas whenever possible. The transects were done following Foster's rapid transect methodology (Foster et al. in prep.). To conduct the transects, a measuring line was carried through a single habitat type; as the line was reeled out, all trees > 10 cm dbh that fell into a 5 m wide path on either side of the line (10 m total width) were identified and herbarium collections were made as necessary. Trees > 10 cm dbh were divided into two size classes, trees 10-30 cm dbh and trees > 30 cm dbh. After 50 trees were counted, measurement of the transect distance was noted. We also collected and took further notes on abundance and density of understory species, epiphytes, and lianas.

Collections

Including the vouchers for the transects, we made 974 collections of plants, the majority of these with 3-5 duplicates. The first set was deposited at the herbarium of the Universidad Nacional de San Marcos (USM), Lima, Peru and the second set at the Marie Selby Botanical Gardens herbarium (SEL), Florida, USA. Additional sets will be distributed to taxonomic specialists and the Field Museum of Natural History in Chicago (F).

To date, approximately only 10% of the collections have been identified to species, and those are mostly herbaceous families such as Bromeliaceae, Orchidaceae, and Araceae. Complete species lists will be posted on the Conservation International website (www.conservation.org).

DESCRIPTION OF THE VEGETATION: CAMP THREE

Bamboo forests (*pacales*)

From overflights of the low, eastern Vilcabamba foothills, one is immediately impressed by the amount of terrain occupied by bamboo of the genus *Guadua* (Poaceae), locally known as *pacales*. Vast colonies covering many square kilometers in expanse can be seen, not only on steep hills where landslides are of common occurrence, but also on gentle slopes, or from the air on what appear to be flat areas. In our study area, *pacales* formed a nearly continuous band around the circumference of the limestone hill, though at places it was interrupted by forest, particularly on more level ground such as the summit or ridges, giving the impression that at least in this area, landslides may have been an important disturbance factor that allowed the bamboo colonization. The important question of how bamboo propagules arrived to these disturbed sites at the appropriate time, given the infrequency of bamboo flowering and fruiting, is not easily answered. It is possible that a reduced number of propagules are all that is needed for establishment of bamboo since it is a rapid grower and since, because of its aggressive vegetative reproduction, it may outcompete other gap colonizers. Some of the *pacales* we observed contained large canopy trees, though these were mostly fast-growing trees such as *Pterygota* (Sterculiaceae) and *Inga* (Fabaceae-Mim.) and may be the rare result of a broadleaf species successfully outcompeting the bamboo. We also observed many instances of bamboo growing well inside forests adjacent to the *pacales*; a long-term study is needed to understand the dynamics between the broadleaf and bamboo forests.

Ridge forests

The ridge forests varied in diversity and composition from site to site, and it was clear that soil depth and type played an important role. Ridge forests also had a fairly high incidence of large tank bromeliads, many of which were night-pollinated species. One site a few hundred meters west of our camp at 1000 m elevation, on a narrow ridge of exposed limestone, was the least diverse of any of the forests we sampled with only 23 species among 50 individuals > 10 cm dbh (Transect 4, Appendix 8). This rocky ridge forest was lower in stature and contained more palms in both the canopy (particularly *Iriartea deltoidea*, as well as the smaller *Wettinia augusta*. and *Geonoma* sp.) and understory (*Aiphanes aculeata*, *Chamaedorea sp.*, *Geonoma* spp., *Bactris* sp., *Desmoncus* sp.). The only small limestone sinkholes in the area were found adjacent to this ridge. The canopy along the rocky ridge was also more irregular than most other areas, possibly due to disturbance by storms. Other common trees on the ridge were *Guarea pterorhachis* (Meliaceae), *Roupala montana* (Proteaceae), and *Coussapoa* cf. *villosa* (Cecropiaceae). Besides the understory palms mentioned above, other shrubs and understory trees were *Clavija weberbaueri* (Theophrastaceae), *Piper* sp. (Piperaceae), *Rinorea* sp. (Violaceae), *Calyptranthes bipennis* (Myrtaceae), *Trichilia* sp. (Meliaceae), *Trophis caucana* (Moraceae), and an unknown Rubiaceae. Common epiphytes were Cyclanthaceae, Gesneriaceae, a fleshy *Psychotria* (Rubiaceae), and miscellaneous pteridophytes and mosses.

This level on the hill marked a transition zone to a more montane flora above and was the last relatively flat area before the steep rise to the summit 450 m above.

At a slightly higher elevation on the same hill (1100 m), on a more eastward-facing slope, we sampled a closed-canopy, open-understory forest on a broad, steep ridge with minimal rock outcropping and generally deeper soils than the rocky ridge near camp. Species diversity on this ridge forest was much higher than on the rocky ridge, with 36 species > 10 cm dbh in 50 individuals counted (Transect 3, Appendix 8). Only about five of the 36 species were also found along the rocky ridge. Understory plants also differed significantly on the broad ridge. Common shrubs and small trees were *Galipea* sp. (Rutaceae), *Tabernaemontana* sp. (Apocynaceae), *Myrcia* sp. (Myrtaceae), *Psychotria* sp. (Rubiaceae), *Clavija* sp. (Theophrastaceae), and *Cordia nodosa* (Boraginaceae). The herbaceous understory was fairly discontinuous, probably as a result of uneven light penetration. Patches of mixed fern species (e.g. *Adiantum, Blechnum*) occurred towards the edge of the forest along the border with a bamboo forest where the light penetrated more readily. Large-bodied epiphytes and hemi-epiphytes such as Cyclanthaceae, *Anthurium* spp. (Araceae), *Billbergia violacea* (Bromeliaceae), *Polybotrya* spp. (Pteridophyte), and several Orchidaceae were abundant in this area.

Continuing farther down the ridge from camp several hundred meters, immediately after a steep drop, an abrupt change in understory vegetation was noticeable at 850 to 950 m elevation. This seemed to be correlated with a change in soil type and depth. The most evident difference between these ridge forests compared to the higher ones was the near understory dominance of a colonial *Calathea* sp. (Marantaceae) that formed dense thickets, along with a common *Olyra* species (Poaceae). We made two transects in this region: Transect 1 (elevation 800 m) located in a medium-height forest of widely separated, fairly small diameter trees; Transect 2 (elevation 950 m) located less than 100 m farther down the ridge in more open understory forest with larger diameter trees (Appendix 8). Neither of the forests, however, had more than a few widely spaced large trees > 30 cm dbh. These two forests shared several important species between them, such as *Faramea* sp. (Rubiaceae), *Myrcia fallax* (Myrtaceae), *Hevea brasiliensis* (Euphorbiaceae), and *Virola sebifera* (Myristicaceae).

Several hundred meters walk farther down the ridge, which is gently sloping at this level, the ridge crest broadens considerably and the soils are deeper, supporting many larger trees. In the transect at this site (Transect 6, elevation 850 m, Appendix 8), 26% of the individuals were > 30 cm dbh, including the most common of the species in the transect, *Gavarettia terminalis* (Euphorbiaceae). Other important canopy trees were Sapotaceae, *Virola* spp. (Myristicaceae), *Hevea brasiliensis* (Euphorbiaceae), Anacardiaceae, and *Neea* sp. (Nyctaginaceae). The dense Marantaceae colonies of the slightly higher elevation sites were replaced in this forest with a more diverse mixture of herbs and shrubs, such as a colonial *Psychotria* (Rubiaceae), *Chamaedorea* sp. (Arecaceae), *Selaginella* sp. (Selaginellaceae), *Costus* sp. (Costaceae), *Miconia* sp. (Melastomataceae), and small, thin-stemmed tree ferns.

Slope forests

We surveyed slope forests at 1000, 900, and 650 m elevation. Other than the summit forest, these were the most species diverse of the transects conducted, yielding between 29 and 37 species out of 50 individuals > 10 cm dbh counted in each transect. Transect 5 (elevation 1000 m) was in a steep, slightly rocky canyon immediately below the Ridge Camp (Appendix 8). The most common tree > 10 cm dbh at this site was *Iriartea deltoidea* (Arecaceae), with 5 individuals. The only other site in the forest where *I. deltoidea* was so common was on the rocky ridge area (Transect 4) immediately adjacent to the transect 5 site. At this site, only one other species had more than two individuals out of 50 trees > 10 cm dbh, *Hevea brasiliensis* (Euphorbiaceae) with three individuals. Species with two individuals were *Inga* sp. (Fabaceae-Mim.), *Ficus* sp. (Moraceae), *Quiina* sp. (Quiinaceae), *Quararibea* sp. (Bombacaceae), *Pourouma cecropiifolia* (Cecropiaceae), *Homalium* sp. (Flacourtiaceae), and two species of unidentified Moraceae. Twenty-six additional species were represented by only one individual in this transect. Pteridophytes virtually covered the rocky ground and rocks at this site and tree ferns were common. A high diversity of other shrubs and herbs were present, the most notable of the families being Araceae, Begoniaceae, Cyperaceae, Marantaceae, Melastomataceae, Piperaceae, Poaceae, and Rubiaceae.

Transects 7 and 8 (elevation 900 m, Appendix 8) were done in similar, tall forest with many large diameter trees on gentle slopes with deep, red soils and little or no rock outcrops. Both transects were rich in species and individuals of Sapotaceae, Fabaceae, Myristicaceae, and Euphorbiaceae. In this same area, several individuals of juvenile *Euterpe* (Arecaceae) were observed, and a single large individual of *Enterolobium cyclocarpum* (Fabaceae-Mim.). The shrub and small tree layer was composed of several Melastomataceae, Rubiaceae, and Arecaceae (*Geonoma*). Common herbs were *Calathea* (Marantaceae), *Olyra* sp. (Poaceae), *Huperzia* (Lycopodiaceae), and epiphytes included diverse genera of Bromeliaceae (*Aechmea* sp., *Guzmania squarrosa*, *Mezobromelia pleiosticha*, and *Tillandsia fendleri*).

The final slope forest we sampled (Transect 10, elevation 650 m, Appendix 8) showed the highest species diversity of any of the sites studied, and 26% of the trees measured were > 30 cm dbh. The transect site was located at the bottom of the limestone hill near the community of Tangoshiari. The topography in this area is of more gently rolling hills and shallow canyons, and soils are deep. *Iriartea deltoidea* (Arecaceae) was the most common tree species at this site, but there was some indication that it is being locally exploited as we saw some stumps in the transect area. Seven of the 37 tree species > 10 cm dbh found in the transect were Moraceae, and the Rubiaceae was also well-represented with four species. The understory layer was rich and included miscellaneous Rubiaceae, Melastomataceae, tree ferns, and *Ischnosiphon* sp. (Marantaceae).

Summit forests

The bamboo forest that forms a band around the eastern and southern slopes of the hill occurs at an elevation between 1000 m and 1300 m elevation. Above 1300 m, the slope becomes less steep and a high-diversity montane forest occurs, showing considerably different species composition from the forests at or below 1100 m. The aspect of the summit forest is also different, with many large diameter trees (26% of the sample > 30 cm dbh), a well-developed understory with vascular and non-vascular epiphytic component, and a greater presence of tree ferns. Of the 50 trees surveyed at the summit (Transect 9), the most common were *Bathysa obovata* (Rubiaceae), *Perebea tessmannii* (Moraceae), *Protium* sp. (Burseraceae), *Pourouma* cf. *tomentosa* (Cecropiaceae), *Hyeronima* sp. (Euphorbiaceae), and a species of tree fern (Appendix 8).

None of these species were found in transects at lower elevations. Several individuals of young *Euterpe* plants were seen outside of the transect area, including one mature tree we saw at a distance with binoculars. Two to four meter tall *Pholidostachys synantheca* palms were dominant in the understory, along with a very common trifoliolate Rutaceae shrub. There were relatively few epiphytic Bromeliaceae, but numerous individuals of Orchidaceae, Araceae, *Blakea* sp. (Melastomataceae) and Hymenophyllaceae. Bamboo appeared to be invading portions of the summit forest, and at one point, at a narrow saddle, the bamboo forest from the slopes on either side of the hill have grown together to form a nearly impenetrable barrier to further exploration of the slightly higher plateau beyond.

Stream forests

We did not run any transects in the forests along the streams, but did substantial general collecting in these very rich habitats. Remarkably speciose families were Piperaceae, Araceae, Gentianaceae, and the pteridophytes. Nearly all of the water courses in the area are in rocky beds. The largest river we explored, the Río Muiti, is a year-round stream that exits the limestone hill series between two fairly equal sized, 1400–1500 m high hills. The canyon that has formed is very deep, with completely vertical, or nearly vertical, walls several hundred meters high in some places. An interesting *Pitcairnia* sp. (Bromeliaceae) formed huge colonies on one of the walls. Most of the riverine vegetation however, was similar to that observed around the community of Tangoshiari. The ruggedness of the deep canyons may help protect this lower-elevation (500–700 m elevation) flora as the forests around the community of Tangoshiari become increasingly fragmented.

CONSERVATION IMPLICATIONS

For the most part, plant communities above 700 m elevation above Tangoshiari are undisturbed by humans, although we did find evidence that *Euterpe* palms have been removed for their edible hearts at least within a radius of one day's walk of Tangoshiari. The forests around the Tangoshirari community at 500 to 700 m elevation have high levels of plant diversity but are also actively being fragmented for agricultural purposes. In addition, the Asháninka are considering growing coffee for export. Broadleaf forests also are fragmented by large patches of bamboo, although it is not clear how the bamboo has become such an important ecological factor in the area. Some timber is being extracted from the forests immediately around Tangoshiari on a small scale, principally to build community structures and facilities for the local church.

From our limited view of the Vilcabamba flora, it is difficult to know if any of the vegetation types that we sampled are rare or threatened. Clearly, the area is very diverse at the local and community level, and in light of the increasing habitat fragmentation at the lower elevations around the massif, further inventory work is critical to understanding regional patterns of habitat and plant species distribution.

The discovery of new orchid species in the area around Tangoshiari highlights the need for conservation activities, including continued inventory work and local education to prevent the loss of such species before they can be adequately studied.

CHAPTER 4

FLORA OF TWO SITES IN THE SOUTHERN CORDILLERA DE VILCABAMBA REGION, PERU

James Comiskey, Alfonso Alonso, Francisco Dallmeier, Shana Udvardy, Patrick Campbell, Percy Núñez, William Nauray, Rafael de la Colina, and Severo Baldeón

INTRODUCTION

The tropical Andes have been identified as one of the most important biodiversity hot spots in the world, hosting about 20,000, or 7.4%, of the world's endemic plant species (Mittermeier et al. 1998). The transition between montane and lowland tropical rainforest is known to be extremely high in diversity, and attention in the past has focused on these forests. In contrast, tropical montane forests are generally poorly described, yet highly endangered (Gentry 1992). Two regions, the northern Andean montane forests and the Andean yungas, have recently been identified among the 200 most biologically valuable ecoregions of the world (Olson and Dinerstein 1998). The southern region of Peru has also been identified as a potential Pleistocene refuge, which would account for the high levels of endemism in the area (Brown 1987, Prance 1982). The range of habitats along elevation and latitudinal gradients means that montane rainforests are more diverse in their total species richness than their lowland counterparts (Henderson et al. 1991, Young and Valencia 1992). More than 40% of the total area occupied by montane forest in South America occurs in Peru, giving rise to an extensive range of habitats (Young and Valencia 1992). The flora of eastern Peruvian montane forests include nearly 2500 species of flowering plants (Young 1992). Montane forests are, however, threatened. Over the past 50 years, the construction of transandean roads has led to extensive settlements, resulting in deforestation and increased agriculture that has created habitat degradation and affected watersheds. These highland environments are less resilient to disturbance than their lowland counterparts (Brack 1992).

Montane rainforest types

Montane rainforest can be classified into three main categories: lower montane, upper montane, and subalpine (Webster 1995). The elevation zones of each type vary with latitude, rainfall patterns, aspect, and degree of exposure. Cloud forest occurs between 1000 meters (m) and 3000 m on most neotropical mountain ranges (Webster 1995) and is determined more by the presence of continuous cloud cover than by the structural or floristic characteristics of the vegetation, although these frequently reflect the increased moisture availability. Diversity tends to decrease above 2000 m to 2500 m (Young 1991).

Lower montane rainforest occurs between 1000 m and 2500 m, with cloud forest becoming more prevalent as altitude increases. The taxa that occur above 1500 m in montane forests rarely overlap those in the lowland forests (Gentry 1992, 1995). Upper montane rainforests tend to occur between 2500 m and 3500 m. This forest type differs not only in floristic composition compared to the lower montane forest, but also in its structure, having a distinctly lower canopy. Pteridophyte diversity increases at these higher elevations (Young 1992). The subalpine tropical forest, also known as timberline forest (Young 1998), elfin woodland (Beard 1955), or locally by the name of "monte chico," usually occurs above 3000 m (Webster 1995). It is predominantly composed of small stature vegetation, with tree ferns and bamboo along the lower margins. The ground is typically carpeted with a dense covering of mosses, and the trees take on a stunted appearance because of the effect of wind (Lawton 1982).

Montane forests in Peru occur at altitudes above 1000 m (Gentry 1992, Young and Valencia 1992). Deep interandean valleys are often dry zones caused by a rain shadow effect, while on the slopes the cloud forests occur between 1500 m and 2500 m in southern Peru (Young 1992).

Young (1991) provides a valuable floristic analysis along elevation gradients in southern Peru.

According to Young (1992), the annual precipitation between 1500 m and 2500 m ranges from 1500 millimeters (mm) to 4000 mm. Mean annual temperatures range from 17° to 21° C. Southern Peru is also subject to the southern climatic oscillations originating from Patagonia during the months of May to October. These "southerlies" have been noted in the Vilcabamba region by other authors, when temperatures may drop 8° C below their normal values (Terborgh 1971). Rainfall originates from Amazonia, occurring mostly between November and April.

METHODS

Site descriptions

The SI/MAB surveys were conducted in the Southern Cordillera de Vilcabamba highlands of the Apurímac region, which are part of the Río Tambo Province (Young 1992). To the west, the Apurímac Valley creates such an important biogeographical barrier that the Urubamba and Madre de Dios regions are the most distinct in Peru in relation to their floristic composition (Berry 1982). Much of this region was composed of dry to moist premontane and montane forests that have been lost to highland agriculture, with many of the remaining forests occurring at lower elevations. At night, fires are clearly visible in the surrounding mountain slopes and valleys, and during the month of September, the whole valley is frequently consumed with smoke from fires started by farmers to clear their land.

Terborgh (1971) first described the vegetation of the Vilcabamba highlands along an elevation gradient not far from the sites of these surveys. He noted that the cloud forest begins at 1380 m, and the elfin forest at 2550 m, being lower on ridges than on the slopes.

SI/MAB surveys were conducted at two sites:

Llactahuaman

The Llactahuaman camp was located on the west side of the mountain range (12° 51'55" S, 73°30'46.0" W, elevation 1710 m) about 4 kilometers (km) northeast of the community of Pueblo Libre on a slope of about 40°, steeper in some areas. The primary aspect was south. Seven study plots were established near the camp at elevations ranging from 1675 m to 1735 m, with slopes between 35° to 70°. Soils of all plots were composed mainly of clay and organic matter. Llactahuaman can be characterized as upper montane forest zone in the cloud forest range, characterized by high epiphyte loading of the trees. The common bamboo of the highlands, *Chusquea* sp., was less dominant on the exposed ridge containing the plots than on the surrounding slopes. Two streams, Bagre Creek and Puma Creek, passed near the camp. The water was transparent and oxygenated, and temperatures ranged from 14° to 16° C. Both streams had permanent water and belong to the watershed that empties into the Río Apurímac (see Acosta et al. this volume).

Wayrapata

The Wayrapata camp was located on the top of a ridge (12° 50'10.1" S, 73°29'42.6" W, elevation 2445 m) about 10 km northeast of Pueblo Libre. Ten study plots were established near the camp, ranging in elevation from 2285 to 2465 m, with slopes from 30° to 65°. Soils of all plots were composed almost entirely of organic matter. The floor structure was composed of a complex vertical web of root systems.

Wayrapata can be described as an elfin forest or sub-alpine tropical forest, and was also located within the cloud forest range. Average temperatures during the day range between 24° and 26° C, falling noticeably during the night to about 8° to 10° C. Fog and wind are characteristic of the site.

At Wayrapata, streams to the north flow into the Río Cumpirusiato and eventually into the Río Urubamba, while streams to the south flow into the Río Apurímac. Both the Río Urubamba and the Río Apurímac, located east and west of the Vilcabamba Mountains, flow north and later join to form the Río Ucayali, eventually flowing into the Río Amazonas. Most streams near the camp had little permanent water and relatively low volume (less than 0.05 m^3/second), with a strong current, high oxygen, and temperatures ranging from 14° to 16° C (see Acosta et al. this volume).

Plot site selection

On arrival at each site, several days were spent scouting the surrounding areas to define the range of vegetation types available for sampling. The botanists conducted floristic studies with the objective of covering as much area as possible within 250 m of base camp. Typically, more than 1000 m were covered in any one direction. Where the terrain allowed, additional trails were cut to facilitate movement east and west of each base camp. The samples from the floristic studies enabled assessment of species composition and vegetation structure revealing the range of vegetation types of the area.

Plot locations were selected to avoid subjective bias by randomly choosing: 1) a trail to follow, 2) a distance from the camp, 3) a side of the trail, and 4) a distance from the trail. These parameters were chosen prior to departing camp, and the locations rejected only if they were in areas that were inaccessible.

Plot establishment

The methodology chosen for the assessment of plant diversity in the region involves the use of modified

Whittaker vegetation plots. These plots have been used extensively in temperate regions (Stohlgren 1995) and provide a simple and quick quantitative method for the assessment of species diversity.

Environmental parameters were measured at each location where a plot was established. These included latitude and longitude, elevation, slope, aspect, and soil characteristics. The modified Whittaker is 0.1 hectares (ha) in size, measuring 20 x 50 m, with several nested subplots (Fig. 4.1). The largest subplot (C) is 20 x 5 m and is centered in the plot. Two 2 x 5-m subplots (B1 and B2) are located in opposite corners of the plot. Ten 2 x 0.5-m subplots (A1-A10) are placed just inside the periphery of the plot.

Field measurements

Different size classes of vegetation were measured in the modified Whittaker plots. In the smallest subplots (A1-A10), all herbs, grasses, and saplings < 40 centimeters (cm) in height were identified and counted. In the two corner subplots (B1 and B2), all trees and shrubs with a diameter at breast height (dbh) of 1 cm were measured. In the central subplot (C), all trees ≥ 5 cm dbh were identified and measured. All trees with a dbh ≥ 10 cm were identified and measured in D, which is the entire 0.1 ha plot. The dbh of all trees was measured following protocols described by Dallmeier (1992). Tree height was measured as accurately as possible using a clinometer.

Voucher specimens

Voucher specimens were sampled for all species encountered in the field. Wherever possible, the botanists identified morpho-species in the field to reduce the need for multiple samples. Most of the specimens sampled were sterile, thus presenting particular challenges for identification. The flora of the highland is less well known than that of the lowlands, and individuals that could be identified only to genus were assigned a species number. Individuals that were determined only to family were likewise assigned a number. Additional random specimens were sampled outside the plots to identify species representative of the area that did not occur in the plots.

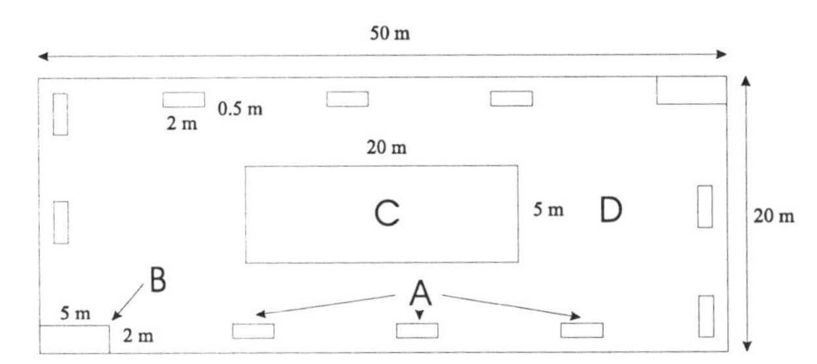

Figure 4.1. Layout of a Whittaker plot and its subplots.

Voucher specimens were deposited in the Museo de Historia Natural de la Universidad Nacional de San Marcos in Lima, the Universidad San Abad del Cusco, and the Smithsonian National Museum of Natural History. The floristic nomenclature follows Brako and Zarucchi (1993) and The Royal Botanic Gardens, Kew (1999).

Data management and analysis

Data were entered at the field camps using BioMon Whittaker, a relational database designed specifically for managing data elicited from the modified Whittaker plots. A master species list for the region greatly reduced the data entry effort and errors. The combined data from all the plots at each of the sites provided quantitative data on the floristic composition and structure of the two habitat types. Using the same methodology, a data set from Pagoreni—a site in southeastern Peru examined earlier in this project (Alonso and Dallmeier 1999)—provides a comparison of the floristic composition and structure of a lowland tropical forest.

RESULTS AND DISCUSSION

The two sites selected for assessment represented two distinct habitats that varied both in their floristic composition and structure (Appendices 9 –12). Structurally, the Wayrapata plots were dominated by trees between 1 cm and 5 cm dbh with few individuals in the large size classes, while Llactahuaman had a higher proportion of larger trees (Fig. 4.2). By comparison, the lowland site at Pagoreni had, on average, twice as many trees > 10 cm dbh per 0.1 ha (Alonso and Dallmeier 1999). The total basal area of trees at Wayrapata was extremely low, again largely because of the low number of individuals > 5 cm. Nevertheless, the basal area occupied by trees with a dbh between 1 and 5 cm at

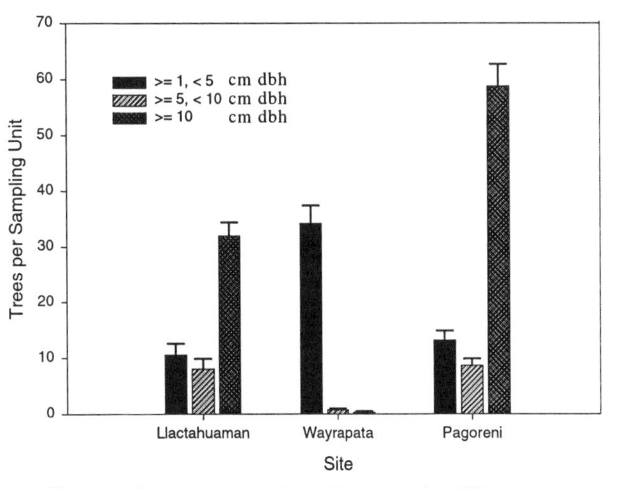

Figure 4.2. Average number of trees in the different subplot units at Llactahuaman, Wayrapata, and Pagoreni, Peru. Data for Pagoreni from Alonso and Dallmeier 1999.

Wayrapata was higher than at either Llactahuaman or Pagoreni. At Pagoreni, the trees > 10 cm dbh accounted for most of the basal area.

The average species richness was higher at Wayrapata than at the lower elevation of Llactahuaman (Table 4.1, Appendices 9-12)). This was mostly because of the higher species richness encountered in the herbaceous subplots and among the smaller tree size classes (Fig. 4.3). In nearby Manu National Park, the high diversity of plants in the montane region was also attributed to the richness of herbaceous plants (Cano et al. 1995). The absence of larger trees at Wayrapata has enabled the establishment of a more prolific understory and herbaceous layer by increasing the potential number of microhabitats for species establishment. In contrast, at Llactahuaman the larger size classes were well represented with an associated poor understory layer. Species richness was generally higher at Pagoreni's lowland forest sites, except among the smaller trees (1 cm to 5 cm dbh).

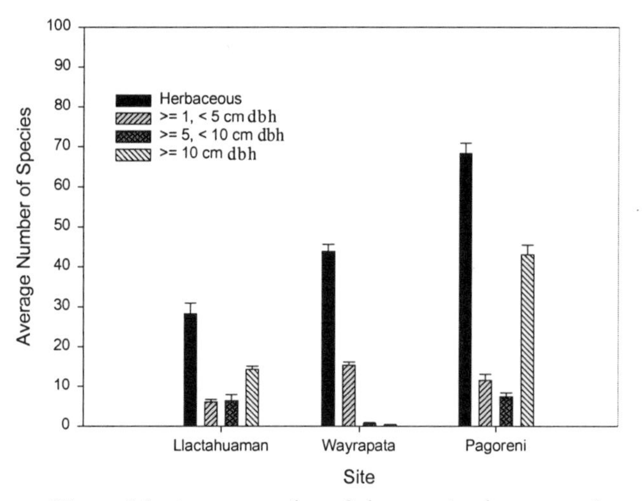

Figure 4.3. Average number of plant species that occurred in the different subplot units at Llactahuaman, Wayrapata, and Pagoreni, Peru. Data for Pagoreni from Alonso and Dallmeier 1999.

The species area curves agreed with these results, Wayrapata having a higher species richness than Llactahuaman in the herbaceous plots (Fig 4.4).

Nevertheless, the latter site accumulated species at a higher rate, and at larger areas we would expect more species than at Wayrapata. In comparison, the Pagoreni lowland forest showed a much higher initial species richness and a considerably higher rate of species accumulation than either Llactahuaman or Wayrapata.

The degree of homogeneity—the number of species overlapping between plots—in each vegetation type appeared to increase as elevation increased (Fig. 4.5). The proportion of species that occurred in only one plot was highest in the lowland forest of Pagoreni (55%) and lowest at Wayrapata (28%). Unlike Llactahuaman and Pagoreni, Wayrapata contained more species in more than five plots (29%) than those that occurred in only one plot.

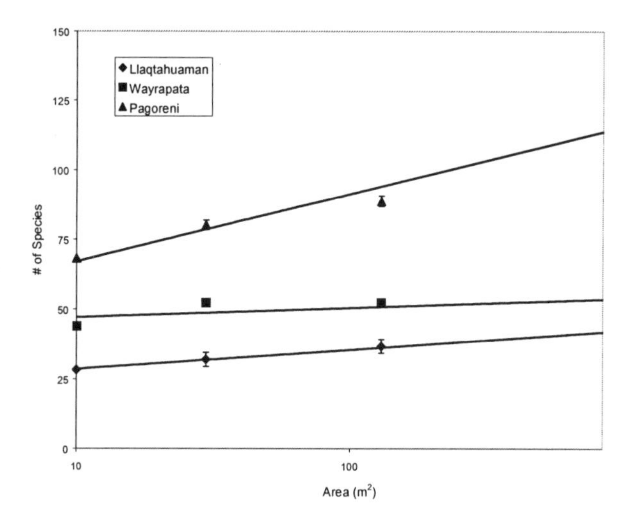

Figure 4.4. Species area regressions for plants at Llactahuaman ($r^2 = 0.99$, slope = 3.1), Wayrapata ($r^2 = 0.48$, slope = 1.5), and Pagoreni ($r^2 = 0.97$, slope = 10.5), Peru. The results are based on the average number of plant species within each subplot (standard error bars shown). The steeper the slope on the line the less likely that an asymptote has been reached. Data for Pagoreni from Alonso and Dallmeier 1999.

Table 4.1. Average number of plant species found in the different size classes in Whittaker plots at Llactahuaman, Wayrapata, and Pagoreni, Peru. Data for Pagoreni from Alonso and Dallmeier 1999.

	Average # Species per Plot		
	Llactahuaman[1]	**Wayrapata[2]**	**Pagoreni[2]**
trees (>=10cm dbh)	47	3	237
trees (>=5cm, <10cm dbh)	33	6	61
trees (>=1cm, <5cm dbh)	21	50	88
trees (>=1cm)	60	51	303
herbaceous	80	115	286
Total	**121**	**130**	**512**

1 Llactahuaman had 7 Whittaker plots

2 Wayrapata and Pagoreni each had 10 Whittaker plots

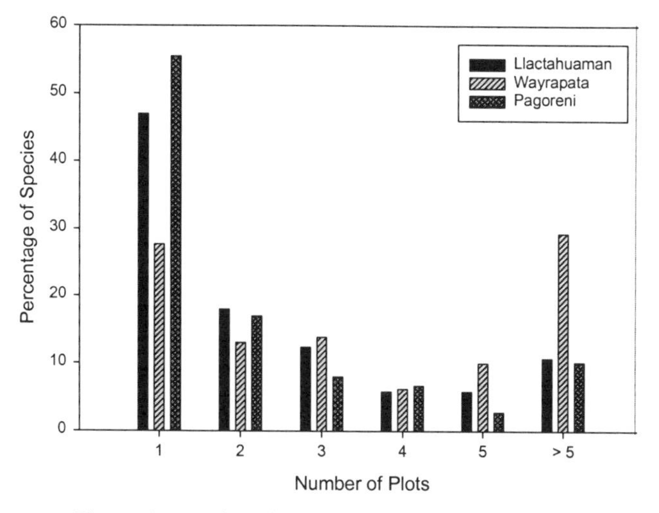

Figure 4.5. Habitat heterogeneity as indicated by the overlap in plant species occurrences between different plots at each site in Llactahuaman, Wayrapata, and Pagoreni, Peru. The graph shows the percent of all plant species at each site that occurred in a different number of plots. Data for Pagoreni from Alonso and Dallmeier 1999.

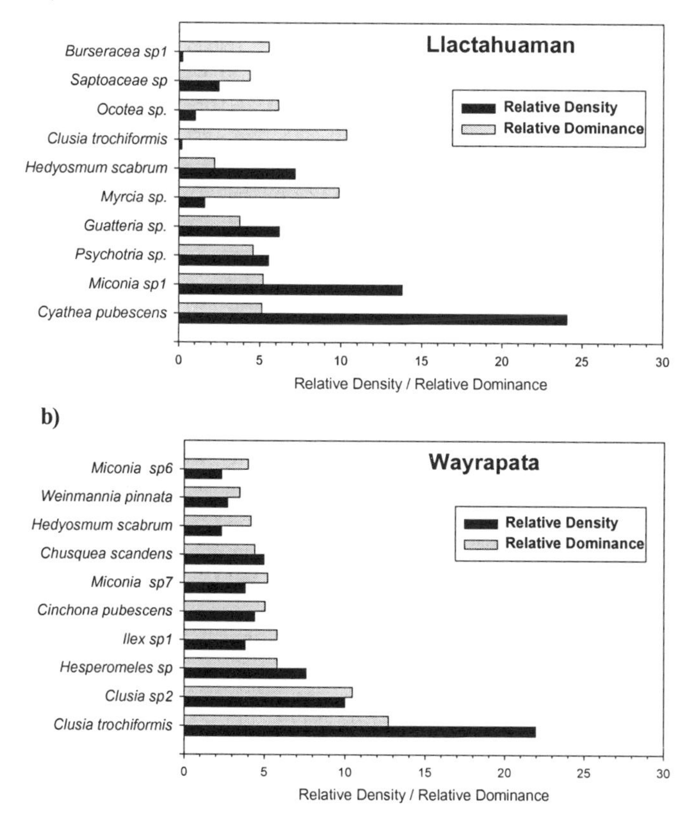

Figure 4.6. Dominant and abundant woody plant species at a) Llactahuaman, and b) Wayrapata, Peru. The relative density is the proportion of individuals of a species in relation to the total number of trees measured in all the plots. Relative dominance is the proportion of the basal area occupied by individuals of a species in relation to the total basal area occupied by all trees in all the plots.

Species richness was low at both Llactahuaman and Wayrapata when compared to other montane sites in Peru at similar elevations. Gentry (1995) reported 39 species >10 cm dbh for 0.1 ha at 1750 m in Chirinos, while at a similar elevation at Llactahuaman an average of just over 14 species per 0.1 ha was registered. Similarly, Wayrapata had an average of only 0.3 species of trees >10 cm dbh while at the same elevation in Montaña de Cuyas, 24 species were registered (Gentry 1995). Species richness at Wayrapata is more comparable to that registered at 3000 m in El Pargo (Gentry 1995). The more southerly location of the sites in the current study increases the susceptibility of the region to strong southerlies that give rise to considerable declines in temperature (Terborgh 1971, Young 1991). As noted in the site description above, the transition between different forest types within our sampling area had previously been reported to occur at lower elevations than in montane regions further to the north (Terborgh 1971).

The most common species at Llactahuaman were *Cyathea pubescens* and *Miconia* sp., with *Clusia trochiformis* and *Myrcia* sp. dominating in relation to the basal area occupied among the woody plants (Fig. 4.6a). On the more protected areas below the ridge, the dominant species was *Chusquea scandens*. At Wayrapata, the most abundant species, *Clusia trochiformis* and *Clusia* sp. 2, dominated in relation to basal area (Fig. 4.6b). *Clusia* also dominated the uplands of the nearby Manu National Park (Cano et al. 1995). *Hedyosmum scabrum* was among the 10 most important species that occurred at both sites. Seven of the top 10 species represented at Wayrapata belonged to genera that have been described as occurring in montane forests at a continental level, from Mexico to Bolivia, including *Clusia, Ilex, Miconia, Hedyosmum*, and *Weinmannia* (Webster 1995). Three of these genera were represented in Llactahuaman. The genus *Cinchona* occurs south of Panama only, while genera such as *Ilex* have not been reported to occur in samples below 1500 m (Gentry 1995).

The Leguminosae is usually represented as the most speciose family of trees > 2.5 cm dbh in lowland tropical forests below 1500 m (Gentry 1988, 1992). At higher elevations, between 1750 m and 2410 m, Lauraceae was represented as the most speciose family in Peru (Gentry 1992, 1995). In this study, Moraceae was represented by 5 species at Llactahuaman, while Lauraceae, Myrtaceae, Melastomataceae, and Euphorbiaceae were each represented by 4 species of woody stems > 1 cm dbh (Appendices 9 and 10). At the higher elevation of Wayrapata, Melastomataceae was represented by 10 species, followed by the Cunoniaceae and Ericaceae with 7 species each and Lauraceae with 2 species (Appendices 11 and 12). Ericaceae occurred at both sites, but was more species rich at higher elevations.

The absence of the Lecythidaceae, Myristicaceae, and Vochysiaceae, all of which occur in the lower montane forests, distinguishes the upper montane forest from the lowland tropical forest (Webster 1995). Several families represented at both the sites are more commonly found in cloud forests, including Clethracea, Ericaceae, and Podocarpaceae at Llactahuaman, and Aquifoliaceae, Ericaceae, and Symplocaceae at Wayrapata. Cyclanthaceae, first described for a montane site in Peru by Cano et al. (1995) in Manu, was encountered at Wayrapata.

CONSERVATION RECOMMENDATIONS

The purpose of the SI/MAB biodiversity surveys was to evaluate the species richness and diversity of two highland sites along a proposed pipeline route across the Southern Cordillera de Vilcabamba. While plans for a pipeline in this area are now on hold, we offer the following recommendations for minimizing impacts on the local flora in the event that a pipeline is built in the general area in the future.

- Vegetation types vary with altitude and aspect, tending to form bands around the mountains. Hence, the least amount of disturbance will occur by traversing the habitats at right angles (directly up slope).
- At higher elevations and along more exposed ridges, plant diversity is fairly uniform over wider scales, creating more homogenous habitats. In these places, it may take a relatively long time for habitats to recover from impacts. Wherever possible, these habitats should be avoided.
- Replanting species native to the area once a pipeline has been installed can minimize long-term disturbance (Mistry 1998).

LITERATURE CITED

Alonso, A., and F. Dallmeier, eds. 1999. Biodiversity Assessment and Monitoring of the Lower Urubamba Region, Perú: Pagoreni Well Site Assessment and Training. SI/MAB Series #3. Smithsonian Institution/ MAB Biodiversity Program. Washington, DC.

Beard, J. S. 1955. The classification of tropical American vegetation types. *Ecology* 36: 89-100.

Berry, P. E. 1982 . The systematics and evolution of Fuchsia Sect. Fuchsia (Onagraceae). *Annals of the Missouri Botanical Garden* 69: 1-198.

Brack, A. 1992. Estrategias nuevas para la conservación del bosque montano. *In* K. R. Young and N. Valencia (eds.) Biogeografía, Ecología y Conservación del Bosque Montano en el Perú. Memorias del Museo de Historia Natural 21. Universidad Nacional Mayor de San Marcos, Lima.

Brako, L., and J. L. Zarucchi. 1993. Catalogue of the Flowering Plants and Gymnosperms of Peru. Missouri Botanical Gardens, St. Louis, Missouri.

Brown, K. S. 1987 Soils. In T. C. Whitmore and G. T. Prance (eds.), Biogeography and Quaternary History in Tropical America. (Clarendon Press, Oxford, UK.

Cano, A., K. R. Young, B. León, and R. B. Foster. 1995. Composition and diversity of flowering plants in the upper montane forest of Manu National Park, southern Peru. *In* S. P. Churchill, H. Balslev, E. Forero, and J. L. Luteyn (eds.) Biodiversity and Conservation of Neotropical Montane Forests. The New York Botanical Garden, New York.

Dallmeier, F. 1992. Long-term monitoring of biological diversity in tropical forest areas. Methods for establishment and inventory of permanent plots. MAB Digest Series, 11. UNESCO, Paris.

Gentry, A. H. 1988. Tree species richness of upper Amazonian forests. *Proceedings National Academy Science* 85: 156-159.

Gentry, A. H. 1992. Diversity and floristic composition of Andean forests of Peru and adjacent countries: implications for their conservation. *In* K. R. Young and N. Valencia (eds.) Biogeografía, Ecología y Conservación del Bosque Montano en el Perú. Memorias del Museo de Historia Natural. 21. Universidad Nacional Mayor de San Marcos, Lima.

Gentry, A. H. 1995. Diversity and floristic composition of neotropical dry forests. *In* S. H. Bullock, H. A. Mooney, and E. Medina (eds.) Seasonally Dry Tropical Forests. Cambridge University Press, New York.

Henderson, A., S. P. Churchill, and J. L. Luteyn. 1991. Neotropical plant diversity. *Nature* 351: 21-22.

Lawton, R. O. 1982. Wind stress and elfin stature in a montane rainforest tree: an adaptaive explanation. *American Journal of Botany* 69: 1224-1230.

Mistry, S. 1998. Depth of plant roots: implications for the Camisea pipelines and flowlines. *In* A. Alonso and F. Dallmeier (eds.) Biodiversity Assessment and Monitoring of the Lower Urubamba Region, Perú: Cashiriari-3 Well Site and the Camisea and Urubamba Rivers. SI/MAB Series #2. Smithsonian Institution/ MAB Biodiversity Program, Washington, DC.

Mittermeier, R. A., N. Myers, J. B. Thomsen, G. A. B. da Fonseca, and S. Olivieri. 1998. Biodiversity hotspots and major tropical wilderness areas: approaches to setting conservation priorities. *Conservation Biology* 12: 515-520.

Olson, D. M., and E. Dinerstein. 1998. The global 200: a representation approach to conserving the Earth's most biologically valuable ecoregions. *Conservation Biology* 12: 502-515.

Prance, G. T. 1982. Forest refuges: evidence from woody angiosperms. In G. T. Prance, (ed.) Biological Diversification in the Tropics. Columbia University Press, New York.

Stohlgren, T. J. 1995. A modified-Whittaker nested vegetation sampling method. *Vegetatio* 117: 113-121.

Terborgh, J. 1971. Distribution on environmental gradients: theory and a preliminary interpretation of distributional patterns in the avifauna of the Cordillera Vilcabamba, Peru. *Ecology* 52: 23-40.

The Royal Botanic Gardens Kew. 1999. Botanical Database. Web Page: www.rbgkew.org.uk.

Webster, G. L. 1995. The panorama of neotropical cloud forests. *In* S. P. Churchill, H. Balslev, E. Forero, and J. L. Luteyn (eds.). Biodiversity and Conservation of Neotropical Montane Forests. The New York Botanical Garden, New York.

Young, K. R. 1991. Floristic diversity on the eastern slopes of the Peruvian Andes. *Candollea* 46: 125-143.

Young, K. R. 1992. Biogeography of the montane forest zone of the Eastern slopes of Peru. *In* K. R. Young and N. Valencia (eds.) Biogeografia, Ecología y Conservación del Bosque Montano en el Perú. Memorias del Museo de Historia Natural 21. Universidad Nacional Mayor de San Marcos, Lima.

Young, K. R. 1998. Composition and structure of a timberline forest in north-central Peru. *In* F. Dallmeier and J. A. Comiskey (eds.) Forest Biodiversity in North, Central and South America and the Caribbean: Research and Monitoring. Man and the Biosphere Series, Vol 21. UNESCO and The Parthenon Publishing Group, Carnforth, Lancashire, UK.

Young, K. R., and N. Valencia. 1992. Introduccíon: los bosques montanos del Perú. *In* K. R. Young and N. Valencia (eds.) Biogeografia, Ecología y Conservación del Bosque Montano en el Perú. Memorias del Museo de Historia Natural 21. Universidad Nacional Mayor de San Marcos, Lima.

CHAPTER 5

AVIFAUNA OF THE NORTHERN CORDILLERA DE VILCABAMBA, PERU

Thomas S. Schulenberg and Grace Servat

INTRODUCTION AND METHODS

(Thomas S. Schulenberg)

In contrast to most other taxa, the avifauna of the Cordillera DE Vilcabamba was relatively well-known prior to the 1997-1998 RAP expeditions, as a result of surveys in the late 1960s and early 1970s by John Terborgh and John Weske. The Terborgh and Weske expeditions were based at Luisiana (ca. 600 m), on the Río Apurímac on the southern flanks of the Cordillera de Vilcabamba, and reached treeline at 3520 m. The purpose of their surveys was to investigate the ecological aspects of bird community structure along an elevational gradient. These expeditions resulted in an important collection, primarily housed at the American Museum of Natural History, New York, and also in a landmark series of papers on bird ecology (Terborgh 1971, 1977, 1985, Terborgh and Weske 1969, 1975).

Terborgh and Weske recorded 405 bird species on the western slopes of the Cordillera de Vilcabamba (data from Appendix 1 of Weske 1972). An additional 158 species were recorded in the floor of the Apurimac Valley, near Luisiana or downstream from there (including areas of as low as 340 m, along the Río Ene). Terborgh and Weske also discovered several previously unknown species (Weske and Terborgh 1974, 1981, Vaurie et al. 1972) and subspecies (Fjeldså and Krabbe 1984: 496, Remsen 1984, Weske 1985, Brumfield and Remsen 1996, Schuchmann and Zuechner 1997) of birds, and found many important range extensions as well (e.g., Terborgh and Weske 1972).

The 1997 and 1998 RAP surveys were the first scientific expeditions to assess biological diversity in areas not surveyed by any other scientific team. For this RAP study, birds were surveyed with binoculars, tape recorders, and directional microphones. We documented the presence of as many species as possible with specimens or tape recordings; recordings by T. Schulenberg will be archived at the Library of Natural Sounds, Cornell Laboratory of Ornithology, Ithaca, New York. Small numbers of specimens were collected at Camps One and Two during the 1997 expedition, and are housed at the Museo de Historia Natural de la Universidad Nacional Mayor de San Marcos, Lima, and at the Field Museum of Natural History, Chicago.

RESULTS

Avifauna of Camp One, 3350 m, headwaters of the Río Pomureni, northern Cordillera de Vilcabamba

(Thomas S. Schulenberg)

RAP Camp One was located one degree north of the Terborgh and Weske transect. This camp was at 3350 m, just below the crest of the highest ridges in this part of the range. The area was a mosaic of *pajonal* (itself a mosaic of different kinds of bogs, some of which supported some grass), shrubbery and forest patches, and taller forest, including bands of *Polylepis* and of taller humid forest. Ornithologists at this site were Schulenberg and Servat. Schulenberg surveyed all habitats, to a radius of several kilometers from the camp; however, only small patches of tall humid forest were readily accessible, and our expedition did not adequately survey this habitat. Servat primarily surveyed birds of *Polylepis* forests (see below).

Forty-three bird species were recorded at Camp One during the survey (Appendix 13). *Schizoeaca vilcabambae* (Vilcabamba Thistletail), a species discovered by Terborgh and Weske (Vaurie et al. 1972), was common in treeline forest and in shrubs on the pajonales, as was an undescribed subspecies of *Ochthoeca fumicolor* (Brown-backed Chat-Tyrant; Fjeldså and Krabbe 1984). The species total at Camp One, was low when compared with some of those reported by Terborgh and Weske. For example, Terborgh and Weske recorded 61 species at an elevation between 3300 and 3500 m, whereas we only recorded 43 species at this elevation. However, most of the species not recorded by us are species typically found in tall humid forest, a habitat we did not fully survey. Thus, a direct comparison is not appropriate. Present even in relatively small patches of tall humid forest were *Coeligena violifer albicaudata* (Violet-throated Starfrontlet; Schuchmann and Zuechner 1997) and *Cranioleuca marcapatae weskei* (Marcapata Spinetail; Remsen 1984), both of which were described based on the Terborgh and Weske collections.

Nonetheless several expected species typical of pajonal and forest edge were not encountered. For example, many sites in the Andes have two species each of *Metallura* hummingbird and *Diglossa* flower-piercer, yet we recorded only a single species per genus. We also had no records of *Gallinago imperialis* (Imperial Snipe), although its discovery in the Vilcabamba was one of the greatest surprises of the Terborgh and Weske surveys (Terborgh and Weske 1972). Since then, this snipe also has been found in several sites in the Andes of Peru and Ecuador (Parker et al. 1985, Stotz et al. 1996), and so it was expected to be found in the northern portions of the Cordillera de Vilcabamba as well.

Two species were found at Camp One that were not reported from the Cordillera de Vilcabamba by Weske (1972): *Buteo polyosoma* (Red-backed Hawk) and *Notiochelidon murina* (Brown-bellied Swallow). Both are widespread Andean species, and their presence here was not surprising.

There were extensive bands of *Polylepis* forests at Camp One. A number of bird species are associated with *Polylepis* woodlands, some of which are almost never found in other habitats (Fjeldså and Kessler 1996). Therefore, perhaps the greatest surprise of the expedition was the apparent absence of these *Polylepis* specialists, as detailed in the following section.

Avifauna of the *Polylepis* forests at Camp One, northern Cordillera de Vilcabamba, Peru

(Grace Servat)

Polylepis (Rosaceae) forests are known to have a number of birds that are narrowly bound to them ("specialists") in addition to many other non-endemic and widespread species that use the forest marginally. As many as 120 bird species can be found in these forests along the Peruvian Andes.

The higher parts of Cordillera de Vilcabamba have been recognized as a priority area for *Polylepis* forest preservation, based on the high concentration of restricted-range bird species. However, nothing is known about the unexplored areas of these mountains due to the inaccessibility of the terrain (Fjeldså and Kessler 1996). The present RAP report contributes information on one of these unexplored areas that contain extensive patches of *Polylepis* forest. The studies were conducted at Camp One (3350 m), in the department of Junín in the northern Cordillera de Vilcabamba.

I censused and mist-netted the *Polylepis* woodland at this locality in order to determine the bird species and specialists present.

1) Censuses. - I conducted four censuses in *Polylepis sericea* woodland, using spot-mapping in transects with fixed distance; in order to do this, I established 3 transects of 100 m each along a 1000 x 3000 m woodland. The terrain was steep and the understory covered with *Chusquea* (Poaceae), which made the transects very difficult to follow. In each transect, there were 6 fixed stations (spaced at regular 20 m intervals) where I tape-recorded and observed birds during 2 minute periods. Three censuses were made between 0800-1000 hr and another one from 1500-1700 hr. In addition, I spent 29 hours in this woodland making general bird observations and mist netting.

2) Mist Nets.- Four mist nets were placed in different locations in the *P. sericea* woodland for two consecutive days and kept open from 0900 to 1600 hr. Later, I changed the location of mist nets from this woodland to a *Polylepis pauta* woodland located in front of the camp.

The censuses indicate that *Mecocerculus leucophrys* (White-throated Tyrannulet; 3 individuals), *Schizoeaca vilcabambae* (2 individuals), *Ochthoeca rufipectoralis* (Rufous-breasted Chat-Tyrant; 2 individuals) and *Metallura tyrianthina* (Tyrian Metaltail) were resident in the *Polylepis sericea* woodland, and probably also *Scytalopus parvirostris* (1 individual), and *Grallaria rufula* (Rufous Antpitta; 1 individual). A mixed flock that included *Margarornis squamiger* (Pearled Treerunner) and *Conirostrum sitticolor* (Blue-backed Conebill) included this woodland as part of their foraging range. The following birds were captured in mist nets at Site 1 (14 hours): *Diglossa mystacalis* (Moustached Flower-piercer) and *Schizoeaca vilcabambae*; at Site 2 (16 hours): *Schizoeaca vilcabambae*, *Ochthoeca fumicolor*, *Uropsalis segmentata* (Swallow-tailed Nightjar), *Zonotrichia capensis* (Rufous-collared Sparrow), *Turdus fuscater* (Great Thrush), *Coeligena violifer*, and *Anisognathus igniventris* (Scarlet-bellied Mountain-Tanager).

No bird species that are narrowly bound to *Polylepis* were found at either of the two woodlands surveyed. All bird species reported are widely distributed near Camp One. A major question that emerges from this survey is: Why are there no narrowly bound bird species associated with these *Polylepis* woodlands? This association is common elsewhere in Peru, including the higher parts of the southern Cordillera de Vilcabamba (around Nevada Sacsarayoc at 5994 m), which was proposed as a priority area for conservation of biological diversity of *Polylepis* woodlands (Fjeldså and Kessler 1996). Some hypotheses that may explain the lack of *Polylepis* specialists are listed here:

1) Low elevation forest.- Is this region too low to contain specialized birds? *Polylepis sericea* and *P. pauta* are among the most primitive *Polylepis* species that usually occur admixed with species from montane areas, and therefore a lower number of specialists is expected to occur given the lack of isolation that has promoted speciation in higher elevation forests. However, *Oreomanes fraseri* (Giant Conebill), one of the most specialized birds in the *Polylepis* system, has been recorded at elevations as low as 2700 m.

2) Isolation.- Are these mountains too isolated to allow colonization? The Cordillera de Vilcabamba is isolated due to geographical barriers that may not have allowed bird species colonization, or perhaps recolonization after local extinction, to occur on these peaks. In order to test this hypothesis, it will be necessary to obtain data from the closest localities where specialized birds occur and determine the distance to the area studied. These data will contribute valuable information on the distance at which colonization of these patchily distributed woodlands becomes unlikely in habitat specialists

3) Forest age.- Is this forest too young to have narrowly bound bird species? No data are available to judge the age of the woodland at this site. However, this must be explored as a possible explanation as to why no narrowly bound bird species occur there. Since no endemics were detected in the other inventoried group either, the case grows stronger. Testing this hypothesis will involve examining old Landsat photos, aging trees, and taking pollen cores.

4) Resources.- Are the food resources (arthropods) the same in terms of abundance and types as in other *Polylepis* forests? To evaluate this hypothesis, I sampled arthropods from *Polylepis* bark. Bark resources are used by *Oreomanes fraseri* and *Cranioleuca baroni* (Baron's Spinetail), two widespread specialists of *Polylepis* woodlands. I compared bark samples from Vilcabamba to those obtained at two other *P. sericea* forests in the Cordillera Blanca: Morococha (3850 m) and Yanganuco (3850 m). An analysis of spider abundance (the most abundant arthropods in the bark and also the main food resource used by these birds, as evidenced by stomach content analyses from my own studies) indicates that no significant differences in spider abundance was found among the three localities ($F = 0.09$, $P > 0.05$). In terms of types of resources, the same arthropod taxa were found at the three localities.

5) Resource accessibility.- Are these resources accessible to foraging birds? Abundant resources may be present, but these may not be accessible to birds due to the amount of moss and epiphytes that will "hide" the resources. Although this may be true for bark foragers such as *Oreomanes* or *Cranioleuca*, this hypothesis may not explain the absence of specialized foliage foragers such as *Lepthastenura*.

Finally, the absence of specialized birds in the area surveyed in the northern Cordillera de Vilcabamba may be due to a combination of factors that would require a more "in depth" study. This might include full documentation on the areas closest to the base of the Cordillera where specialized birds have been reported, and a subsequent comparison of these sites with the area surveyed during the RAP expedition. Nonetheless, the information obtained in this previously unsurveyed region of the Cordillera de Vilcabamba contributes new information about a distinctive and highly threatened ecosystem.

Avifauna of Camp Two, 2050 m, headwaters of the Río Poyeni, Northern Cordillera de Vilcabamba

(Thomas S. Schulenberg)

Camp Two was located on a small plateau in the headwaters of the Río Poyeni, bordered to the west by a sharp narrow ridge, and to the north and east by the gorge of the Poyeni. Ornithologists present at this camp were T. Schulenberg and L. López. The dominant habitat of the area was tall humid forest, although there were several open *Sphagnum* bogs and areas covered by tall bamboo. Initial survey efforts were concentrated on trails established to the west of camp (leading part way up the ridge, to about 2100 m), and in forests along the edge of the escarpment across the *Sphagnum* bog. In the final days of the survey, there also was trail access part way down the north face of the Poyeni gorge, so that areas as low as about 1800 m were sampled (although on only a few occasions). In addition, López erected two mist nets in humid forest between our camp and the largest *Sphagnum* bog.

We recorded 115 bird species at this site (Appendix 13). Bird abundance and bird species diversity at this site was typical for forests at this elevation; for example, Terborgh and Weske reported 90 species between 2000-2100 m farther south in the Vilcabamba (Weske 1972), and Davis recorded 95 species during a two week period m in a disturbed (logged) forest at 2200 m in the Department of Amazonas, northern Peru (Davis 1986). Of the species recorded during this survey, nine were not reported from the Cordillera de Vilcabamba by Weske (1972): *Buteo leucorrhous* (White-rumped Hawk), *Pionus tumultuosus* (Plum-crowned Parrot), *Ciccaba albitarsus* (Rufous-banded Owl), *Lurocalis rufiventris* (Rufous-bellied Nighthawk), *Myiotheretes striaticollis* (Streak-throated Bush-Tyrant), *Ampelion rufaxilla* (Chestnut-crested Cotinga), *Catharus fuscater* Slaty-backed Nightingale-Thrush, *Basileuterus signatus* (Pale-legged Warbler), and *Carduelis olivacea* (Olivaceous Siskin). Although some of these species are relatively uncommon or can be difficult to detect, all of them are widespread Andean species known from sites both to the north and to the south of the Cordillera de Vilcabamba, and their presence here was expected.

Species only recorded below the lip of the plateau (i.e., below 2000 m) include *Columba plumbea* (Plumbeous Pigeon), *Herpsilochmus motacilloides* (Creamy-bellied Antwren), *Formicarius rufipectus* (Rufous-breasted Antthrush), *Scytalopus bolivianus* (Southern White-crowned Tapaculo), *Rupicola peruviana* (Andean Cock-of-the-Rock), *Cyanocorax yncas* (Green Jay), *Myioborus miniatus* (Slate-throated Redstart), *Basileuterus signatus*, and *Chlorochrysa calliparaea* (Orange-eared Tanager).

Several species, including the White-throated Screech-Owl (*Otus albogularis*), and the Citrine Warbler (*Basileuterus luteoviridis*) were found at lower elevations (2050-2100 m) than is typical for most other Peruvian localities where they are found.

Camp Two was approximately situated in the middle of the known elevational range (1700-2200 m) of *Otus marshalli* (Cloud-Forest Screech-Owl), a species first discovered in the Cordillera de Vilcabamba (Weske and Terborgh 1981). Therefore, it was disapointing not to have found this poorly-known bird during our expedition. *Otus ingens* (Rufescent Screech-Owl) was found at the upper lip of the Poyeni gorge, near its upper elevational limit. At the type locality, farther south in the Cordillera de Vilcabamba, *Otus marshalli* occupied an elevational band between these two species, and so the (local) absence of *O. marshalli* presumably is related to the factors that allow *O. albogularis* to occur so low. *Otus marshalli* is otherwise known only from a single record from the Cordillera Yanachaga, central Peru (Schulenberg et al. 1984). Schulenberg et al. (1984) reported that "Both the Yanachaga and Vilcabamba are semi-isolated spurs of the main Andean cordillera, suggesting that *Otus marshalli* may have a relictual distribution restricted to such outlying ranges." Actually, it is more likely that *Otus marshalli* is relatively widespread in montane forests in central Peru.

Grallaria erythroleuca (Red-and-White Antpitta) was fairly common at Camp Two; this species has a very restricted distribution, known only from the Cordilleras de Vilcabamba and Vilcanota. Although detailed comparisons have not yet been made, the population in the Cordillera de Vilcabamba appears to differ subtly from birds of the Vilcanota in both voice and plumage, and so may represent an undescribed taxon.

Avifauna of the upper Río Pangoreni, Northern Cordillera de Vilcabamba: RAP Camp Three

(Thomas S. Schulenberg)

The camp was located on a small saddle at about 1000 m, on a low ridge just west of Tangoshiari. Much of the area above the camp, up to the summit of the ridge, was dominated by bamboo, whereas most of the areas at and below camp were in tall humid forest. Ornithologists present at this site were T. Schulenberg and A. Valdes. We regularly worked between about 750 and 1150 m, both along the trail returning to Tangoshiari and along a trail (old hunting trail?) following a side ridge south towards the headwaters of the Río Muiti. From 30 April-7 May there was rain daily, often for much of the day, which made field work more difficult.

We recorded a total of 150 bird species at Camp Three (Appendix 13). We found at least 36 bird species here that were not recorded by Terborgh and Weske in the upper Apurimac valley or on the western slopes of the Vilcabamba (based on Appendix 1 in Weske 1972). The majority of these are widespread Amazonian species, and their apparent absence in the upper Apurimac, an area with a clearly Amazonian avifauna, is a little surprising. It is possible that the drier, deciduous forests of the Río Ene, downstream from the Apurímac, have acted as a "filter" for movements of Amazonian species.

More notable was the presence near Camp Three of a number of characteristic "upper tropical" species, such as *Phaethornis koepckeae* (Koepcke's Hermit), *Heliodoxa branickii* (Rufous-webbed Brilliant), *Hemitriccus rufigularis* (Buff-throated Tody-Tyrant), the recently described *Phylloscartes parkeri* (Cinnamon-faced Tyrannulet; Fitzpatrick and Stotz 1997), *Pipreola chlorolepidota* (Fiery-throated Fruiteater), *Ampelioides tschudii* (Scaled Fruiteater), *Lipaugus subularis* (Gray-tailed Piha), and *Oxyruncus cristatus* (Sharpbill). Of these, only two (*Heliodoxa branickii* and *Phylloscartes parkeri*) had been reported from the western side of the Cordillera de Vilcabamba, and most of them are known in Peru from only a few localities.

Our records of *Pipreola chlorolepidota*, however, are of some interest. Until recently this species was known south only to central Peru, in the department of Pasco; in 1992, however, Ted Parker found this species in the Cerros del Távara, department of Puno, in far southern Peru (Foster et al. 1994). Our records from the Cordillera de Vilcabamba are the first records for southern Peru since Parker's, and help to fill the "gap" established by this huge range extension.

Although there are a number of "bamboo specialists" in southwestern Amazonia (Kratter 1997), only few such species were recorded during our survey. That we did not record more bamboo specialists was a little surprising, in view of the abundance of bamboo in this region (visible from the air and from satellite image). Our site probably was at the upper elevational limit for most of these bird species, however, and that may have had an attenuating effect on this portion of the avifauna.

No cracids were detected during the survey. However, the men from Tangoshiari who hunted clearly were familiar with *Penelope jacquacu* (Spix's Guan) and *Crax tuberosa* (Razor-billed Curassow), and explained that the scarcity of the two species was a local effect due to hunting pressure. These hunters were well aware of the deleterious effects of hunting with shotguns on game populations, a common practice in this community for the past few years. It is not clear that these hunters had modified their practices in any way, however, other than traveling farther to hunt. They did not seem to be familiar with *Crax unicornis* (Horned Curassow), a rare and local species that is expected to occur in this region.

CONSERVATION RECOMMENDATIONS

The area around Camp Three (ca. 1000 m) is in the elevational zone that represents the upper limit of the Amazonian avifauna, and the lower limit of Andean species. Consequently, these "upper tropical" forests typically have very high levels of forest bird species diversity; many of the Andean species found in this habitat also are restricted to narrow elevational bands. Given the high levels of diversity and endemism, and rates of forest loss, protection of such forests ranks as one of the most urgent conservation priorities in South America (Stotz et al. 1996).

Several of the bird species observed at Camp Three apparently are restricted to low outlying ridges along the base of the Andes. Consequently, their distributions are exceptionally narrow and naturally fragmented, making these species of particular interest for conservation (Stotz et al. 1996:44-45).

The apparent absence of bird species specializing on *Polylepis* forests deserves further study. Polylepis forests and their inhabitants have been identified as a priority for conservation in Peru (Fjeldsa and Kessler 1996). The abundance of this habitat type in the higher elevations of the Cordillera de Vilcabamba provides the ideal opportunity for its conservation. A potentially new species of Antpitta discovered at Camp Two highlights the potential species richness of the Cordillera and its conservation importance.

LITERATURE CITED

Brumfield, R. T., and J. V. Remsen, Jr. 1996. Geographic variation and species limits in *Cinnycerthia* wrens of the Andes. Wilson Bulletin 108: 205-227.

Davis, T. J. 1986. Distribution and natural history of some birds from the departments of San Martín and Amazonas, northern Peru. Condor 88: 50-56.

Fitzpatrick, J. W., and D. F. Stotz. 1997. A new species of tyrannulet (*Phylloscartes*) from the Andean foothills of Peru and Bolivia. Pages 37-44 *in* Remsen, J. V., Jr. (editor). Studies in Neotropical ornithology honoring Ted Parker. Ornithological Monographs Number 48.

Fjeldså, J., and M. Kessler. 1996. Conserving the biological diversity of *Polylepis* woodlands of the highland of Peru and Bolivia: a contribution to sustainable natural resource management in the Andes. Copenhagen: NORDECO.

Fjeldså, J., and N. Krabbe. 1984. Birds of the high Andes. Copenhagen: Zoological Museum, University of Copenhagen.

Foster, R. B., J. L. Carr, and A. B. Forsyth (eds.) 1994. The Tambopata-Candamo Reserved Zone of southeastern Perú: a biological assessment. RAP Working Papers Number 6. Washington, D.C.: Conservation International.

Kratter, A. W. 1997. Bamboo specialization by Amazonian birds. Biotropica 29: 100-110.

Parker, T. A., III, T. S. Schulenberg, G. R. Graves, and M. J. Braun. 1985. The avifauna of the Huancabamba region, northern Peru. Pages 169-197 *in* Buckley, P. A., M. S. Foster, E. S. Morton, R. S. Ridgely, and F. G. Buckley (editors), Neotropical ornithology. Ornithological Monographs Number 36.

Remsen, J. V., Jr. 1984. Geographic variation, zoogeography, and possible rapid evolution in some *Cranioleuca* spinetails (Furnariidae) of the Andes. Wilson Bulletin 96: 515-523.

Schuchmann, K.-L., and Zuechner, T. 1997. *Coeligena violifer albicaudata* (Aves, Trochilidae): a new hummingbird subspecies from the southern Peruvian Andes. Ornitología Neotropical 8: 247-253.

Schulenberg, T. S., S. E. Allen, D. F. Stotz, and D. A. Wiedenfeld. 1984. Distributional records from the Cordillera Yanachaga, central Peru. Gerfaut 74: 57-70.

Stotz, D. F., J. W. Fitzpatrick, T. A. Parker III, and D. K. Moskovits. 1996. Neotropical birds: ecology and conservation. Chicago: University of Chicago Press.

Terborgh, J. 1971. Distribution on environmental gradients: theory and a preliminary interpretation of distributional patterns in the avifauna of the Cordillera Vilcabamba, Peru. Ecology 52: 23-40.

Terborgh, J. 1977. Bird species diversity on an Andean elevational gradient. Ecology 58: 1007-1019.

Terborgh, J. 1985. The role of ecotones in the distribution of Andean birds. Ecology 66: 1237-1246.

Terborgh, J., and J. S. Weske. 1969. Colonization of secondary habitats by Peruvian birds. Ecology 50: 765-782.

Terborgh, J., and J. S. Weske. 1972. Rediscovery of the Imperial Snipe in Peru. Auk 89: 497-505.

Terborgh, J., and J. S. Weske. 1975. The role of competition in the distribution of Andean birds. Ecology 56: 562-576.

Vaurie, C., J. S. Weske, and J. W. Terborgh. 1972. Taxonomy of *Schizoeaca fuliginosa* (Furnariidae), with description of two new subspecies. Bulletin of the British Ornithologists' Club 92: 142-144.

Weske, J. S. 1972. The distribution of the avifauna in the Apurimac Valley of Peru with respect to environmental gradients, habitats, and related species. Unpublished Ph. D. thesis. Norman, Oklahoma: University of Oklahoma.

Weske, J. S. 1985. A new subspecies of Collared Inca hummingbird (*Coeligena torquata*) from Peru. Pages 41-45 *in* Buckley, P. A., M. S. Foster, E. S. Morton, R. S. Ridgely, and F. G. Buckley (editors), Neotropical Ornithology. Ornithological Monographs Number 36.

Weske, J. S., and J. Terborgh. 1974. *Hemispingus parodii*, a new species of tanager from Peru. Wilson Bulletin 86: 97-103.

Weske, J. S., and J. W. Terborgh. 1981. *Otus marshalli*, a new species of screech-owl from Peru. Auk 98: 1-7.

CHAPTER 6

BIRDS OF THE SOUTHERN VILCABAMBA REGION, PERU

Tatiana Pequeño, Edwin Salazar, and
Constantino Aucca

INTRODUCTION

In the Neotropics, the greatest bird diversity usually occurs on cold and humid western slopes and along the lower mountains facing the lowland forest (O'Neil 1992). These relatively unexplored areas may contain many bird species with restricted and patchy distributions, factors that motivated our study.

Montane forests are stands of mature forest usually found on mountain slopes and valleys, especially on the western slope of the Andes. Trees are usually covered with epiphytes such as bromeliads, orchids, ferns, and mosses. Arboreal ferns and bamboo (*Chusquea* sp.) are also characteristic of this type of forest (Parker et al. 1982). Most authors consider the montane forest's lower elevation limit to be 1000 meters (Young and Valencia 1992).

Montane forests occupy a very important place in the biogeography of Peru because so much of the characteristic Peruvian flora and fauna is found in the Andean region. The complex topography of the region presents biogreographic barriers for several species, thus creating narrow distributions that are further limited by the steep slopes.

Montane habitats are among the most diverse in Peru. Approximately 930, or 55%, of the 1700 bird species for Peru listed by Parker et al. (1982) are found above 1000 m. But because the high elevation zones are not well studied, it is possible that up to 60% of the Peruvian avifauna is distributed in the montane regions. Furthermore, of the 88 known bird families in Peru, 71 are found in montane forests (Franke 1992, O'Neil 1992).

The Andes mountain range lies across six countries (Venezuela, Colombia, Ecuador, Peru, Bolivia, and Chile). Peru contains the greatest concentration (40%) of this tropical Andean region, which makes the Peruvian montane forest a very complex habitat. Despite the large area encompassed by this habitat, research efforts and funding support have been minimal, making this region the least known in the country.

The elfin forest is usually found on the crest of mountains and resembles humid montane forest with abundant arboreal epiphytes (Parker et al. 1982). Essentially, it is a specialized kind of cloud forest, abundant with moss and practically impenetrable because of its tangled root system that lies above ground. Many bird species have been discovered recently in this kind of environment (Ridgley and Tudor 1989). The elfin forest is also characterized by patches of humid arboreal vegetation found above protected places and frequently associated with impenetrable layers of vegetation with abundant *Chusquea* sp. along the edges of the forest. This tangled formation is also found in forests close to mountain tops that are continuously exposed to low temperatures, wind, thick fog, and mist, earning them the name "mist forests."

METHODS

The assessment of the avifauna was carried out around the Llactahuaman camp (elevation 1710 m above sea level) between July 14 and 28, 1998, and around the Wayrapata camp (elevation 2400 m above sea level) between August 1 and 13, 1998, in the province of La Convención, Cusco department. The coordinates for the Llactahuaman camp were 12°51'55.5" S, 73°30'46.0" W. From the camp, we developed a trail system with an elevation range between 1500 m and 1900 m. The coordinates for the Wayrapata camp were 12°50'10.1" S and 73°29'42.6" W. This camp was located on the highest point of a mountain crest (2445 m in elevation), and the trails used reached as high as 2000 m in elevation.

Although the assessment was carried out during the dry season, the weather was usually very humid, especially during the first hours of the day. Temperatures varied between 16° and 30° C. Rains are infrequent during this time of the year, although occasional light precipitation can be recorded, especially during the morning in the form of morning dew. We experienced heavy rains and gusty winds on a few occasions at the Wayrapata camp.

Llactahuaman

Floral components of the area include bamboo (*Chusquea* spp.), which is usually patchily distributed in large clumps throughout the study area, thus indicating its high density. Another type of bamboo, known as "paca" (*Guadua* spp.) forms smaller patches next to *Cecropia* sp. We also observed arboreal ferns of the genera *Cyathea* and *Alsophila* as well as bromeliads, orchids, palm trees (*Dyctyocaryum*), and trees belonging to several genera (*Miconia, Clusia, Podocarpus*). The altitude and climate characteristics, as well vegetation components, place this area in the macro habitat category of humid montane forest (Parker et al. 1982, Walker and Ricalde 1988).

We set up four mist netting stations parallel to the Hananpata, Hurinpata, and Brichero trails, and along the Bagre ravine. The names of the trails are for reference purposes and were used only during this study. Records from each mist net station were made for a period of 4 to 5 effective days, at the end of which the capture rate tended to decline. We chose to maintain two active stations simultaneously, which allowed us to maximize capture efficiency. To identify species without capturing them, we recorded vocalizations or songs and used direct observation. References materials included Hilty and Brown (1986), Fjeldsa and Krabbe (1990), and Ridgley and Tudor (1989). Later, we compared our specimens with the ornithological collection at the Museo de Historia Natural de la UNMSM, Lima.

We noted four important bird habitats in the study area: 1) Mature montane forest with an abundance of *Chusquea* sp. and thick undergrowth accessible through the Hurinpata trail, 2) An area where *Chusquea* spp. dominated and accessible through the Hananpata trail, which also led to a dwarf forest, 3) The *pacal*, with *Guadua* spp. and *Cecropia* spp. as the predominant genera, accessible through the Brichero trail, and 4) the Bagre ridge, about nine meters wide and ending at the Río Apurímac, which was accessible through a steep trail called Tahuachaqui.

Wayrapata

Shrubby patches occurred along mountain crests and around the camp. These were composed mostly of Ericaceae, Araceae, and terrestrial orchids. Patches of *Chusquea* sp. were not as extensive as at Llactahuaman. Also, because this site was a transition zone between dwarf forest and a plateau at higher elevation, there were very few trees. The East Trail ran along a dwarf forest, where Podocarpaceae, Clusiaceae, arboreal ferns, and epiphytes (bromeliads and orchids) were the dominant vegetation types (Comiskey et al., Chapter 4).

We set up three sampling stations with mist nets along the three principal trails: West Trail, East Trail, and North Trail. The name of the trails describes the approximate direction they followed. The trails ran along the crest of a mountain chain that was a major thoroughfare for birds. Thus, mist net stations remained open during a period of time longer than usual without the capture rate decreasing. We also tape-recorded bird songs and vocalizations and recorded the presence of any bird species observed in the area. To identify species, we used the same guides as used at Llactahuaman and the ornithological collection at the Museo de Historia Natural de la UNMSM, Lima.
The area next to the trail and along the top of the mountain was composed of thick shrubby vegetation. At the lower elevations the area was less exposed to the inclement weather, and the vegetation changed to small trees and large arboreal ferns.

Sampling protocols

To document the bird community in the study sites, we used a combination of methods that included mist netting and audio and visual records. We emphasized mist netting because the density of the undergrowth, dominated by large patches of *Chusquea* spp., made visual records very difficult. Furthermore, mornings were characterized by low visibility due to the thick fog. However, we were experienced in identifying bird by calls or songs.

Llactahuaman

We used ATX mist nets 12 m long and 2.6 m high with a 36 millimeter (mm) mesh. The nets were distributed at 4 sampling stations

- Station 1: Hurinpata Trail

Along the main trail (the proposed pipeline route), down from the camp to about 1500 m in elevation, at the edge of a forest with thick *Chusquea* spp. patches.

- Station 2: Hananpata Trail

Also along the main trail, but up from the camp to about 1800 m in elevation. Similar habitat to Station 1, but with many more *Chusquea* clumps and with shorter vegetation toward the end of the trail.

- Station 3: Quebrada Barge (ridge)

The nets were parallel to the ridge and over the same body of water, at about 1550 m in elevation.

- Station 4: Brichero Trail

Trail ran diagonally to the main trail with its lowest point at a pacal (*Guadua* spp.), reaching up to 1520 m in elevation. This *pacal* also had *Cecropia* spp. trees that provided food for several fruit-eating bird species.

The thick undergrowth made access very difficult. It was possible to open only 6 nets on the first day of the study. The following days, we used 10 nets per station. To maximize the rate of capture, we opened nets at 2 stations simultaneously, with nets open 3 to 5 days per station.

This was done because birds learn to avoid the nets, especially the territorial species. Since the Hananpata Trail nets were placed on a hilltop, a main thoroughfare for birds, the nets at this station remained open for a longer period of time. There, the capture rate did not experience a considerable drop, and only 2 nets were relocated.

Wayrapata camp

We used ATX mist nets, 12 m long by 2.6 m high, with a 36 mm mesh. These nets were distributed at 3 sampling stations.

- Station 1: West Trail

This station ran along the main trail (the path of the proposed pipeline) from the camp to as low as 1800 m. The first part of the trail was characterized by thick, shrubby vegetation that continued to the helipad, about 350 m from the camp. This was a very humid zone with Ericaceae ferns. The second part of the trail was much dryer, with the ground covered by *Sphagnum* sp. moss.

- Station 2: East Trail

This station also ran along the main trail (pipeline path) from the camp to about 1800 m in elevation. At about 200 m from the camp, this trail descended rapidly and crossed a dwarf forest along steep and dry cliffs. At 500 m from the camp, conditions were very dry, although a very tall forest was located at about 650 m from camp. The source of the river began at 1800 m in elevation, some 750 m to 800 m from the camp along this trail. These were the only bodies of water that we found in the area.

- Station 3: North Trail

We placed mist nets along the trail, which ran along the crest of the hill.

Mist nets reduce the source of error in sampling caused by the observer and allow the collection of reliable quantitative information in a short amount of time. They also make it easier to repeat the study (Karr 1981).

The inaccessibility of the sampling area (due to the thick shrubby vegetation) precluded us from setting more than 6 nets along the West Trail on the first sampling day. The 10 mist nets eventually reached as far as 600 m away from the camp. We gradually placed several nets along the other 2 trails, but we were not able to reach our goal of 10 nets because of the difficult terrain. Also, because the trail ran along the edges of cliffs, we were limited in the amount of space to open all 10 nets. To complete the required number of net-hours, we left the nets open for a longer period of time (up to 7 days). The fact that we kept the nets open for a longer period of time did not reduce the capture rate, possibly because many bird species are forced to cross the mountain at the top. The number of captures for these birds was greater than that registered at the Llactahuaman camp.

Sampling with nets is one of the most reliable methods of documenting the presence of birds in an area, while identifying birds through visual or vocalization records requires much more experience. However, the latter method usually yields higher numbers of species recorded. Therefore, we tape recorded birds' songs and calls, noting habitat characteristics, time of record, and height of calling individual for later identification and comparison.

Birds captured in mist nets were identified in the field. The following data were obtained for each capture: mass, culmen length, wing chord, tarsus length, tail length, molt score, presence of brood patch, and time and location of capture. We photographed each species as visual evidence of its presence. We also marked each individual by making an incision for evaluation of density using the mark-recapture technique from other studies.

RESULTS

Llactahuaman

Species list

We recorded a total of 111 species belonging to 26 families at Llactahuaman (Appendix 14). The family with the largest representation was Emberezidae with 27 species (23.9%), followed by Tyrannidae with 17 species (15.0%), Formicaridae with 12 species (10.6%) and Trochilidae (8.0%).

Species richness and species accumulation curves

As in previous sampling studies at San Martín, Cashiriari, and Pagoreni (Dallmeier and Alonso 1997; Alonso and Dallmeier 1998, 1999) mist nets are a standardized method that allows for a comparison of species richness among bird communities. This is especially true for communities in understory habitats. Although mist nets did not provide a complete sampling of the bird community in the forests of Llactahuaman, the results for each area are comparable because each of the samples obtained in each area are proportional to the population size of that area.

The species accumulation curve for birds is shown in Figure 6.1. The curve seems to level off, possibly indicating that most of the species present in the habitat were captured. However, since sampling took place during the cold and dry season, it is likely that many species (except for the resident species) migrated toward lower, warmer, and more favorable areas. If the sampling were to take place during the warm and rainy season, it is possible that the curve would increase significantly, not only because of an increase in species with elevation migrations, but also because of an increase of migrants from the Northern Hemisphere.

The more abundant species were *Basileuterus coronatus* (9.5%), *Mionectes olivasceus* (7.1%), *Anabacerthia striacollis* (5.0%), and *Conopophaga castaneiceps* (4.7%). While we sampled only one specimen

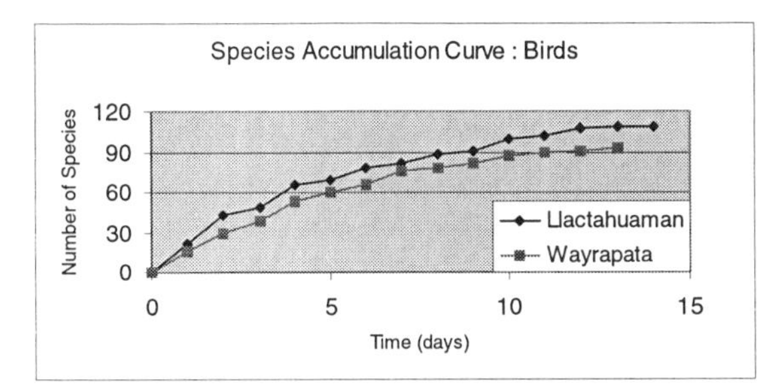

Figure 6.1 Species accumulation curve for Birds at Llactahuaman (1710 m) and Wayrapata (2445 m), Peru.

each for 24 of the 86 sampled species (27.9%), the top 12 most abundant species represented 48.9% of all the captured individuals (423 captures, 30 recaptures).

Based on their relative abundance (Stotz et al. 1996), the uncommon species were *Eutoxeres condaminii, Malacoptila fulvogularis, Eubucco versicolor, Colaptes atricollis, Campephilus pollens, Lochmias nematura, Chamaeza campanisona, Thamnophilus aethiops, Rupicola peruviana, Pipra pipra, Phylloscartes poecilotis, Rhynchocyclus fulvipectus, Knipolegus poecilurus, Cacicus leucorhamphus, Chlorochrysa calliparaea,* and *Creurgops dentata.*

Reproduction and molt

To determine which individuals were reproducing in the study area, we recorded the presence or absence of the brood patch for individuals sampled in mist nets. Of the 396 captured individuals (427 total captures minus 31 recaptures), two *Drymophilia caudata* individuals, one *Grallaricula flavirostris*, one *Conopophaga castaneiceps*, one *Trogon personatus*, and one *Rhynocyclus fulvipectus* showed old brood patches, indicating that they had brooded chicks some time ago. The rate of reproduction for this time of the year at Llactahuaman was very low, with only 1.5% of the captured individuals showing an old brood patch. No captured individuals had a recent or active brood patch. Nor did we observe nests, chicks, or individuals displaying typical reproductive or copulatory behavior.

Capture rates

Capture rates for each trail are shown in Table 6.1. Capture rates allow the data to be compared with greater ease between two or more study sites. This is done by calculating the total number of captures for each site for every 100 net-hours. Net-hours are obtained by multiplying the number of open nets on each trail times the number of hours the nets remained open. As Table 6.1 shows, the nets cannot always be open the same number of hours for several reasons, including rain, wind, and other factors.

The capture rate, the number of captured individuals, and number of recorded species were highest for the Hananpata station. This was because the trail ran along a hilltop, which is a major thoroughfare for birds going across the hill. It is important to note the capture rate for the first day of mist netting, which was much higher on this trail despite having only 5 open nets (40 net-hours).

Wayrapata

Species list

At Wayrapata, we recorded a total of 92 bird species belonging to 25 families (Appendix 15). Of these, the most speciose family—as at Llactahuaman—was Emberezidae with 22 species (23.7% of the total), followed by Trochilidae with 12 species (12.9%) and Tyrannidae with 10 species (10.8%).

With the exceptions of parrots, swifts, swallows, and vultures, the majority of species records at Wayrapata were obtained by mist net captures. We had relatively few records based on direct observation because of the low visibility in the first hours of the day, which is precisely the time of greatest bird activity in this area.

Species richness and species accumulation curves

The species accumulation curve in this area did not level off completely, indicating the possibility of obtaining new species (Fig. 6.1). This may be due to the fact that sampling took place at the only overpass crossing a mountain that separates the Río Apurímac and the Río Cumpirusiato. Furthermore, this hilltop allowed us to make observations and record species such as large vultures, swifts, and swallows that were not captured in our nets or that we did not see at Llactahuaman.

Table 6.1. Capture rates for mist nets and number of bird species captured at Llactahuaman (1710 m).

Location	Habitat	Net-Hours	Number of Captures	Capture Rates (Captures/ 100 net-hours)	Capture Rates (1st day)	Number of Species
Hurinpata	Forest and *Chusquea*	350	73	21	25*	24
Hananpata	*Chusquea*	852	202	24	77	62
Bagre	Ridge	448	76	17	24*	33
Brichero	Bamboo	356	72	20	27*	34
Total		**2006**	**423**	**21**		**85**

* Captures calculated only for the first effective day (i.e., when all 10 nets were open simultaneously.

The eleven most abundant species represent half of all captures (50.1% of total captures). The most abundant species were *Anisognatus lacrimosus* (8.8 %), *Iridosornis jelski* (6.3%), *Diglossa cyanea* (6.1%), and *D. brunneiventris* (4.1%). Of 92 species recorded in Wayrapata, 77 species (84%) were captured using the mist nets.

It is important to mention that we captured five individuals of *Notichelidon flavipes*, a species that is usually difficult to capture with mist nets because of its foraging and flying behavior. This bird constantly forages for insects in the air at high altitudes and almost never perches.

Reproduction and molting

Unlike at Llactahuaman, some bird species at Wayrapata showed evidence of reproductive activity:

- *Uropsalis lyra*. We found three nesting sites along the West Trail. The first nest was 600 m from the camp and had only one egg that hatched just a few days after it was found. The second nest was approximately 1150 m away and contained a hatchling. The third nest was abandoned but contained egg remains belonging to *U. lyra*.
- *Trogon personatus*. We found a nest of this species in a log on the East Trail, about 875 m from the camp. The male and female took turns defending the nest.
- *Cacicus cela*. We found several nests in the lower area of the Bagre cliff.
- *Chloromis riefferii*. We observed several adults displaying courtship rituals, consisting of flying upward for several meters above the tree crowns and then plunging downward while giving a characteristic call.
- *Myadestes ralloides*. One captured female was in the early stages of egg development (largest egg diameter = 3.5 mm) and showed well-developed oviducts.

A brood patch was observed on 8.7% of all the captured birds in the following species: *Anisognathus lacrymosus, Colibri coruscans, Catamblyrhynchus diadema, Diglossa albilatera, Hemispingus atropileus, Piculus rivolii,* and *Phrygilus unicolor*. About 31% of the captures showed total or partial molt. It is important to mention that the incubation brood in the hummingbird cannot be accurately identified as such because of a highly vascularized ventral area that appears to be (normally) pinkish blue.

Capture rates

Capture rates for each trail are shown in Table 6.2. Capture rates (number of captures per 100 net-hours) allow for data comparison among sites.

DISCUSSION OF AVIFAUNAL PATTERNS

Llactahuaman

We determined an order of abundance for all bird species recorded for Llactahuaman and Wayrapata based on their feeding habits. At Llactahuaman we found the following distribution of feeding habits: 57 (50.0%) species of insectivores; 34 (29.8%) frugivores species; 13 (11.4%) nectivores species; 7 (6.1%) granivores speceis; and 3 (2.6%) species of carnivores (Fig. 6.2).

We frequently observed canopy and sub-canopy flocks of mixed species composition, with most of the species belonging to the Thraupidae and Parulidae families: *Myioborus melanocephalus, Tangara xantocephala, Tangara parzudakii, Iridosornis analis, Iridosornis jelskii, Anisognathus flavinuchus, Thraupis cyanocephala*, and *Chlorornis riefferii*. Among the undercanopy mixed flocks, we observed the following species: *Premnoplex brunescens, Henicorhyna leucophrys, Grallaricula flavirostris, Anabacertia striaticollis*, and *Sindactyla subalaris*.

We recorded single-species flocks of *Carduellis magellanicus* that we observed feeding on a palm tree in a group of 8 to 10 individuals. *Columba plumbea* and *Chlorothraupis carmioli* were frequently heard in the upper canopy of the forest. A group of three *Campylorhamphus trochilirostris* individuals was captured in a mist net. We also observed a flock of *Cacicus cela* flying over the canopy.

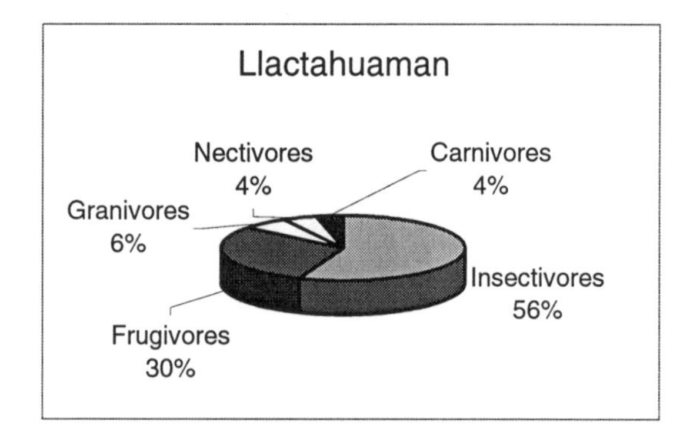

Figure 6.2. Bird feeding guilds at Llactahuaman.

Table 6.2. Bird capture rates and number of captures per net at Wayrapata (2400 m).

Location	Habitat	Net-hours	Number of captures	Captures/ 100 net-hours	Number of Species
West Trail	Shrubs	1230	267	21.71	58
East Trail	Forest	480	83	17.29	40
North Trail	Shrubs	300	35	11.67	26
Total		**2010**	**385**	**19.15**	**77**

Species that we heard singing in the same territories included *Drymophila caudata* (always in pairs), *Scytalopus unicolor* (difficult to see but heard constantly and frequently), *Grallaria flavirostris* (apparently alone), *Thamnophilus aethiops*, *Henicorhina leucophrys* (also seen and heard in pairs), and *Euphonia xanthogaster* (feeding on the same palm tree every day).

As for terrestrial species, the tinnamid *Crypturellus obsoletus* was captured in one the mammalogists' snap traps. Among the gralarids, we recorded *Chamaeza nobilis, Grallaria guatimalensis, Grallaricula flavirostris*, and *Conopophaga castaneiceps*.

Among the habitat specialists, we recorded the following species.

- Aquatic specialits: *Lochmias nematura* is a terrestrial bird associated with aquatic habitats. It was captured with the nets located on the Bagre ridge.
- Bamboo specialists: *Drymophila caudata, Thryothorus einsenmanni*, and *Cacicus holocericeus*.
- Montane forest habitat specialists: *Doryfera ludoviciae, Eutoxeres condaminii, Adelomyia melanogenis, Coeligena coeligena, Trogon personatus, Malacoptila fulvogularis, Campephilus pollens, Xiphorhynchus triangularis, Premnoplex brunescens, Anabacerthia striaticollis, Xenops rutilans, Lochmias nematura, Creurgops dentata, Conopophaga castaneiceps, Grallaria guatimalensis, Grallaricula flavirostris, Chamaeza campanisona, Myrmotherula schiticolor, Dysithamus mentalis, Pipreola pulchra, Rupicola peruviana, Chloropipo unicolor, Myiotriccus ornatus, Platyrinchus mystaceus Rhynchocyclus fulvipectus, Pseudotriccus pelzelni, Leptopogon superciliaris, Phylloscartes ventralis, Phyllomyias cinereiceps, Henicorhina leucophrys, Basileuterus coronatus, Basileuterus tristiatus, Hemispingus frontalis, Chlorospingus ophthalmicus, Iridosornis analis*, and *Anisognatus flavinucha*.
- Several species with patchy distributions or about which little is known in relation to their distribution were recorded. Such is the case of *Amazilia viridicauda*, which is known from this part of the Andes between Pasco and Cusco, but only from a few records. *Grallaricula flavirostris* has a patchy distribution east of the Andes, but it occurs frequently in this area. *Thryothorus einsenmanni* is an endemic species restricted to the Vilcanota and Vilcabamba mountain ranges in Cusco. While the majority of the recorded species are distributed along the Andes mountain range, other species such as *Crypturellus obsoletus, Oroaetus isidori, Pionus tumultuosus, Eutoxeres condamini, Trogon personatus, Aulacorhynchus prasinus, Premnoplex brunescens, Drymophila caudata, Grallaria guatimalensis*, and *Pseudotriccus ruficeps* are restricted mostly to the eastern slopes of the Andes. Some species from lower elevations are quite rare, including *Piaya cayana, Trogon collaris*, and *Momotus momota*.

Wayrapata

Fifty-eight of the 92 species recorded at this site fell into two feeding categories: frugivorous (29 species) and insectivorous (29 species). These two groups represented more than 33.0% of all the species in the community (Fig. 6.3).

- We observed mixed flocks in the canopy and just below the canopy. Among the species observed in these flocks were *Margarornis squamiger, Myioborus melanocephalus, Tangara vassorii, Tangara nigroviridis, Iridosornis analis, Iridosornis jelskii, Anisognathus lacrymosus, Anisognathus igniventris, Thraupis cyanocephala*, and *Chlorornis riefferii*.
- We also observed single-species flocks. The most notable species was *Chloromis riefferii*, which traveled along the forest canopy while calling loudly and sometimes displaying courtship behavior. Large large flocks of *Columba fasciata* flew above the forest canopy, while *Streptoprogne zonaris* flocks foraged all over the area, especially on clear days. Smaller groups of *Turdus serranus, T. chiguanco*, and *Iridosornis jelskii* were also seen flying in the forest. Three individuals, perhaps a family group, of *Buteo polyosoma* were observed flying over the valley near the Mamachuaycco cliff.
- We observed several species only within the forest. *Drymophila caudata* and *Scytalopus unicolor* are diffcult to see but can be heard singing throughout the day. A male and female of *Trogon personatus* were observed defending their nest in a log near a cliff, as was a female of *Urospalis segmentata*.
- We also captured several non-arboreal species, and *Grallariculla flavirostris* was heard singing.

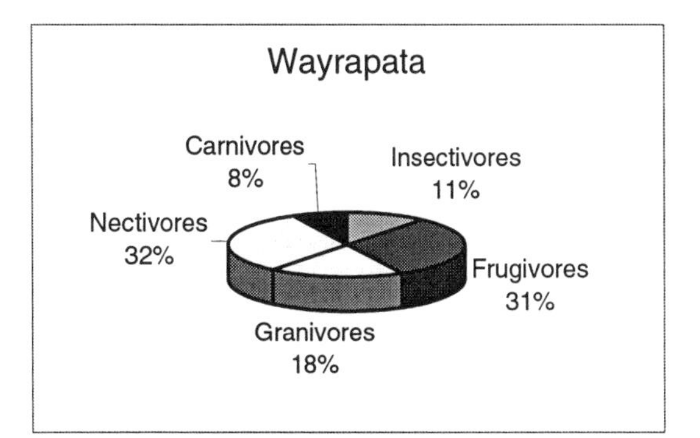

Figure 6.3. Bird feeding guilds at Wayrapata.

We classified the recorded species according to habitat (following Stotz et al. 1996):

- Bamboo specialists: *Bolborhynchus lineola, Synallaxis unirufa Drymophila caudata, Poecilotriccus ruficeps, Cinnycerthia peruana, Thryothorus einsenmanni, Haplospiza rustica, Catamblyrhynchus diadema, Hemispingus atropileus*, and *Cacicus holosericeus*.
- Shrubby montane humid forest species: *Aglaeactis cupripennis, Catamenia inornata, Catamenia homochroa*, and *Diglossa bruneiventris*.
- Elfin forest species: *Notiochelidon flavipes, Catamblyrhynchus diadema*, and *Dubusia taeniata*.
- Montane forests species: *Pipreola pulchra. Iridosornis analis, Basileuterus coronatus, Doryfera ludoviciae, Adelomyia melanogenys, Coeligena coeligena, Dubusia taeniata, Cinnycerthia peruana, Hemispingus atropileus, Pseudotriccus ruficeps, Margarornis squamiger, Andigena hypoglauca, Grallaricula flavirostris, Hemispingus frontalis, Iridosornis analis, Trogon personatus, Ochthoeca pulchella*, and *Ochthoeca cinnamomeiventris*.

Species such as *Bolborynchus lineola*, which show a distribution that includes the area where this study was conducted, are rarely seen, and there are only a few specimens. We observed two individuals of this species flying over the Wayrapata camp. As at Llactahuaman, we frequently saw individuals of *Amazilia viricauda* and *Grallaricula flavirostris*. Other species with a patchy distribution on the eastern slope of the Andes, and for which there are few records, are *Myotheretes fuscorufus* (we captured a breeding pair), *Knipolegus poecilurus* (very difficult to observe), *Notiochelidon flavipes* (difficult to sample using a mist net and seldom reported, although they may be common in this area), and *Diglossa albilatera* (distributed east of the Andes with its southernmost limits in this area).

LITERATURE CITED

Alonso, A. and F. Dallmeier (Eds.). 1998. Biodiversity assessment and monitoring of the Lower Urubamba Region, Perú: Cashiriari-3 well site and the Camisea and Urubamba Rivers. SI/MAB Series #2. Washington, D.C.: Smithsonian Institution/MAB Biodiversity Program.

Alonso, A. and F. Dallmeier (Eds.). 1999. Biodiversity assessment and monitoring of the Lower Urubamba Region, Perú: Pagoreni well site assessment and training. SI/MAB Series #3. Washington, D.C.: Smithsonian Institution/MAB Biodiversity Program.

Dallmeier, F. and A. Alonso (Eds.). 1997. Biodiversity assessment of the Lower Urubamba Region, Perú: San Martin-3 and Cashiriari-2 well sites. SI/MAB Series #1. Washington, D.C.: Smithsonian Institution/MAB Biodiversity Program.

Fjeldsa, J., and N. Krabbe. 1990. Birds of the High Andes. Zoological Museum, University of Copenhagen and Apollo Books, Svendorg, Denmark.

Franke, I. 1992. Biogeografía y Ecología de las Avers de los Bosques Montanos del Peru Occidental. Memorias del Museo de Historia Natural UNMSM (Lima) 21: 181-188.

Hilty, S., and W. Brown. 1986. A Guide to the Birds of Colombia. Princeton University Press, Princeton.

Karr, J. 1981. Surveying birds with mist nets. Studies in Avian Biology 6: 62-67.

O´Neill, J. 1992. A general overview of the montane avifauna of Peru. Memorias del Museo de Historia Natural UNMSM (Lima) 21: 47-55.

Parker, T., S. Parker, and M. Plenge. 1982. An Annotated Checklist of Peruvian Birds. Buteo Books, Vermillion, South Dakota.

Ridgely, R., and G. Tudor. 1989. The Birds of South America. Vol. 1 and Vol. 2. University of Texas Press, Austin.

Stotz, D., J. Fitzpatrick, T. Parker, and D. Moskovits. 1996. Neotropical Birds: Ecology and Conservation. The University of Chicago Press, Chicago and London.

Walker, B., and D. Ricalde. 1998. Aves de Machu Picchu y alrededores. Boletín de Lima 58: 69-79.

Young, K., and N. Valencia. 1992. Los Bosque Montanos del Peru. Memorias de Museo de Historia Natural UNMSM (Lima) 21: 5-9.

CHAPTER 7

MAMMALS OF THE NORTHERN VILCABAMBA MOUNTAIN RANGE, PERU

Louise H. Emmons, Lucía Luna W., and Mónica Romo R.

OVERVIEW AND CONSERVATION RECOMMENDATIONS

The 1997-1998 RAP expeditions were the first to survey the northern Vilcabamba mountain range at higher elevations for non-flying mammals. However, some bat collections were made as bycatch from bird studies in the 1960's by Terborgh and Weske (Koopman 1978). Its large geographical extent and isolation have marked this region as one of the potentially most interesting of the unexplored areas of South America. In the framework of the very short survey times we had at the higher elevations (9 -11 days per camp), the Cordillera de Vilcabamba abundantly justified its potential, with perhaps half a dozen new mammal taxa among the 86 species we recorded at the three camps during both years. As we are continuing to work on the collections, species identifications and numbers given here may change slightly in the future.

As would be predicted from the geography, the fauna of the range becomes more distinctive and divergent with increasing elevation because within the range we surveyed, the upper elevations are increasingly isolated from faunas at similar elevations of the Andean chain. From the perspective of conservation, a protected area limited to elevations above about 1500 m would preserve the mammal species which are either endemic to the Cordillera de Vilcabamba, or with small Andean geographic ranges. Because lowland mammals, including primates and other game species, reach to above 1000 m, a reserve protecting the fauna above this level would also have some function as a reservoir and refuge for species pressured by subsistance hunting below.

INTRODUCTION

Machu Picchu and the snow-capped peaks of the Cordillera de Vilcabamba lie at the southern base of a giant ridge of the Andes which is sandwiched between the canyons of the Ene and Urubamba rivers. This ridge, which we call the Northern Vilcabamba, is largely isolated from the main Andean mountain range by the 5000-6000 m snow-capped peaks, and by the deep canyons of the major rivers. The canyon of the Río Costreni nearly cuts across the base of the ridge at elevations of 1000-1500 m. The only place where the fauna of the 2000-3000 m level can pass to the main Andean chain along a contour line at this elevation is a tiny pass above the upper Río Apurímac; but all passage from there to the main Andes to the north is blocked by the deep canyon of the Apurimac, while to the south the contour is blocked by the Apurímac and high mountain passes. The montane faunas of the Northern Vilcabamba are thus almost isolated from the faunas at similar elevations of the main Andean Cordillera. The island-like nature of the Vilcabamba ridge, its unusual limestone formations, and its enormous tracts of undisturbed forest, have long inspired the interest of naturalists, but it has remained almost unexplored because of difficulty of physical access and terrorist activity.

Following his discovery of the ruins of Machu Picchu in 1911, in 1912 Hiram Bingham of Yale University organized a large multidisciplinary expedition to the Upper Urubamba basin around Machu Picchu, during which some mammal remains were found in association with human burials in caves, including two new taxa (Eaton 1916). Another Bingham expedition in areas surrounding Machu Picchu, from April-November 1915, included the zoologist Edmund Heller, who made extensive collections of birds and mammals on a complete elevational transect that included sites on two tributaries of the Urubamba from the Northern Vilcabamba (Río Comberciato, Río Cosireni). These collections largely reside in the United States National Museum (Smithsonian Institution), but were sent for identification to Oldfield Thomas (British Museum of Natural History), who described 12 new taxa from the 65 species collected (885 specimens, now diagnosed as 72 species; Thomas 1920, Appendix 18). We know of no other significant mammal collections from this mountain range until Terborgh and Weske collected about 900 bats on an altitudinal transect, incidental to studies of the avifauna in 1966 -68 (Koopman 1978; Appendix 17; see Chapter 5). We append to our report lists of mammals collected on both of these important expeditions (Appendices 17 and 18).

Our mammal list bears a close resemblance to the list of species collected by Heller at similar elevations in cloud forest less than 150 km from our sites (Appendix 16). However, many of his localities were either completely deforested by humans, or above the treeline, and thus he discovered open-habitat and high elevation taxa which we did not encounter (Chapman 1921; Appendix 18). Nonetheless, almost every small rodent and marsupial we collected shows minor to important divergences from its likely nearest relatives, largely found in Heller's collections, to the point that for about six of the species, we find it difficult to place them with confidence in known taxa. We are currently studying the collections and we will eventually describe several new species, but as of this writing, all results are preliminary. A number of identifications from DNA derived from tissues have kindly been made for us by Dr. James N. Patton, of the Museum of Vertebrate Zoology, University of California at Berkeley. Several identifications are based largely on his results.

THE MAMMAL FAUNA OF CAMP ONE (3350 M)

(Louise Emmons and Mónica Romo)

Methods

At Camp One we surveyed small mammals by trapping in the open *Pajonal* and *Sphagnum* bogs, in the *Polylepis/Weinmannia/Chusquea* forest, and in the sparsely wooded *Weinmannia* open woodlands. We trapped with Sherman, Tomahawk, and Victor snap traps for a total of 1,039 trap/nights (789 forest; 233 pajonal), and set a few additional Conibear traps. A line of pitfall traps across the bog was unsuccessful because the water table was too high and the buckets floated. At this site we experienced cold drizzling rain and wind almost every night, and our bat-netting efforts (a total of 12 net/nights) yielded poor results, as did 8.6 h of night observation walks, when no mammals were seen.

Results

We captured a total of 85 individuals and collected 57 voucher specimens of nine species, and identified three additional species from tracks and other signs (Appendix 16). The species accumulation curve from trapping (Figure 7.1c) shows that we had probably captured most species on our traplines, but we think it likely that within the short survey time we did not discover all species present; for example, we did not survey the pure *Polylepis* forests at some distance from the camp.

At this elevation the mammal fauna was numerically dominated by the grass mouse, *Akodon torques*, which we captured in high numbers in all habitats. In the forest we also captured three species of Thomas' paramo mice, *Thomasomys*, two of them possibly undescribed. The one individual marsupial collected, a *Gracilinanus* cf. *aceramarcae*, is only the second record of this species for Peru, and represents a considerable range extension of 400 km from the first (collected by J. L. Patton in Puno Dept.), and it is the most northerly record by this distance. Our specimen shows some divergences from the holotype, to which it was compared.

The most exciting mammalian find from Camp One was a large arboreal rodent of the family Abrocomidae, which was found dead, apparently just killed by a weasel (we collected longtailed weasels at both 2,000 and 3,000 m in 1997). This rat bears close resemblance to a species known only from two skulls and some other bones excavated from pre-Columbian (Inca?) burial sites at Machu Picchu, *Abrocoma oblativa* (Eaton 1916), which was thought to be extinct (Thomas 1920). Our specimen, externally very unlike other *Abrocoma*, shows that these specimens, collectively, represent a new genus and our individual a new species, which has recently been described as *Cuscomys ashaninka* (Emmons 1999).

Along the forest edges there were abundant signs of a tiny deer, which had created runways and open arenas marked with many pellets and damaged vegetation. We made tracings of footprints, and T. Schulenberg found a place where a predator (likely a puma) had killed one, where we collected large tufts of hair. At 2,000 m we found feces of a large cat, which included the foot bones of a small deer. We compared this hair and bones to specimens of small Andean deer and found that it closely matches specimens (including the type) of *Mazama chunyi*, a rare species known from only a few individuals from Cusco south to N. Bolivia. Interestingly, a scapula of this deer was also found in one of the Machu Picchu burials (Hershkovitz 1959). The open pajonal, especially near water-filled sinkholes, had good populations of guinea pigs (*Cavia tschudii*). There were a few old signs of Andean bears, which had uprooted and eaten the hearts of some terrestrial bromeliads or *Puya* in the pajonal, but bears were evidently uncommon or rare.

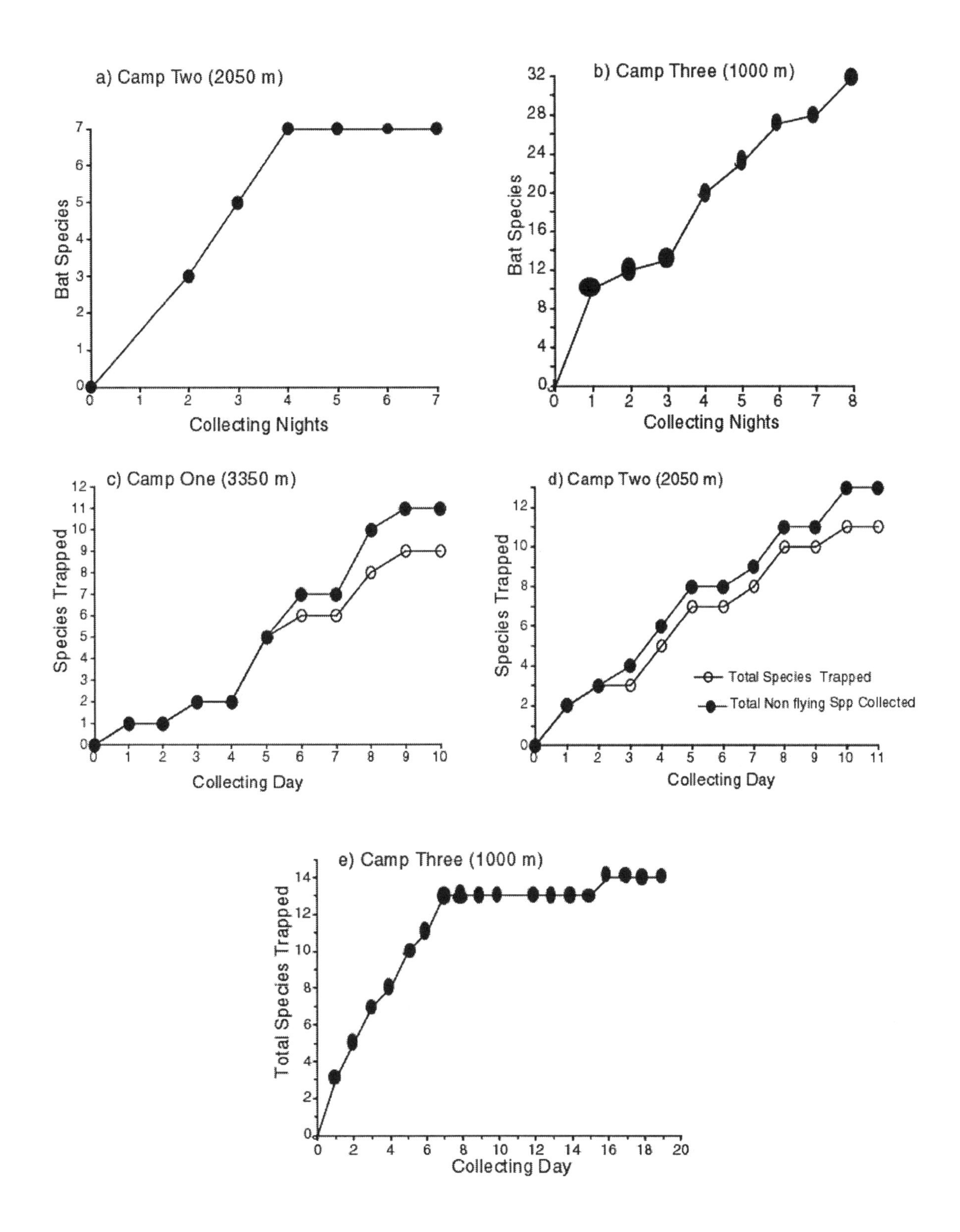

Figure 7.1. Species Accumulation curves for small mammals.

THE MAMMAL FAUNA OF CAMP TWO (2000 M)

(Louise Emmons, Lucía Luna and Mónica Romo)

Methods

At the second camp, we sampled a variety of forested habitats, with only a few traps and a line of pitfalls in the *Sphagnum* bog. We trapped for a total of 1,159 trap-nights and 26 mist net-nights, and made 23.43 h of night observation/collecting walks.

Results

The mammal fauna was richer at this elevation than at Camp One. We recorded 28 species, four of which were also found at Camp One (Appendix 16). All the genera collected were typical of this elevation, and the same rodent species, *Akodon torques*, was numerically dominant at both camps.

The most speciose and interesting genus was of the forest murid rodents *Thomasomys*, with four species at Camp One (3300 m), three species at Camp Two (2000 m) and five species total. Possibly as many as three of these are undescribed. While the species accumulation curve for bats leveled off after 4 nights (Figure 7.1a), indicating that we had captured most nettable species, the curve for trapping continued to rise, suggesting that we likely had not captured all non-flying small mammal species in the area (Figure 7.1d).

Five marsupials were captured at 2000 m, including two short-tailed opossums, *Monodelphis*, for one of which (possibly an undescribed species), we obtained the third and fourth specimens from Peru. All of the short-tailed opossums were captured at the edge of the *Sphagnum* bog near our camp, where the bog was overgrown with shrubs and bamboo.

Dense stands of impenetrable bamboo had completely covered another old bog near our camp, and a tall, canopy-forming *Chusquea* sp. formed patches within the forest. Where these had large stem diameters, the bamboo thickets were occupied by montane bamboo rats (*Dactylomys peruanus*), which called throughout the night. Our specimens seem to be the most northerly records and a new range extension for this rarely collected species. Probably because of its small size, the bog/pajonal at 2000 m lacked *Cavia* and other open-habitat species.

Our campsite was on a flat ledge or bench on the Urubamba "wall" of the Vilcabamba ridge. A kilometer or so west of the campsite, where this ledge joined the base of a vertical cliff, there was a narrow band of tall forest, at an elevation of about 2500 m. This included 30-40 m high, straight-trunked trees of an architecture similar to trees in the lowlands. This anomalous forest strip was used by at least two species of primates, spider (*Atelus belzebuth*) and night monkeys (*Atelus* sp.), while capuchin monkeys (*Cebus* sp.) were seen on the lower rim of our ledge. Because primates have been severely reduced by hunting in the inhabited lowlands, the higher elevations of the Vilcabamba are an important local refuge. The tall forest, which included many *Ficus* trees, is probably nurtured by the accumulation of nutrients and moisture from above at the base of the cliff, and it is an edaphic anomaly.

MAMMAL FAUNA OF CAMP THREE (1000 M)

(Louise Emmons, Lucía Luna, Mónica Romo)

Methods

We spent a longer time in Camp Three than at the other camps and thus had greater survey effort. Our traplines ranged in elevation from about 850 to 1200 m and included all the local habitat types of ridge top, moist ridgeside and valley bottom, drier ridge side, and bamboo forest. The whole surveyed area was forested. We trapped for 2,143 trap-nights; mistnetted for 36 net-nights, and spent 23.58 h collecting at night with a shotgun.

Results

We preserved 188 specimens, including about 46 species, and identified another 12 species from observations, tracks, or calls, for a total of 58 species (Appendix 16). Three of the observed species were found at lower elevation, near the village of Tangoshiari; *Lagothrix lagothricha* (Woolly monkey) and *Cuniculus paca* (Paca) were captured by hunters for food, and tracks of *Dasyprocta* sp. were observed near the village. The species accumulation curves for mistnetting (Figure 7.1b) and trapping (Figure 7.1e) suggest that we did not survey all of the bat fauna, but we probably sampled most of the small mammal fauna present around our traplines during that month.

As expected at 1000 m, the mammal fauna was largely composed of lowland genera and species. The only overlap between the faunas recorded at 1000 and 2000 meters were two bats and the three primates. In contrast to that at higher elevations, the lowland fauna at 1000 m on the eastern slope of the Cordillera de Vilcabamba is not isolated from similar faunas in the Amazon lowlands of Peru, so that it is not differentiated from them. Because we could not access forest other than bamboo at higher elevations above our Tangoshiari camp, we were unfortunately unable to sample the elevational transition between 1000 and 2000 m.

At this elevation the interesting aspect of the fauna was in the quantitative representation of species, rather than in their identities. Among murid rodents, the common species of the lowlands, *Oryzomys megacephalus*, which dominates lowland faunas, was rare at our camp, with only three specimens collected, while *Oryzomys macconnelli*, which is rare in the lowlands, was exceedingly common, with over twenty captures. The spiny rat, *Proechimys simonsi*, numerically dominated the small mammal fauna, especially in the bamboo forests above our camp, where it was the only rodent captured in the highest traps at about 1200 m (all in bamboo).

Thickets of large bamboo (*Guadua sarcocarpa*), which dominates enormous areas of landslides and (probably human-disturbed) habitats in nearly monspecific stands throughout this region, includes dense populations of the bamboo rat, *Dactylomys boliviensis*, which occurred within the *Guadua* bamboo up to at least 1400 m elevation. At the higher camp the bamboo *Chusquea* c.f. *picta* was similarly used by the montane bamboo rat, *D. peruanus*. Bamboo rats are herbivores that feed on the bamboo itself.

The bat fauna was similarly distinctive in that the most common lowland genera, *Artibeus* and *Carollia*, were here rare, while *Platyrrhinus* was the most common taxon in both numbers, along with the nectar feeder *Anoura*, and in species, with four.

The Asháninka of Tangoshiari hunt game for subsistance, and the populations of all large game species seemed extremely low. Repeating 12 gauge shotguns are used for civil defense against terrorists and these increase the effectiveness of hunting, especially for animals which live in troops, such as monkeys and peccaries. We saw no peccaries or deer during our stay, and only few other large mammals. The local hunters stated that when they wished to hunt for a good quantity of meat, they traveled to another valley about three days distant.

LITERATURE CITED

Chapman, F. M. 1921. The distribution of bird life in the Urubamba valley of Peru. United States National Museum Bulletin 117.

Eaton, G. F. 1916. The collection of osteological material from Machu Picchu. Memoirs of the Connecticut Academy of Arts and Sciences 5: 1-96 (plus 39 plates).

Emmons, L. H. 1999. A new genus and species of a brocomid rodent from Peru (Rodentia: Abrocomidae). American Museum Novitates, Number 3279.

Hershkovitz, P. 1959. A new species of South American brocket, genus *Mazama* (Cervidae). Proceedings of the Biological Society of Washington 72: 45-54.

Koopman, K. F. 1978. Zoogeography of Peruvian bats with special emphasis on the role of the Andes. American Museum Novitates Number, 2651.

Thomas, O. 1920. Report on the Mammalia collected by Mr. Edmund Heller during the Peruvian Expedition of 1915 under the auspices of Yale University and the National Geographic Society. Proceedings of the United States National Museum 58: 217-249.

CHAPTER 8

SMALL MAMMALS OF THE SOUTHERN VILCABAMBA REGION, PERU

Sergio Solari, Elena Vivar, Paul Velazco, and Juan José Rodríguez

INTRODUCTION

The diversity of small mammals in montane forests along the eastern slopes of the Andes is poorly known, especially when compared to studies of neotropical rainforests (Voss and Emmons 1996, Emmons and Feer 1997). The information that does exist reveals interesting patterns of diversity and distributions, with a high level of endemism for some groups of species (Thomas 1920, Koopman 1978, Leo and Romo 1992, Pacheco et al. 1993, Patterson et al. 1996). During the SI/MAB surveys, our primary objectives were to 1) determine the diversity of small mammals (bats, rodents, and opossums) in the montane forests along a proposed natural gas pipeline route from the Lower Urubamba region in southeastern Peru to the pacific coast, 2) obtain baseline information for measuring changes if a pipeline is established and for monitoring how the community of small mammals may react to reforestation, and 3) identify the factors underlying species distribution within a general comparative framework. Small mammals are a main component of tropical rainforest ecosystems (Voss and Emmons 1996), and their diversity appears to vary at different rates for each group in relation to altitude (Patton 1986, Pacheco et al. 1993, Patterson et al. 1996, 1998). To discern the patterns of diversity associated with altitudinal change, we compared our findings to those from other montane forests on the eastern slope, following the suggestions of Young (1992).

METHODS

Study Sites

Two camps were established during the SI/MAB surveys, both of them northeast of Pueblo Libre, a small town in the Kimbiri district, and southwest of the gas and condensate activity in Camisea. The first camp, Llactahuaman, was located on the western side of the range (12°51'55.5" S, 73°30'46.0" W) at 1710 meters (m) in elevation and approximately 4 kilometers (km) from Pueblo Libre. The second camp, Wayrapata, was established on the top of the ridge (12°50'10.1" S, 73°29'42.6" W) at 2445 m in elevation and about 10 km from Pueblo Libre.

Llactahuaman

Vegetation in the vicinity of Llactahuaman is typical of lower montane forest (Frahm and Gradstein, 1991; Comiskey et al., Chapter 4), acting as an elevation transition between montane and sub-montane forests (Patton 1986, Young 1992). Slopes are moderate (less than 45°), and large, entangled woody roots cover the ground. The understory is dense and it is dominated by *Chusquea* sp. (bamboo). The taller trees are covered by epiphytes. During our visit, the palm *Dyctiocaryum* sp. was fruiting, evidenced by the accumulation of seeds at the base. Palms in many forest types are considered keystone species since their seeds are the preferred food for many animals, including several bats (e.g. Stenodermatinae) and large rodents (e.g., *Dasyprocta* sp.).

At this camp, we studied small mammals from July 15-28, 1998, at the end of the dry season. Rainfall was minimal, and the morning clouds generally cleared out by noon. The nights were clear under a new moon phase. The weather was constant, with maximum temperatures of 24° C and lows of 15.5° C.

We sampled along four trails, all of them departing from the base camp:

1. Hananpata trail followed a northeast direction to the area's highest point; the primary understory vegetation was *Chusquea* sp. interspersed with palms and small trees covered by mosses and epiphytes.
2. The Hurinpata trail ran south to the Apurimac valley on a continual descent; the vegetation consisted of *Chusquea* sp., ferns, palms, and medium-sized trees

(about 40 centimeters [cm] in diameter at breast height [dbh]).

3. The Brichero trail led west through a transition from *Chusquea* sp. to a young forest of small trees (20 cm dbh) and then to a bamboo forest.
4. Tawachaqui trail was rather steep (45^0) and led east to a small creek where *Chusquea* sp. was absent and riverine vegetation dominated. Bagre creek was the nearest water source to Llactahuaman camp (see Acosta et al, Chapter 13).

Wayrapata

Vegetation at Wayrapata was of an elfin forest type (referred as "Monte Chico"by Terborgh 1971). It is characterized by the absence of tall trees (higher than 15 m) and is found on the tops of ridges, but is different in species composition from tropical subalpine forest in species composition (Frahm and Gradstein 1991, Comiskey et al, Chapter 4). According to Frahm and Gradstein (1991), elfin forests are dominated by small plants with large leaves and few ground bryophytes; due to strong wind, epiphyte and moss cover is scarce. At our study site, however, many epiphyte mosses (e.g., *Sphagnum*) grew over the exposed and entangled roots, resulting in a spongy "false ground" up to 0.5 m thick. On both sides of the ridge, the vegetation was like that of other montane forests at similar elevations, although *Chusquea* sp. was quite abundant, especially at the highest portions. Descending to the eastern side, the forest changed to a more typical montane forest with taller trees (15 to 20 m), more epiphytes, and a greater sensation of moistness. Slopes on the eastern side reached 45^0.

We sampled at Wayrapata between August 2-12, 1998. Rainfall was infrequent; it rained on the first few days and then in the middle of the sampling period only during the afternoons and nights. The mornings were cloudy and cold, but the sky was clear by early afternoon in the first week (August 4-7). The nights were generally clear after some early rain, and the full moon was usually visible. Temperatures ranged from 9^0 C at night to 28^0 C at noon; the mean was 20^0 C. Wet winds blew during most of the sampling period.

We sampled along three trails leaving the base camp:

1. The West trail, which ran in a southwest direction along the top of the ridge through typical elfin forest vegetation influenced by *Chusquea* sp.
2. The East trail, the steepest trail, leading down slope to an upper montane forest.
3. North trail, including the north part of the ridge but in contrast to the West trail, also including flat areas with palms and ferns. Some small creeks were located on the East trail, but these were not included in the sampling.

Sampling Methods

The methods used are described in detail in a previous papers concerning this project (Solari et al. 1998). The standardized protocols were designed for comparisons among several sampling locations. The sampling effort—measured by number and type of traps and nets employed—varied at each site.

We set an arrangement of trap lines combining snap (Victor) and live (Sherman and National) traps for sampling non-volant small mammals. At Llactahuaman, the trapping effort was primarily focused on the Hananpata and Hurinpata trails, where we used four parallel transects to each trail. Transects included 10 Victor traps, 10 Sherman traps, and three National traps, for a total of 92 trap-nights per trail. At Wayrapata, we used a single line trap for each of the three trails, incorporating 40 Victor traps, 40 Sherman traps, and four National traps totaling 84 trap-nights per trail. None of the trap lines extended more than 600 m from the camps. We checked them every day, in the morning and afternoon, to record activity patterns for the captured species. Traps were baited every day.

We used mist nets to sample bats and devoted most of the effort to the dominant habitats in each locality. At Llactahuaman, we set mist nets for 13 nights along the Hananpata trail, six along Bagre Creek, four along the Hurinpata trail, and three along the Brichero trail, with an average effort of 18 nets per night. At Wayrapata, we sampled 11 nights on the West trail, seven on the East trail, four on the North trail, with an average effort of 15 nets each night. Nets were open all night and were checked between 1930 and 2300 every night.

We compared the diversity and abundance of the small mammals among sampled habitats at each camp, but because of differences in the sampling effort, we emphasized analysis of species composition. We obtained the abundance of bat species from the ratio of the number of captures to netting effort (Wilson et al. 1996). A similar index was obtained for non-volant species using the trap-night results.

Although different habitats were sampled at each locality, we combined the small mammal lists from Llactahuaman and Wayrapata and compared the result with other montane forests in Peru. The comparison was based only on the number of species recorded within the elevation range of our two study locations (between 1700 and 2500 m).

Specimens were preserved as skins and skeletons or fixed in a 10% formaldehyde solution and stored in 70% alcohol. We recorded external measurements, reproductive status, and data related to each capture. These specimens and data are housed in the Museo de Historia Natural, Universidad Nacional Mayor de San Marcos (MUSM), Lima. Because the systematics of montane small mammals, especially rodents, is poorly known (Patton 1986, Carleton and Musser 1989, Patton and Smith 1992), our identifications are preliminary and must be compared with larger museum series.

RESULTS AND DISCUSSION

We recorded a combined total of 28 different species from both camps (19 at Llactahuaman and 20 at Wayrapata, Table 8.1). Eleven species were common to both localities, while eight were restricted to the lower montane forests of Llactahuaman and nine to the elfin and transitional forests of Wayrapata. The species list comprises 16 genera from 5 families and 3 orders (Table 8.1). Most of the species were typical of montane forests, but some are from pre-montane forests or have a wide altitudinal range that includes several types of montane forests (Patton 1986, Pacheco et al. 1993, Patterson et al. 1998).

Even though the sampling effort at Llactahuaman was more intense than that at Wayrapata—2400 trap nights and 230 net nights at Llactahuaman compared to 1800 trap nights and 160 net nights at Wayrapata—the success rate for non-volant small mammals (marsupials and rodents) was higher at Wayrapata (1.47%; 26 captures) compared to Llactahuaman (0.96%; 23 captures). For bats the success rate was 0.45 individuals per net-night at both camps (103 captures at Llactahuaman and 73 at Wayrapata). Eight species of non-volant mammals and 14 species of bats were recorded at Wayrapata while five species of non-volant mammals and 12 species of bats were recorded at Llactuahuaman (Table 8.1, Figs. 8.1 and 8.2).

Table 8.1. Comparative diversity of small mammals of Llactahuaman (LLH) and Wayrapata (WAY), two montane forest localities in the Southern Cordillera de Vilcabamba, Peru.

		Sampling Locality	
Species	**Family**[1]	**LLH (1700 m)**	**WAY (2500 m)**
Didelphimorphia (marsupials)			
Caluromys lanatus	cal		X
Marmosops noctivagus	mar	X	
Micoureus demerarae	mar	X	
Monodelphis osgoodi	mar		X
Monodelphis theresa	mar		X
Chiroptera (bats)			
Anoura caudifer	phy	X	X
Anoura geoffroyi	phy	X	X
Anoura latidens	phy		X
Carollia brevicauda	phy	X	
Carollia perspicillata	phy	X	
Carollia sp. nov.	phy	X	
Dermanura cinerea	phy	X	
Dermanura glauca	phy	X	X
Enchistenes hartii	phy		X
Platyrrhinus cf. *dorsalis*	phy	X	X
Platyrrhinus cf. *nigellus*	phy	X	X
Platyrrhinus vittatus	phy	X	X
Sturnira bidens	phy		X
Sturnira erythromos	phy	X	X
Sturnira oporaphilum	phy	X	X
Sturnira tildae	phy	X	
Myotis keaysi	ves	X	X
Rodentia (rodents)			
Akodon aerosus	mur	X	X
Akodon torques	mur		X
Microryzomys minutus	mur		X
Oecomys superans	mur	X	
Oligoryzomys destructor	mur		X
Oryzomys keaysi	mur	X	X
Total species		**19**	**20**

[1] cal = Caluromyidae; mar = Marmosidae; phy = Phyllostomidae; mur = Muridae; ves = Vespertilionidae

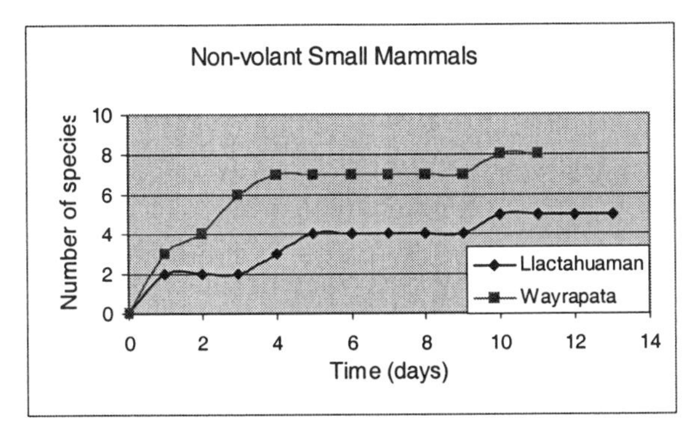

Figure 8.1 Species accumulation curve for Non-volant small mammals at Llactuahuaman and Wayrapata, Peru.

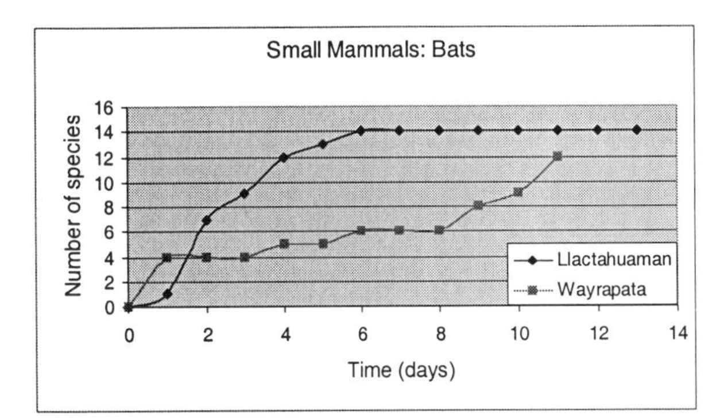

Figure 8.2 Species accumulation curve for Bats at Llactuahuaman and Wayrapata, Peru

The species accumulation curves for each camp were compared (Figs. 8.1 and 8.2). Capture of non-volant small mammals leveled off after 10 days of sampling at both sites (Fig. 8.1). In contrast, capture of new bat species at Wayrapata continued throughout the study while captures at Llactahuaman leveled off after six days (Fig. 8.2). Noteworthy records from these montane forests included the sympatry (i.e., species from the same ecological or systematic group that live in the same location) within several genera such as the marsupials *Monodelphis* (two species at Wayrapata), the rodent Akodon (two species at Wayrapata), and the bats *Anoura* (three species at Wayrapata) and *Platyrrhinus* and *Sturnira* (three species of each at each camp; Table 8.1).

Diversity and Abundance

Non-volant Small Mammals: Marsupials

We recorded four species of marsupials (Didelphimorphia) of the family Marmosidae, including *Marmosops noctivagus* and *Micoureus demerarae* at Llactahuaman and *Monodelphis* cf. *osgoodi* and *Monodelphis* cf. *theresa* at Wayrapata (Table 8.1). A fifth record was the arboreal *Caluromys lanatus* (Caluromyidae) sighted on a tree in the transitional forest along the East trail at Wayrapata (feces were also found but the record was not included in the analysis).

Non-volant Small mammals: Rodents

Among the rodents, we recorded two tribes of the subfamily Sigmodontinae (Muridae)—the Akodontini, represented by two species of *Akodon* (one at Llactahuaman, both at Wayrapata), and the Oryzomyini, with four species within four genera (two of them at Llactahuaman, three at Wayrapata; Table 8.1). T he Oryzomyini species at Llactahuaman were *Oryzomys keaysi* and *Oecomys superans*; at Wayrapata, *O. keaysi*, *Oligoryzomys destructor*, and *Microryzomys minutus* (Table 8.1). The most abundant species were *O. keaysi* and *Akodon aerosus* (11 and 10 individuals, respectively) at Llactahuaman and *A. aerosus* (16 individuals) at Wayrapata (Fig. 8.3). Some species typical of montane forests, including *Oligoryzomys destructor* and *Microryzomys minutus* (Carleton and Musser 1989), were scarce at Wayrapata and absent at Llactahuaman. *Akodon* maintained similar abundance at both camps, but *Oryzomys* showed large differences between the camps. Forest structure, weather, and food availability may all play roles in shaping rodents communities in montane forest.

Another trait of the forests at our camps was the absence of the highly diverse native genus *Thomasomys*, which is generally present in eastern-slope montane forests in Peru (Appendix 19). It could be that our camps represent the lower and upper limits to the distribution of this genus, or it could be that the genus has highly diminished populations during the time of the year when our study was conducted. Whatever the cause, this phenomenon has not been manifested in other studies of cloud and montane forests (Patton 1986, Leo and Romo 1992, Pacheco et al. 1993, Emmons et al. Chapter 7).

Trapping success for the most abundant rodent species was less than 1% (one individual/100 night traps), with 0.90% for *Akodon aerosus* at Wayrapata, and 0.46% for *Oryzomys keaysi* and 0.42% for *A. aerosus* at Llactahuaman.

Volant Small Mammals: Bats

Only two bat families were recorded at the camps. The Phyllostomidae was the most diverse family, with 16 species in six genera (Table 8.1); the subfamilies Stenodermatinae (genera *Platyrrhinus* and *Sturnira*) and Glossophaginae (genus *Anoura*) were the most abundant groups at Llactahuaman and Wayrapata, respectively (Fig. 8.4). These three genera have a wide elevation range and high species diversity in the montane forest of nearby Manu National Park (Patterson et al. 1996). A new species of *Carollia* (Phyllostomidae) was discovered at Llactahuaman. The other bat family, Vespertilionidae, was represented only by *Myotis keaysi* at both camps (Table 8.1).

Abundance among genera differed significantly between the camps (Fig. 8.4). At Llactahuaman *Platyrrhinus* and *Sturnira* accounted for nearly 60% of the captures, while at

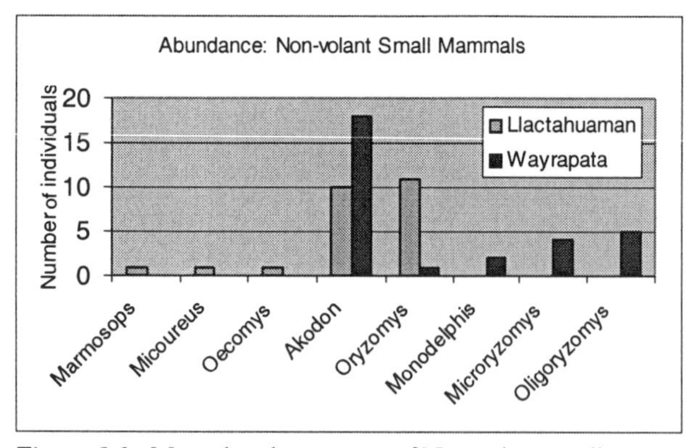

Figure 8.3. Most abundant genera of Non-volant small mammals at Llactuahuaman and Wayrapata, Peru.

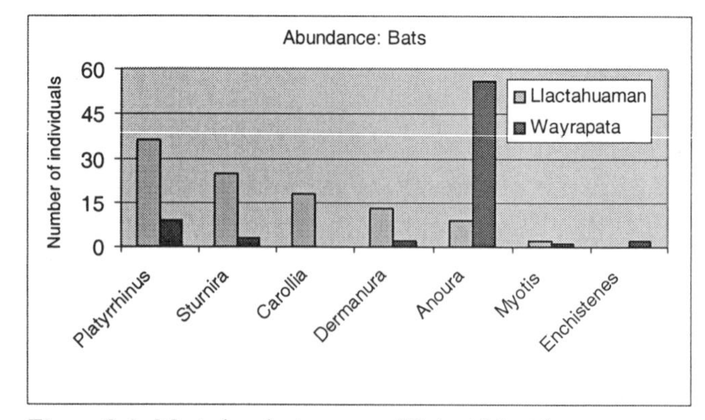

Figure 8.4. Most abundant genera of Bats at Llactahuaman and Wayrapata, Peru.

Wayrapata *Anoura* species comprised 75% of the captures. The explanation may lie in foraging strategies. Patterson et al. (1996) place these genera in different trophic guilds, which show differences in their diversity and abundance patterns along the elevation gradient. The abundance of *Anoura* at Wayrapata could be attributed to their broad food niche (ranging from insects to nectar) which allows them to find sufficient food resources even in the dry season. The abundance of *Platyrrhinus* and *Sturnira* at Llactahuaman could be due to the presence of a comparatively more diverse vegetation (Comiskey et al., Chapter 4) where these animals find fruits in different strata (canopy and understory, respectively).

We have much to learn about the influence of seasonal changes on the structure of bat communities in montane forests. Still we suggest that a reduction of fruit availability in forests with fewer plant species could be critical to frugivore (e. g., Stenodermatinae) and omnivore (e.g., Glossophaginae) groups.

Small Mammals from Other Eastern Slope Montane Forests

We compared our results to other studies and defined diversity for the habitat by the presence of the species, regardless of their abundance (Appendix 19). Most non-volant mammals (Didelphimorphia and Rodentia) are exclusive to one altitudinal range. All of the marsupials found at Llactahuaman and Wayrapata were typical of lowland (three species) and upper (six species) montane forests. Rodents are known to be more restricted in altitude changes than marsupials (Patton 1986, Patterson et al. 1998). However, two species of rodents were sampled at both sampling locations; these two species are known to have wide altitudinal ranges (Patton 1986, Pacheco et al. 1993).

Among the 17 bat species recorded in this study, 9 were sampled at both locations and 8 were exclusive to one. Species in highly diverse genera are usually segregated in highland and lowland forms (e.g., *Anoura* and *Sturnira*), while others have a wide range (*Platyrrhinus* spp.). Four species were restricted to the lowland and three to the upper montane forests. Our survey was not extensive enough to record all potential species (see Patterson et al. 1996), but the sampling covered most of the habitats in the area, and comparisons were based on this procedure.

Our records of nearly 20 species of small mammals at each camp during a sampling period of less than two weeks were close to records from other sites in Peru such as Manu (Pacheco et al. 1993, Patterson et al. 1998, S. Solari pers. obs.) and Yanachaga Chemillen (E. Vivar and S. Solari pers. obs.), but different from Abiseo (Leo and Romo 1992, M. Leo and M. Romo pers. com.), which is at a more northern latitude (Patton 1986, see Appendix 19).

We also compared our findings to the results of bat samples studied by J. Terborgh and J. Weske in montane forests of the Cordillera de Vilcabamba (see Koopman 1978), and results of small mammal research undertaken by Heller at Machu Picchu (see Thomas 1920). The resulting list included 42 species in 28 genera, 8 families, and 4 orders (Appendix 19). Some species were common throughout the latitudinal gradient, primarily bats such as *Micronycteris megalotis, Anoura* spp., *Carollia brevicauda, Dermanura glauca, Platyrrhinus* spp., *Sturnira* spp., and *Myotis keaysi*, as well as the rodent *Microryzomys minutus*.

The Didelphimorphia were encountered much less often, since only a few of these species are considered "montane," including *Marmosops impavidus*, *Micoureus demerarae*, and *Monodelphis* cf. *theresa*. Very few (e.g., *Marmosops noctivagus*) reach higher elevations. As mentioned by Patton (1986), the latitudinal gradient has a significant influence on some highly diverse rodent genera such as *Akodon*, *Oryzomys*, and *Thomasomys*. This pattern may be expressed as a low similarity at a specific level, but with the species richness maintained. The few species with wide latitudinal ranges corresponded to mono-specific groups such as *Microryzomy minutus* or to species tentatively grouped under a single name (e.g. *Thomasomys aureus*; V. Pacheco pers. com.).

The greatest similarity among sites was found in bat species, an expected result because of their high mobility. The genera *Anoura*, *Platyrrhinus*, and *Sturnira* were found most frequently, followed by *Micronycteris megalotis*, *Carollia brevicauda*, *Dermanura glauca*, and *Myotis keaysi*.

DISCUSSION AND CONSERVATION IMPLICATIONS

In summary, the most common species in our study were in the genera *Platyrrhinus* and *Sturnira* (Phyllostomidae), and *Oryzomys* and *Akodon* (Muridae) at Llactahuaman, and *Anoura* (Phyllostomidae) and *Akodon* at Wayrapata. These species accounted for more than 50% of the captures at each camp, indicating the dominance of their respective trophic guilds. Because *Platyrrhinus* and *Sturnira* are considered frugivorous (Patterson et al. 1996) and *Oryzomys* is probably granivorous, forests at Llactahuaman would be the main source of food for that community. At Wayrapata, insects could be responsible for the dominance of the omnivorous *Anoura* (Patterson et al. 1996) and the insectivorous *Akodon*. Climatic conditions could also be important in determining the availability of food and/or microhabitat resources for some species as shown by greater bat activity during the first half of the night, sympatric presence of co-generic species at the upper or lower altitudinal limit of their ranges, and scarce use of rodent "roosts" on the roots covering the ground.

Peculiarities of montane forests of the Apurimac Valley include the absence of Thomasomyini rodents, especially the genus *Thomasomys*, which could be considered an indicator of montane forests (Patton 1986). Another noteworthy absence was the shrew opossum, *Lestoros inca*, known from the southeastern Peruvian forests and adjacent forests in Bolivia (N. Bernal pers. com.). Among the Didelphimorphia, we recorded two species of the short-tailed opossum *Monodelphis* at Wayrapata.

Several genera recorded in these forests included species from lowlands (below 1000 m in elevation), some of them exhibiting high diversity (*Monodelphis*, *Platyrrhinus*, *Sturnira*, *Myotis*, *Oligoryzomys*, and *Oryzomys*; see Pacheco et al. 1993, Voss and Emmons 1996). Other genera such as *Anoura*, *Sturnira*, and *Akodon* were most diverse in montane forest, while some species such as *Caluromys lanatus*, *Anoura caudifer*, *Dermanura glauca*, *Enchistenes hartii*, and *Myotis keaysi* occurred over a wide altitudinal range. A few of these (e.g., *A. caudifer*), however, could be complexes of two or more species with narrower altitudinal ranges or subject to a climate variation along the gradient.

Our estimates of the small mammal fauna at the Apurimac Valley exhibited similarities to other Peruvian sites, including the Northern Cordillera de Vilcabamba, Machu Picchu, and Manu National Park—despite the absence of some representative groups (Appendix 19). The biogeographic importance of this study lies with its use in delimiting zones of high endemism and/or diversity. This is significant in relation to conservation of biodiversity, especially in the currently threatened Peruvian montane forests (Young 1992).

LITERATURE CITED

Carleton, M. D., and G. G. Musser. 1989. Systematics studies of Oryzomyine rodents (Muridae: Sigmodontinae): a synopsis of *Microryzomys*. Bull. Am Mus. Nat. Hist. 191: 1-83.

Emmons, L. H., and F. Feer. 1997. Neotropical rainforest mammals: A field guide. 2nd Ed. The University of Chicago Press, Chicago.

Frahm, J. P., and S. R. Gradstein. 1991. An altitudinal zonation of tropical rain forests using bryophytes. J. of Biogeog. 18: 669-678.

Koopman, K. F. 1978. Zoogeography of Peruvian bats with special emphasis on the role of the Andes. Am. Mus. Novitates 2651: 1-33.

Leo, M., and M. Romo. 1992. Distribución altitudinal de roedores sigmodontinos (Cricetidae) en el Parque Nacional Río Abiseo, San Martín, Peru. Mem. del Mus. de Hist. Nat., UNMSM 21: 105-118.

Pacheco, V., B. D. Patterson, J. L. Patton, L. H. Emmons, S. Solari, and C. F. Ascorra. 1993. List of mammal species known to occur in Manu Biosphere Reserve, Peru. Publ. del Mus. de Hist. Nat., UNMSM, Zool. (A) 44: 1-12.

Patterson, B. D., V. Pacheco, and S. Solari. 1996. Distribution of bats along an elevational gradient in the Andes of southeastern Peru. J. of Zool. (London) 240: 637-658.

Patterson, B. D., D. F. Stotz, S. Solari, J. W. Fitzpatrick, and V. Pacheco. 1998. Contrasting patterns of elevational zonation of vertebrates in the Andes of SE Peru. J. of Biogeog. 25: 593-607.

Patton, J. L. 1986. Patrones de distribución y especiación de fauna de mamíferos de los Bosques Nublados andinos del Peru. Anal. del Mus. de Hist. Nat. de Valparaíso 17: 87-94.

Patton, J. L., and M. F. Smith. 1992. Evolution and systematics of the akodontine rodents (Muridae: Sigmodontinae) of Peru, with emphasis on the *genus Akodon. Mem. del Mus. de Hist. Nat., UNMSM* 21: 83-103.

Solari, S., E. Vivar, J. J. Rodríguez, and J. L. Mena. 1998. Small mammals: biodiversity assessment in the Lower Urubamba region. *In* A. Alonso and F. Dallmeier, (eds.) Biodiversity Assessment and Long-term Monitoring, Lower Urubamba Regon, Peru: Cashiriari-3 Well Site and the Camisea and Urubamba Rivers. SI/MAB Series #2 Smithsonian Institution/MAB Biodiversity Program, Washington, DC.

Terborgh, J. 1971. Distribution on environmental gradients: theory and a preliminary interpretation of distributional patterns in the avifauna of the Cordillera Vilcabamba, Peru. Ecology 52: 23-40.

Thomas, O. 1920. Report on the mammalia collected by Mr. Edmund Heller during the Peruvian expedition of 1915 under the auspices of Yale University and the National Geographic Society. Proc. USNM 58: 217-249.

Voss, R. S., and L. H. Emmons. 1996. Mammalian diversity in neotropical lowland rainforests: a preliminary assessment. Bull. of the Am. Mus. Nat. Hist. 230: 1-115.

Wilson, D. E., C. F. Ascorra, and S. Solari. 1996. Bats as indicators of habitat disturbance. *In* (D. E. Wilson and A. Sandoval (eds.) Manu: The Biodiversity of Southeastern Peru Smithsonian Institution Press, Ed. Horizonte, Lima.

Young, K. R. 1992. Biogeography of the montane forests zone of the eastern slopes of Peru. Mem. del Mus. de Hist. Nat., UNMSM 21: 119-154.

CHAPTER 9

MEDIUM AND LARGE MAMMALS OF THE SOUTHERN VILCABAMBA REGION, PERU

Juan José Rodríguez and Jessica M. Amanzo

INTRODUCTION

The objective of this study was to evaluate the diversity and abundance of medium and large mammals along a proposed natural gas pipeline route across the Southern Cordillera de Vilcabamba as part of the SI/MAB biodiversity surveys (see Alonso and Dallmeier 1998, 1999). This information will be used to develop a baseline for future assessments and monitoring efforts aimed at detecting changes in species populations and abundance.

Surveys of medium and large mammals were focused on montane and dwarf forest areas. Montane forests are biogeographically important because they serve as connectors for north-south species interactions. Montane forests also exhibit important elevation gradients and latitudinal differences (Young 1992). These characteristics allow study of the speciation of selected groups, based on past and present environmental factors, as well as their phylogeny and evolution.

Montane forests are defined as forest vegetation above 1000 meters (m) in elevation. In the study region, they are either continuous forests along the western slopes and the northern part of the eastern slopes or fragmented forests occupying areas above the tree line on the western slopes (Young and Valencia 1992).

Dwarf forests are characterized by the absence of trees taller than 15 m (Terborgh 1971). This forest type is usually present on sharp cliffs at lower elevations than the subalpine montane forests (Frahm and Gradstein 1991). The Ericaceae and Asteraceae families are dominant in relation to diversity and abundance. Many epiphytic mosses (e.g., Sphagnum) grow over tangled roots, creating a spongy "false floor" up to 0.5 m deep. With decreasing elevation, this forest type resembles a more typical montane forest, with tall trees (15 to 20 m), more epiphytes, and greater humidity (Comiskey et al. Chapter 4).

METHODS

Site descriptions

Two camps were established in the montane forest near the helipads built during the topographic survey for the proposed pipeline. Llactahuaman (12°51'55.5" S, 73°30'46.0" W; elevation 1710 m) was located about 4 kilometers (km) from the Matsigenka community of Pueblo Libre. Sampling at this site occurred during the dry season, in July 1998. The second camp, Wayrapata (12°50'10.1" S, 73°29'42.6" W; elevation 2445 m), was about 10 km northeast of Pueblo Libre. Sampling at Wayrapata took place in August 1998.

The Llactahuaman site contained areas of abundant bamboo (*Chusquea* spp.) and an irregular forest floor covered mostly with *Chusquea* leaf litter 10 to 20 centimeters (cm) deep. The site was on a 40° slope that is steeper in some areas. Ravines approximately 1 km from the camp were very rocky at their edges and along their bottoms. Vegetation in the ravines was thick, and *Chusquea* spp. were not dominant.

Wayrapata lay within the transition zone between dwarf forest and interandean valley forests. Cloud forests, which have very different floristic structure and composition compared to montane forests, were found at lower elevations from the site. Daytime temperatures averaged between 24° and 26° C, dropping noticeably during the night to 8° - 10° C. Fog and wind were characteristic of the site.

Trail system

We established a trail system that extended into three habitat types. One was the ecotone between dwarf forest and interandean valley forest. This ecotone is primarily found on hilltops between 2200 m and 2450 m in elevation. The vegetation was composed of small shrubs and some trees shorter than 10 m. The large grass (*Arundinella* sp.) was patchily distributed.

The dwarf forest type is found below the dwarf forest and interandean valley ecotone just described. Vegetation is thick and generally short (< 1.5 m). The more abundant tree genera belong to *Clusia, Befaria, Wemmannia, Brunellia*, and *Cinchona*. There were also several orchid species and colonizing epiphytic bromeliads. In some areas, *Chusquea* was fairly common.

The cloud forest type, occurring at about 2280 m, is composed of tall trees (20 to 25 m), ferns, and many *Macgravia* sp. *Chusquea* sp. was also found in some areas.

At Llactahuaman, we cut trails perpendicular to the proposed pipeline route, with the camp as the reference point. For this study, we cut four trails and chose three of them for sampling. The pipeline trail was divided into two sections: the North Trail, or Hurinpata, and the South Trail, or Hananpata. Two additional trails, the Tawachaqui Trail heading east and the Brichero Trail heading west, divided the North and South trails transversally. Trails sampled included:

1. The Hananpata Trail – with gradient from 1710 m at the camp to 1800 m at the last sampling point for this trail and a total length of 1 km. It encompassed areas abundant with Chusquea spp. and containing thick vegetation with difficult access. There were few tall (> 25 m) trees along this trail.
2. Brichero Trail – a 500 m trail that began at the camp elevation of 1710 m, descending to 1540 m in a bamboo thicket. The most common species were *Chusquea* spp., the soil was spongy, and tall trees were scattered throughout the area.
3. The Hurinpata Trail – this trail descended from 1710 m at the camp to 1550 m at the last sampling point for this trail. The forest showed characteristics similar to those found on the Hananpata Trail.

At Wayrapata, we divided the pipeline trail into the West and East trails and cut a new trail, the North Trail. For sampling purposes, we used lengths of 1600 m, 1600 m, and 900 m, respectively, for these trails. The trails sampled were:

1. The West Trail – this trail ran along a hilltop in a southwester direction, descending from 2445 m to 2290 m over the sampling length through areas of dwarf forest and the ecotone between dwarf forest and interandean valley forest. The vegetation along this trail was short and thick without any significant changes along the way.
2. The North Trail – this trail also ran along a hilltop dropping from 2445 m in elevation to 2370 m over the sampling length primarily across typical dwarf forest. Strong westerly winds were common on the trail.
3. The East Trail – this trail dropped from 2445 m in elevation to 2190 m over the sampling length. It wound through a variety of habitats, starting at the ecotone between dwarf forest and interandean valley, then moving into the cloud forest 650 m from the camp, and continuing through areas with trees taller than 20 m along a cliff side (used as a transect). The trail ended at a 30-m waterfall.

Sampling Methods

To evaluate the diversity of medium and large mammals (those mammals as big as squirrels or bigger), we used visual methods (direct observation), vocalizations, tracks (e.g., paths, dens, footprints, eaten fruits, feces, etc.), scent stations, and photographic records (Boddicker 1998). We also interviewed local guides, who provided some information regarding the presence of animals around the study area.

Direct observations and vocalizations

We conducted day and night walks along the selected trails for approximately 9 hours each day (from 0800 to 1200 hours, 1500 to 1800 hours, and 2000 to 2200 hours). We registered each mammal species and described the habitat in which the species was found. Observations by other researchers were also noted. During each sampling period, we stopped occasionally for 3 to 5 minutes to listen for animal sounds. We noted species and the approximate distance between an animal and researcher.

Track records

We defined tracks as any path created by animals, footprints along paths, dens, half-eaten fruits or plants, feces, and hair. Recorded data included family, genus, or species (when the record was fairly evident), type of record, and location with respect to the trail. Footprint records were confirmed using identification guides (Aranda 1981, Emmons and Feer 1997).

Scent stations

Scent stations are used to record footprints. Each station consists of 4 to 6 circles (also known as "points") on the ground, each about one m in diameter and cleared of vegetation. Cotton balls are soaked with olfactory attractants based on animal gland secretions, urine, or food smells (Boddicker 1998) and placed in the circles. We located our stations 200 m to 250 m apart and checked them daily between 0800 and 1600 hours.

Two scent stations were placed along each of the selected trails at Llactahuaman for a total of six scent stations. Stations 1, 4, 5, and 6 consisted of 4 points each. Station 2 had 6 points, and Station 3 had 5 points. Variation in the number of points per station depended on whether ground conditions were favorable (Table 9.1).

The locations of scent stations at Llactahuaman were as follows:

- Station 1: 200 m from the camp on the Hurinpata Trail; spongy forest floor; *Chusquea* spp. very abundant.

Table 9.1. Olfactory attractants used in scent stations at Llactahuaman (1710 m).

Trail	Station #	Scent Type	Preference
Hurinpata	1	Canine Call	Carnivores – Canids
		Trophy Deer	Herbivores - Cérvids
		Bobcat Gland	Carnivores - Felins
		Trail's End	Herbívores in general
	2	Canine Call	Carnivores – Canids
		Trophy Deer	Herbivorss - Cervids
		Bobcat Gland	Carnivores - Felins
		Trail's End	Herbivores in general
		Wind River	Aquatic Carnivores
		Last Call Marmot Bit	Herbivores ingeneral
Hananpata	3	Canine Call	Carnivores - Canids
		Trophy Deer	Herbivores - Cervids
		Bobcat Gland	Carnivores - Felids
		Trail's End	Herbivores in general
		Muskrat Fig	Herbivores
	4	Canine Call	Carnivores – Canids
		Trophy Deer	Herbivoers - Cervids
		Bobcat Gland	Carnivores – Felids
		Trail's End	Herbivores in general
Brichero	5	Canine Call	Carnivores – Canids
		Trophy Deer	Herbivores - Cervids
		Mink Gland Lure	Herbivores
		Muskrat Fig	Herbivores
	6	Canine Call	Carnivores – Canids
		Trophy Deer	Herbivores - Cervids
		Bobcat Gland	Carnivores – Felids
		Trail's End	Herbivores in general

- Station 2: 250 m from Station 1; forest floor is more spongy and irregular with small dirt mounds covered by *Chusquea* spp. leaf litter; short, twisted vegetation; vine-like lianas present.
- Station 3: On the Hananpata trail, 100 m from camp; narrow, steep entrance road; thick vegetation; thin trees; irregular forest floor.
- Station 4: 400 m from the camp at the highest point (1850 m) of the Hananpata Trail; *Chusquea* spp. dominate; irregular forest floor.
- Station 5: On the Brichero Trail about 200 m west of the camp; trail ends at a bamboo forest.
- Station 6: At the edge of the botany plot #3 about 200 m from station 5; patches of *Chusquea* spp. Occur (see Comiskey et al, Chapter 4).

At Wayrapata, we established 7 stations on the West Trail, 4 on the East Trail, and 4 on the North Trail (total of 15 stations) in an attempt to test the different attractants and to observe the animals' reactions to each one (Table 9.2).

The locations of the scent stations at Wayrapata were:

- Station 1: 200 m from the camp on the West Trail in the dwarf forest; spongy, moss-covered forest floor.
- Station 2: On a hilltop 250 m from Station 1; *Arundinella* spp. very abundant; spongy forest floor.
- Station 3: 600 m from the camp on the West Trail; habitat similar to Station 2, but with some trees 2 to 3 m tall.
- Station 4: 250 m from Station 3; forest characteristics similar to Stations 2 and 3.
- Station 5: On a hilltop covered with shrubs on the West Trail, about 1100 m from the camp.
- Station 6: 50 m from Station 5; vegetation mostly shrubs; a few trees 2 to 3 m tall occur.
- Station 7: 1670 m from the camp on one of the heliports along the pipeline route; open area; no vegetation within a 10-m radius.
- Station 8: On the East Trail in dwarf forest 150 m from the camp; irregular, spongy soil; *Chusquea* spp. present.
- Station 9: 250 m from Station 8; *Arundinella* spp. abundant; mossy; very spongy soil.
- Station 10: 750 m from the camp; more humid than the other locations; dominated by arboreal vegetation such as *Macgravia* sp.
- Station 11: Near a cliff about 1000 m from the camp; cloud forest abundant with *Chusquea* spp.
- Station 12: On the North Trail 10 m from the camp; *Arundinella* sp. very abundant.
- Station 13: 250 m from Station 12 on a hilltop in open area; numerous shrubs and trees < 3 m in height.

- Station 14: 500 m from the camp on the North Trail in dwarf forest; arboreal species covered with moss dominate; spongy forest floor.
- Station 15: 200 m from Station 14, also on a hilltop; area similar to Station 13.

Photographic record

Photographic recording registers the occurrence of animals at a particular place through use of an infrared beam emitter and a corresponding receptor connected to a camera. When the continuity of the beam is interrupted, the camera takes a picture of whatever caused the interruption. We programmed the receiver to record all interruptions > 0.2 seconds (such as an animal walking by) so that falling leaves or other non-relevant interruptions would not be recorded. We used this method only at Wayrapata on the East and West trails for a total of 8 days.

Interviews

We interviewed people from the town of Pueblo Libre, who worked with us as guides, showing them pictures from a guide to mammals (Emmons and Feer 1997) to help in identifying the animals. The guides also provided information on hunting frequency and animal abundance.

RESULTS

We recorded a total of 19 species of medium and large mammals, or 68 % of the 27 species believed to exist in the area, belonging to 5 orders and 12 families (Table 9.3). Twelve species were reported at Llactahuaman and 12 at Wayrapata. Only five species were found in common between the two sampling sites (Table 9.3).
Scent stations proved to be the most effective method for recording species at Llactahuaman. The stations yielded 5 records. The second most effective method, observation, yielded 3 records. At Wayrapata, the most effective method was tracks (footprints and feces), which yielded 6 records. The scent station at Wayrapata yielded 5 records (Table 9.3).

We did not obtain any records using the photographic equipment, possibly because the climatic conditions affected the system. Thick fog, humidity, and solar radiation caused false activation of the mechanism.

We used collections of species from Machu Picchu (Thomas 1920) as a reference for identifying species that we expected to find because of that site's similarity to the environmental conditions of areas sampled in this study. Scientific names followed Wilson and Reeder (1993). Some identifications were based on distribution and elevation ranges reported (Patton 1986). English common names were obtained from Emmons and Feer (1997).

Table 9.2. Olfactory attractants used in scent stations at Wayrapata (2445 m).

Trail	Station #	Scent	Preference
West	1 - 6	Canine Call	Carnivores - Canids
		Trophy Deer	Herbivores - Cervids
		Bobcat Gland	Carnivores - Felids
		High Noon	Herbivores
	7	Canine Call	Carnivores – Canids
		Trophy Deer	Herbivores - Cervids
		Bobcat Gland	Carnivores – Felids
		High Noon	Herbivores
		Predator Call	Carnivores
East	8 - 11	Canine Call	Carnivores – Canids
		Trophy Deer	Herbivores - Cervids
		Bobcat Gland	Carnivores – Felids
		High Noon	Herbivores
North	12 - 15	Canine Call	Carnivores - Canids
		Trophy Deer	Herbivores - Cervids
		Bobcat Gland	Carnivores - Felids
		High Noon	Herbivores

Species recorded at Llactahuaman

Dasypus sp. (Long-nosed armadillo) was recorded by noting its trails and dens and by the discovery of footprints on two days along the Brichero Trail about 100 m to 200 m from the camp. Two dens were found on the same trail.

The record for *Tremarctos ornatus* (Andean bear) was made through footprints and half-eaten plants (*Calatola* sp.: Zingiberaceae). We found the footprints on slopes next to cliffs, which made following the track very difficult because of the inaccessibility of the area. *T. ornatus* is known as a good climber (Peyton 1980, 1983, 1986). *Bassaricyon gabbii* (Olingo) was observed 15 m from the ground in a tree by a researcher on the Hurinpata Trail between 1700 and 1800 hours. The animal was passive. *B. gabbii* is a nocturnal, arboreal, solitary species that feeds on fruits and invertebrates, although it occasionally will drink the nectar of certain flowers (Emmons and Feer 1997). *Nasua nasua* (South American coati) was observed between 1300 and 1400 hours hanging from a tree 10 m off the ground near the camp on the Hurinpata Trail. We observed 2 individuals of *Potos flavus* (Kinkajou) between 2000 and 2100 hours in a *Dyctyocaryum* sp. palm (Arecaceae) about 20 m from the ground feeding on the tree's fruits. The observation lasted about an hour before one of the individuals jumped to the ground when we approached the base of the palm tree.

Leopardus pardalis (Ocelot) was a more common species. We found its footprints (4.5 x 4 cm) at each scent station baited with the Canine Call attractant. *Puma concolor* (Puma) was also common. Its footprints (7 x 8) were found at stations with the Canine Call attractant as well as near cliffs and inside the forest proper. *Mazama americana* (Red brocket deer) was recorded on one occasion through identification of its footprints (3 x 6 cm) at Station 2, where we used the Trophy Deer attractant.

Table 9.3. Large mammal species recorded during this study and expected to occur at Llactahuaman (1710 m) and Wayrapata (2445 m), Southern Cordillera de Vilcabamba, Peru (Ll = Llactahuaman; W = Wayrapata).

Order Family	Species Expected to Occur	English common name	Recorded during this study	Site	Type of Record
Didelphimorphia					
Didelphidae (Opossum family)	*Caluromys lanatus*	Western woolly opossum	X	W	Observation
	Didelphis albiventris	White-eared opossum	X	W	Scat
	Didelphis marsupialis	Common opossum			
Xenarthra					
Myrmecophagidae (Anteater family)	*Cyclopes didactylus*	Silky (pygmy) anteater	X	W	Vocalization
Dasypodidae (Armadillo family)	*Dasypus* sp.	Long-nosed armadillo	X	Ll - W	Dens, tracks
Primates					
Atelidae	*Lagothrix lagothricha*	Common woolly monkey			
	Ateles belzebuth	White-bellied spider monkey			
Carnivora					
Canidae (Dog family)	*Pseudalopex culpaeus*				
Ursidae (Bear family)	*Tremarctos ornatus*	Andean bear	X	Ll - W	Tracks, feeding marks, den, hairs
Procyonidae (Racoon family)	*Bassaricyon gabbii*	Olingo	X	Ll	Observation
	Nasua nasua	South American coati	X	Ll	Observation
	Potos flavus	Kinkajou	X	Ll	Observation
Mustelidae (Weasel family)	*Conepatus chinga*				
	Eira barbara	Tayra	X	W	Tracks
	Mustela frenata	Long-tailed weasel	X	W	Tracks
Felidae (Cat family)	*Herpailurus yaguarondi*	Jaguarundi	X	W	Scent station: tracks
	Leopardus pardalis	Ocelot	X	Ll	Scent station: tracks
	Leopardus tigrinus	Oncilla	X	W	Scent station: tracks
	Puma concolor	Puma	X	Ll - W	Scent station: tracks
Artiodactyla					
Cervidae (Deer family)	*Mazama americana*	Red brocket deer	X	Ll	Scent station: tracks
	Mazama rufina				
Rodentia					
Sciuridae (Squirrel family)	*Sciurus ignitus*	Bolivian squirrel	X	Ll	Collected
Erethizontidae (New World Porcupine family)	*Coendu bicolor*	Bicolor-spined porcupine	X	Ll	Hairs
Agoutidae (Paca family)	*Cuniculus* cf. *taczanowskii*	Paca	X	Ll - W	Tracks, trails; Scent station: tracks
Dasyproctidae (Agouti family)	*Dasyprocta* cf. *kalinowskii*	Agouti	X	Ll - W	Buried seeds; Scent station: tracks
Echimyidae (Spiny and Tree Rat family)	*Dactylomys peruanus*	Montane bamboo rat			
TOTAL	**26**		**19**	**Ll – 12 W - 12**	

Sciururs ignitus (Bolivian squirrel) was sampled when an individual became tangled in a mist net located 1 m above the ground near the bamboo thicket on the Brichero Trail. We recorded the presence of *Coendu bicolor* (Bicolor-spined porcupine) based on hairs (spines) found on a sticky trap used for lizards. On another occasion, we saw a *C. bicolor* individual at 2000 hours. *Cuniculus* cf. *taczanowskii* (Paca) was recorded from small paths it left near the one of the main trails. We also identified the presence *of Dasyprocta* sp. (Agouti) by digging up buried seeds at the base of palm trees and other trees.

Species recorded at Wayrapata

The highest number of individuals for a single species at Wayrapata was recorded for *Tremarctos ornatus* (Andean bear). On the North Trail we found 4 places with leftover food (*Xeroxylum* sp. and *Chamaedorea* sp. leaves). The palms showed scratches and torn leaves. We also found an area 900 m from the camp made up of palm leaves with some hairs. All the evidence was located in dwarf forest and in the transition zone between dwarf forest and interandean valley forest.

We found feces of *Didelphis albiventris* (White-eared opossum) on the East Trail 100 m from the camp in the transition zone of dwarf forest to interandean valley forest. We saw one adult of *Caluromys lanatus* (Western woolly opossum) 3 m above the ground in a *Chusquea* sp. tree during a day trek on the East Trail in the cloud forest. We heard vocalizations of a *Cyclopes didactylus* (Silky or pygmy anteater) individual about 15 m away in the cloud forest on the East Trail during a night walk (between 2000 and 2100 hours). The animal stood near a cliff, which made access to it difficult. We found footprints of *Dasypus* sp. (Long-nosed armadillo) on the West Trail in the cloud forest. This individual had dug a hole in the ground in search of food. We found footprints of *Mustela frenata* (Long-tailed weasel) that crossed 300 m of the cliff on the East Trail in the cloud forest. We also recorded footprints (4 x 4 cm) at Station 4 near the point containing the High Noon attractant. We identified the presence of *Eira barbara* (Tayra) through footprints (5 x 7 cm) near the cliff in the cloud forest about 950 m from camp.

Leopardus tigrinus (Oncilla) visited the scent station near camp in the transition zone of dwarf forest and interandean valley forest at the points with the Canine Call and Bobcat Gland attractants. Identification was based on footprints (3.3 x 4 cm). We also recorded *Herpailurus yaguarondi* (Jaguarundi) through footprints (4 x 4.5) at Station 8 with the High Noon attractant. The footprints were on the East Trail about 1350 m from camp. *Puma concolor* (Puma) was identified by its footprints (7.5 x 8

Table 9.4. Mammals observed and reported to occur in the Pueblo Libre area. Names in Matsigenka, Quechua, and Spanish are local to the Pueblo Libre area.

		Common Name		
Species	**Abundance***	**Matsigenka**	**Quechua**	**Spanish**
Didelphis marsupialis	c		Carachupa	
Tamandua tetradactyla	r			
Priodontes maximus	r	Quintero		Vaca quirquincho
Dasypus sp.	a	Iteri		Quirquincho
Saimiri sciureus	a	Chiguiri	Aguaro	
Aotus sp.	c	Pitoni		
Alouatta seniculus	a	Aguaro		
Ateles belzebuth	c	Oshito	Aguaro	
Nasua nasua	a	Kapis		
Potos flavus	r			Gato montés
Eira barbara	r	Huatari		
Lontra longicaudis	r		Yacu león	
Leopardus pardalis	c			Tigrillo
Herpailurus yaguarondi	r			
Puma concolor	r			León
Tapirus terrestris	c			Sacha vaca
Tayassu tajacu	a	Centori	Monte cuchi	
Mazama americana	c			Manero
Sciurus sp.	c			Ardilla
Coendu bicolor	r		Kishca añas	Zorrillo
Hydrochaeris hydrochaeris	r	Ibito		
Dinomys branickii	a	Oturi		
Cuniculus paca	a	Samani		Majaz
Dasyprocta sp.	c	Sharoni	Siwa	

*Relative abundance based on accounts of local residents: a = abundant, c = common, r = rare.

cm) at a cliff by the waterfall where the East Trail ended in the mist forest. We also found a trail used by this animal on the higher part of that cliff. We identified footprints of *Cuniculus taczanowskii* (Paca) at Station 6 (on the point with Trophy Deer attractant) along the West Trail near the dwarf forest, and interandean valley transition zone. We saw footprints of *Dasyprocta* sp. (Agouti) at Station 1 (point with the Canine Call attractant) on the West Trail in the transition zone between dwarf forest and interandean valley forest. We also recorded footprints of this genus about 850 m from the camp in the cloud forest habitat on the East Trail.

Interviews

We interviewed Santiago López, Crisóstomo Yucra, and Hugo Gutiérrez, our guides from Pueblo Libre, concerning the occurrence of certain mammal species near their community. They, like other inhabitants of the area, frequently hunt the red brocket deer (*Mazama americana*) and the rodent *Cuniculusi paca*, which indicates that these are very common species. The guides noted that *Hydrochaeris hydroachaeris* (Capybara) was abundant in the area, but because of excessive hunting its population had diminished, and it is now considered rare by the locals. Table 9.4 shows the mammals reported to occur in this area with some of the common names in English, Matsigenka, Quechua, and local Spanish.

Table 9.5. Calculating Occurrence and Abundance Indices for Large Mammals.

Occurrence Index

The Occurrence Index is based on assigning numerical values to each piece of evidence found for a species and adding them for a total occurrence value. A minimal value is necessary to confirm that the species occurs in the area of study. This value is arbitrary and depends on the experience and knowledge of the researcher (Boddicker, 1998). In this study, we considered a value of 8 points as the minimum for a confirmed occurrence.

Type of evidence	Points
Collected species	10
Observed species	10
Footprints	5
Feces	3
Hair	3
Bones	5
Food remains	2
Dens and trails	3
Vocalizations, smells	5
Identification by local residents	5
Total possible points	53

Abundance Index

The Abundance Index is obtained by multiplying the value for a type of evidence times the number of times it was found, assuming each sample is independent. This index gives relative values and may depend, among other factors, on the type of habitat where sampling is conducted. For the montane forests sampled in this study, we considered a high value (> 25) to mean that the animal was very active in the sampled area, whereas a low value (< 25) meant lower levels of activity.

Example Calculation of Abundance Index

Species	Type of Evidence	Occurrence Index	Abundance Index
Leopardus pardalis	Collected	0	0
	Observation	0	0
	Tracks	5	5(6) = 30
	Feces	0	0
	Hair	0	0
	Bones	0	0
	Food	0	0
	Dens, trails	0	0
	Vocalizations, odors	0	0
	Local identification	0	0
	TOTAL	**5**	**30**

Table 9.6. Occurrence and abundance values for recorded species from Llactahuaman. Values are based on the number of times evidence of presence was recorded (see Table 9.5). A value of 8 or greater is considered a confirmed occurrence record for this study.

Species	Occurrence Index	Abundance Index*
Xenarthra		
Dasypodidae		
Dasypus sp.	5 + 3 = 8	5(2) + 3(2) = 16
Carnivora		
Ursidae		
Tremarctos ornatus	5 + 2 = 7	5(3) + 2(2) = 19
Procyonidae		
Nasua nasua	10	10(1) = 10
Bassaricyon gabbii	10	10(1) = 10
Potos flavus	10	10(1) = 10
Felidae		
Leopardus pardalis	5	5(6) = 30
Puma concolor	5	5(5) = 25
Artiodactyla		
Cervidae		
Mazama americana	5	5(2) = 10
Rodentia		
Sciuridae		
Sciurus ignitus	10 + 10 = 20	10(1) + 10(1) = 20
Erethizontidae		
Coendu bicolor	10 + 3 = 13	10(1) + 3(1) = 13
Agoutidae		
Cuniculus cf. *taczanowskii*	5 + 3 = 8	5(3) + 3(4) = 27
Dasyproctidae		
Dasyprocta cf. *kalinowskii*	5	5(1) = 5

*Abundance value of > 25 indicates that the animal was very active in the sampled area; a value of <25 indicates lower levels of activity.

Occurrence and Abundance

The occurrence and abundance of mammal species recorded at Llactahuaman and Wayrapata are based on the numerical values assigned to each piece of evidence (footprints, hair, trails, observation, food, etc, see Table 9.5). Any species with a total occurance or abundance value of 8 or more was considered as a confirmed record for this study. At Llactahuaman, *Dasypus* sp., *N. nasua, B. gabbii, P. flavus, S. ignitus, C. bicolor,* and *A. taczanowskii* had values of 8 or more (Table 9.6). At Wayrapata, only *C. lanaus* and *T. ornatus* had values of 8 or more (Table 9.7). The more abundant species at Llactahuaman were *L. pardalis, P. concolor,* and *A. taczanowskii*. At Wayrapata, *T. ornatus* was relatively more abundant.

DISCUSSION

In this study, we found that the most effective methods for recording medium and large mammals were tracks and scent stations. Effective use of the track method requires that researchers have some experience in tracks and knowledge of the behavior of the species in the study area. Experienced researchers are able to tell the difference between a rodent trail left by *Cuniculus paca* (Paca) and a trail left by larger animals such as *Puma concolor* (Puma). They know how to determine which type of mammal is responsible for various kinds of leftover food and for marks and incisions made by teeth or scratches made by paws. They know how to "read" feces (size and shape of dropping, hair and bone fragments in the dropping). These skills are essential in identifying many medium and large mammal species that stay out of the way of humans and therefore are difficult to detect visually.

The scent station method proved very useful for recording footprints of animals attracted to food smells. It is important to note in using this method that different animals react differently to each smell. Some may defecate, some may regurgitate, and others may roll around on the ground or scratch at the ground. Evidence of these behaviors should be noted on a daily basis.

To compare the results over time and space, all the aspects that can influence an animal's behavior and therefore cause variation in the number of species recorded should be standardized. Some of these aspects are the space between transects and stations, the substratum around each station, the frequency with which this method is used, and the type and presentation of the attractant (Roughton and Sweeny 1982, Conner et al. 1983). Also, these methods can be optimized if local guides are used.

Topographical and climate conditions are important factors for obtaining results. The geography of the montane forests is very different to the geography of the lowland forest. The forest floor of the latter is much better for recording tracks, since it is a firmer type of soil where footprints are clearly visible. The forest floor in our study areas was unstable and covered with many shrubs, which made finding footprints or other kinds of tracks much more difficult. Also, the accessibility into the interior of these forests in the Llactahuaman and Wayrapata camps was another factor influencing the results. In some cases the sharp cliffs and thick vegetation restricted sampling to hilltops.

The occurrence and abundance indices, based on assigning numerical values to each piece of evidence, are proposed as a standardized method for monitoring without the need for collecting. This method will make the monitoring of large and diverse groups, such as large mammals, much easier, especially for short-term projects where observations, collecting, and marking of individuals does not yield sufficient information. The values assigned to each piece of evidence (Table 9.5) are based on the field experience of the authors and other investigators. The proposed indices and assigned values are open to discussion (Boddicker 1998).

CONSERVATION IMPLICATIONS

Interviews of local residents of Pueblo Libre allowed us to find out more about the status of some species based on presence, abundance, and hunting frequency accounts. The large mammal species revealed during our interviews include *Mazama Americana* (Red Brocket Deer), *Cuniculus paca* (Paca), and *Nasua nasua* (South American Coati). The first two species are a frequent source of food for communities. Other species such as *Tapirus terrestris* (Brazilian tapir), *Hydrochaeris hydrochaeris* (Capybara), *Lontra longicaudis* (Neotropical otter), and *Puma concolor* (Puma) were reported to be common in the area ten years ago, but due to excessive hunting and habitat degradation (agricultural activities, logging, slash and burn), their populations have

Table 9.7. Occurrence and abundance values for recorded species from Wayrapata. Values are based on the number of times evidence of presence was recorded (see Table 9.5) A value of 8 or greater is considered a confirmed record in this study.

Species	Occurrence Index	Abundance Index
Didelphimorphia		
Didelphidae		
Caluromys lanatus	10	10(1) = 10
Didelphis albiventris	3	3(1) = 3
Xenarthra		
Myrmecophagidae		
Cyclopes didactylus	5	5(1) = 5
Dasypodidae		
Dasypus sp.	5 + 2 = 7	5(1) + 2(1) = 7
Carnivora		
Ursidae		
Tremarctos ornatus	5 + 3 + 2 + 3 = 13	5(2) + 3(1) + 2(4) + 3(1) = 24
Mustelidae		
Eira barbara	5	5(2) = 10
Mustela frenata	5 + 2 = 7	5(3) + 2(1) = 17
Felidae		
Herpailurus yaguarondi	5	5(2) = 10
Leopardus tigrinus	5	5(1) = 5
Puma concolor	5	5(2) = 10
Rodentia		
Agoutidae		
Cuniculus cf. *taczanowskii*	5	5(1) = 5
Dasyproctidae		
Dasyprocta cf. *kalinowskii*	5	5(2) = 10

*Abundance value of > 25 indicates that the animal was very active in the sampled area; a value of <25 indicates lower levels of activity.

declined drastically. These data can be compared to the occurrence of large mammals in both lowland and upper (montane) forests. If species near areas like Pueblo Libre are in decline, other sensitive species in areas of higher elevation such as the Andean Bear and Ocelot could also be displaced. Many of these species are listed in CITES (Convention on International Trade in Endangered Species) Appendix I as currently threatened with extinction in parts of their range, including the Andean bear, Puma, Jaguarundi, Oncilla, Ocelot (Emmons and Feer 1997). These species are primarily threatened by hunting, loss of habitat due to deforestation, and overhunting of their prey.

Significant alteration of habitat occurs at elevations ranging from 500 m to 2000 m where deforestation is prevalent. The Apurimac Valley and Convención province in Cusco are prime examples of this degradation. This suggests that species restricted to elevations below 2000 m are particularly threatened due to the apparent incompatibility of human activities with the needs of wild flora and fauna (Young 1992). Protection of habitat both within this range and at higher elevations (such as those sampled in this study) is important in order to provide a refuge for species that lose their lowland habitat.

LITERATURE CITED

Alonso, A., and F. Dallmeier, eds. 1998. Biodiversity Assessment and Monitoring of the Lower Urubamba Region, Peru: Cashiriari-3 Well Site and the Camisea and Urubamba Rivers. SI/MAB Series #2. SI/MAB Biodiversity Program, Washington, D.C.

Alonso, A., and F. Dallmeier, eds. 1999. Biodiversity Assessment and Monitoring of the Lower Urubamba Region, Peru: Pagoreni Assessment and Training. SI/MAB Series #3. SI/MAB Biodiversity Program, Washington, D.C.

Aranda, J. M. 1981. Rastros de los Mamíferos Silvestres de México. Insttituto Nacional de Investigaciones sobre Recursos Bióticos. México.

Boddiker, M. 1998. Medium and large mammals: Biodiversity assessment in the Lower Urubamba region. Pages 219-244 in Biodiversity Assessment and Monitoring of the Lower Urubamba Region, Peru: Cashiriari-3 Well Site and the Camisea and Urubamba Rivers. SI/MAB Series #2. (A. Alonso and F. Dallmeier, eds.) Smithsonian Institution/MAB Biodiversity Program, Washington, D.C.

Conner, M. C., R. F. Labisky, and D. R. Progulske. 1983. Scent-station indices and measures of population abundance for bobcats, raccons, gray foxes and opposums. *Wildlife Soc. Bull.* 11(2): 145-152.

Emmons, L. H., and F. Feer. 1997. Neotropical Rainforest Mammals: A Field Guide. Second Edition. The University of Chicago Press, Chicago.

Frahm, J. P., and S. R. Gradstein. 1991. An altitudinal zonation of tropical rainforests using bryophytes. *J. of Biogeography* 18: 669-678.

Patton, J. L. 1986. Patrones de distribución y especiación de la fauna de mamíferos de los Bosques Nublados Andinos del Perú. *Ann. Mus. Hist. Nat. Valparaiso* 17: 87-94.

Peyton, B. 1980. Ecology, distribution, and food habits of Spectacled Bears, *Tremarctos ornatus*, in Peru. *J. Mamm.* 61 (14): 639-652.

Peyton, B. 1983. Uso de hábitat por el Oso Frontino en el Santuario Histórico de Machupicchu y zonas adyacentes en el Perú. *Symp. Conserv. Manejo de Fauna Silv. Neotrop. (IX CLAZ). Perú*: 23-31.

Peyton, B. 1986. A method for determining habitat components of the Spectacled Bear (*Tremarctos ornatus*). *Vida Silvestre Neotropical* 1(1): 68-78.

Roughton, R. D., and M. W. Sweeny. 1982. Reffinements in scent-stations methodology for assessing trends in carnivore populations. *J. Wildl. Manage.* 46(1): 217-229.

Terborgh, J. 1971. Distribution on environmental gradients: theory and a preliminary interpretation of distributional patterns in the avifauna of the Cordillera Vilcabamba, Peru. *Ecology* 52: 23-40.

Thomas, O. 1920. Report on the mammalia collected by Mr. Edmund Heller during the Peruvian expedition of 1915 under the auspices of Yale University and the National Geographic Society. *Proc. U. S. Nat. Mus.* 58 (2333): 217-249.

Young, K. 1992. Biogeography of the montane forest zone of the eastern slopes of Peru. In Biografía, ecología y conservación del Bosque Montano en el Perú. Young y Valencia (eds). *Mem. Mus. Hist. Nat. UNMSM* (21) 119-154 p.

Yound, K. R., and N. Valencia, eds. 1992. Biografía, ecología y conservación del Bosque Montano en el Perú. *Mem. Mus. Hist. Nat. No. 21 UNMSM*, 227 pp.

Wilson, D., and D. M. Reeder, eds. 1993. Mammal Species of the World: A Taxonomic and Geographic Reference (2d Edition). Smithsonian Institution Press, Washington, D. C.

CHAPTER 10

THE HERPETOFAUNA OF THE NORTHERN CORDILLERA DE VILCABAMBA, PERU

Lily Rodríguez

INTRODUCTION

The herpetofauna of the slopes of the Andes at high elevations Andes are known to be relatively species poor but rich in endemic species. Some groups of lizards, such as *Proctoporus* (Gymnophtalmidae), or frogs such as *Phrynopus* (Leptodactylidae) are restricted to the highlands and can be good indicators of areas of endemism.

Until the 1960s, the Cordillera de Vilcabamba remained one of the largest unexplored mountain ranges of the Andes. Rising to 3500 m, the northern Vilcabamba range was explored by a National Geographic expedition in 1963 (Baekeland 1964) and intensively surveyed for birds and, to a lesser extent, bats in the following years, in the Apurímac (western) slope of the range (Terborgh 1971, 1977, 1985, Terborgh and Weske 1975, Koopman 1978). Also in the 1960s the Organization of American States proposed the establishment of the Cutivereni National Park, while a large area already was set as the Apurímac National Forest, covering 2,071,700 ha. In April 1988, a new regulation established an area of 1,669, 300 ha as the Reserved Zone of Apurímac, a category that preserves areas, enabling studies for the settlement of a protected area. This area did not include any of the Río Ene forest.

Because of its isolation, geographic characteristics, and uniqueness of the avifauna reported by Terborgh and associates, and despite the lack of information on levels of biodiversity for other taxa, the Cordillera de Vilcabamba has been recognized as one of the priority areas for conservation in Peru (Rodríguez 1996).

In this report we present information on the species richness and levels of endemism of the herpetological fauna of the northern ridges of the Cordillera de Vilcabamba gathered during the RAP expedition.

METHODS

L. Rodríguez collected amphibians and reptiles at Camps One and Two (1997), and C. Rivera collected at the Ridge Camp and near Tangoshiari (1998). Rodríguez was primarily responsible for preliminary identifications of all material from both expeditions.

Surveys for reptiles and amphibians were made by visual contact or by listening for frog calls on walks of 3 h at night or about 5 h during the day. While searches at 3350 m took several directions from the campsite, most searches at the 2050 m and 1000 m sites were made along trails. Frogs were collected and photographed before preservation. Rodríguez recorded frog calls with a SONY WM D6C tape recorder. Specimens are deposited at the Museo de Historia Natural de la Universidad Nacional Mayor de San Marcos, Lima, and the American Museum of Natural History, New York.

At Camp One (3350 m), three different habitats were sampled: patchy short forest of *Polylepis*; grasslands mixed with deep mossy wet areas; and mixed forest. Rocks were calcareous and soils were poor (to the eye). Frogs and reptiles were searched mostly during the day from 7-17 June 1997, across different types of habitats.

At Camp Two (2050 m), the dominant vegetation was a standing forest, about 18-20 m tall, with a high number of epiphytes and terrestrial Araceae, ferns and orchids. Weather was more temperate than at Camp One, with warmer diurnal temperatures. Calling activity was apparently regulated by sun and humidity. Sampling across the forest and along streams was conducted both at day and night, from 21 June-2 July 1997.

At Camp Three (elevation range 550 - 1200 m), collections were made between 28 April and 24 May 1998, near Tangoshiari. Most of this period was spent along the crest of a narrow ridge, which had little standing water, and thus there were few available sites for frog reproduction.

RESULTS

Camp One (3350 m)

Four species, three frogs and one lizard, were found at 3350 m and all appear to be new to science (Appendix 20). All of these genera (*Gastrotheca, Phrynopus, Eleutherodactylus* and *Proctoporus*) are restricted to or are most diverse at high elevations. The most abundant species at this site was a small frog, *Phrynopus* sp., which was less than 25 mm in snout-to-vent length (SVL), and displayed a typical *Phrynopus* call. This species was widespread throughout the ground vegetation and inside the forest patches of *Polylepis*. Eggs were found in the moss and hatched small frogs (18 eggs, direct development).

All three frog species found at this site have terrestrial eggs and direct development, which probably is the most effective way in which to reproduce at this elevation. An unidentified species of *Eleutherodactylus* represents the southernmost report of the genus at these altitudes. Only one specimen (a male) was found, on the long leaves of a terrestrial bromeliad in forest.

The *Proctoporus* lizard usually was associated with bromeliads or *Sphagnum* vegetation. It is not too surprising to discover a new species in this poorly known genus of Andean lizard, as members of this genus often have highly restricted distributions.

Camp Two (2050 m)

We collected 13 species, 11 frogs and two snakes, from a montane forest site near Camp Two (Appendix 20). The most common species were *Gastrotheca* sp. (very likely an undescribed species) and *Eleutherodactylus* cf. *rhabdolaemus*. The absence of any *Hyla* of the *armata* or *pulchella* groups possibly could be explained by a lack of adequate habitat; streams were rather scarce and usually subterranean. However, *Telmatobius* and centrolenids were present along exposed running streams.

The Centrolenidae (glass frogs) were represented by two species at the expected elevation (2000 m). The *Centrolene* sp. is more likely to be new, although related to *C. hesperia* (from the western humid northern forest) and to *C. lemniscatum* (from the eastern northern cordillera). This find represents a southern extension of the distribution of this genus, previously known south only to the Cordillera del Sira. Proper identification of the *Cochranella* requires further comparisons, but it also could be an undescribed species.

Of the *Eleutherodactylus*, *E. rhabdolaemus* was described from the Apurimac valley (Duellman 1978a) and must be closely related to the one found at the Cordillera de Vilcabamba. However, the population found at Camp Two exhibits some consistent morphological differences. Another species described from Manu, *Eleutherodactylus pharangobates* (Duellman 1978b), was synonymized with *E. rhabdolaemus* by Lynch and McDiarmid (1987); these populations are being carefully compared to resolve this taxonomic question. This species was common in the area, and showed calling activity throughout the day. *Eleutherodactylus* sp. gr. *unistrigatus* is a small frog that was fairly abundant in the lower vegetation, especially at night. *Eleutherodactylus* sp. gr. *conspicillatus* is known from only one specimen, similar to *E. danae*, but apparently represents a new species. Also, only one specimen was found of *Eleutherodactylus* sp. gr. *lacrimosus*, which requires further comparisons.

Telmatobius sp. is a large aquatic frog that was found calling under rocks in water. It is similar to (but not the same as) a new species of the highlands from Manú; it might be also a new species.

Chironius monticola is a common snake species of *yungas*, found from Venezuela south to Bolivia. The identification to species of the snake *Dipsas* is difficult to determine on the basis of just one specimen; it may be related to *D. catesby*, but it is likely to be an undescribed species.

Camp Three

The number of species (38) recorded in this area (elevation range 550 - 1200 m) was low. We collected only 24 frog species, five species of toad, no salamanders, five lizards, and four snakes in approximately four weeks of field work in an area 2 km southwest of Tangoshari (see Appendix 20).

Two records are of great interest. A marsupial frog, *Gastrotheca* sp., probably is an unnamed species. Furthermore, this record might be one of the lowest elevation localities for members of this genus in Peru. The other frog species of particular interest is a large *Colostethus* (SVL= 34 mm), and maybe a new species of the poison arrow frogs family (Dendrobatidae). The genus *Colostethus* has a few large species in the Andean foothills, but its taxonomy is poorly known and systematic studies remain to be done; however, this group clearly represents the lower Andean foothills.

The distribution of *Epipedobates macero*, a poison arrow frog known from Manu, the Camisea area and the Alto Purus, is extended west of the Urubamba river by our collection. The geographical range of *Eleutherodactylus variabilis* is extended from Panguana (9° 37S) in Pasco, to the south. *Hemiphractus johnsoni*, a horned frog that feeds on other frogs, and *Eleutherodactylus toftae* were two common species at this camp.

DISCUSSION

Four species at Camp One (3 frogs, 1 lizard) and 13 at Camp Two (11 frogs and 2 snakes) fit well under the expected number of species for these elevations, as reported from samples at similar altitudes from Manu National Park (13? S), Abiseo National Park (8? S) (Rodríguez, unpublished) and other Andean sites (Péfaur and Duellman 1980, Duellman 1988, Duellman and Wild 1993). For example, the frog fauna at 3500 m in Abiseo consists of seven species, including species of *Phrynopus* and *Eleutherodactylus*. The fauna at 3500 m in Manu has levels of species diversity intermediate between those of Abiseo and the Cordillera de Vilcabamba, with five frog species representing three genera. Population densities seemed low at the higher elevations in the Cordillera de Vilcabamba. Insect populations (an amphibian food source) also seemed to be low, which could explain the lower densities of amphibians and reptiles, and the smaller sizes of frogs, in the Cordillera de Vilcabamba as compared to comparable elevations in Manu. Further studies are needed to test this hypothesis.

At least twelve or thirteen undescribed species of amphibians and reptiles were found during the RAP expeditions. In view of the isolation of the Cordillera de Vilcabamba and the fact that there were no previous herpetological collections (other than a few specimens collected opportunistically by Terborgh and Weske) from the region, such a number of new species found was not surprising. However, there are some collections of herpetological material from other Andean sites, relatively close to the Cordillera de Vilcabamba, such as from Machu Picchu (near 13° S), Manu (near 13° S) and from Yanachaga-Chemillen (near 11° S), that will need to be compared to our material. For example, studies of faunal collections on the road from Abra Tapuna and the western side of the Río Apurimac, directly across the Cordillera de Vilcabamba, have found a very different amphibian fauna, characterized by such species as *Phrynopus pereger* and *P. lucida* (Leptodactylidae); a representative of a new genus in the toad family (Bufonidae: *Truebella tothastes*); the salamander *Bolitoglossa digitigrada* (Plethodontidae); *Cochranella phenax, C. ocellata, C. spiculata, C. siren*, and *Hyalinobatrachium bergeri* (Centrolenidae); and *Eleutherodactylus rhabdolaemus* and *E. scitulus*. Differences in the fauna probably indicate a long divergence of the Cordillera de Vilcabamba, which apparently is more related biogeographically to the faunas of the eastern and southern slopes, as in Manu.

CONSERVATION RECOMMENDATIONS

Preliminary comparisons of the herpetofauna of the Cordillera de Vilcabamba show that most taxa differ among the three sites sampled, indicating that isolation has led to differentiation among populations. This is consistent with studies suggesting that vicariance is responsible for more than 50% of speciation events in some groups (Lynch 1989, Patton et al. 1990).

The large percentage of new species discovered during these expeditions points once again to the uniqueness of the biota of the Cordillera de Vilcabamba. Further studies will undoubtedly increase the number of endemic species and new records from this area.

Establishing new protected areas in montane regions, such as the Cordillera de Vilcabamba, might be the only way to ensure the survival of a large percentage of the amphibian and reptile species endemic to the country. The Andean mountains contain many species with specialized ecological, reproductive, behavioral, and morphological adaptations that restrict their distribution.

LITERATURE CITED

Baekeland, G. B. 1964. By parachute into Peru's lost world. National Geographic 126: 268-296.

Duellman, W. E. 1978a. Two new species of *Eleutherodactylus* (Anura: Leptodactylidae) from the Peruvian Andes. Transactions of the Kansas Academy of Science 81: 65-71.

Duellman, W. E. 1978b. New species of leptodactylid frogs of the genus *Eleutherodactylus* from the Cosnipata Valley, Peru. Proceedings of the Biological Society of Washington 91: 418-430.

Duellman, W. E. 1988. Patterns of species diversity in anuran amphibians in the American tropics. Annals of the Missouri Botanical Garden 75: 79-108.

Duellman, W. E., and E. R. Wild. 1993. Anuran amphibians from the Cordillera de Huancabamba, northern Peru: systematics, ecology, and biogeography. Occasional Papers of the Museum of Natural History, University of Kansas Number 157.

Koopman, K. F. 1978. Zoogeography of Peruvian bats with special emphasis on the role of the Andes. American Museum Novitates Number 2651.

Lynch, J. 1989. The gauge of speciation: on the frequencies of modes of speciation. Pages 527-553 *in* Otte, D., and J. A. Endler (editors). Speciation and its consequences. Sutherland, Massachusetts: Sinauer Associates.

Lynch, J. D., and R. W. McDiarmid. 1987. Two new species of *Eleutherodactylus* (Amphibia: Anura: Leptodactylidae) from Bolivia. Proceedings of the Biological Society of Washington 100: 337-346.

Patton, J. L., P. Myers, and M. F. Smith. 1990. Vicariant versus gradient models of diversification: the small mammal fauna of eastern Andean slopes of Peru. Pages 355-371 *in* Peters, G., and R. Hutterer (editors). Vertebrates in the tropics. Bonn: Museum Alexander von Koenig.

Péfaur, J., and W. E. Duellman. 1980. Community structure in high Andean herpetofaunas. Transactions of the Kansas Academy of Sciences 83: 45-65.

Rodríguez, L. O. 1996. Diversidad biológica del Perú: zonas prioritarias para su conservación. Lima: Proyecto Fanpe (GTZ and INRENA): Deutsche Gesellschaft für Technische Zusammenarbeit (GTZ) and Ministerio de Agricultura, Instituto Natural de Recursos Naturales (INRENA).

Terborgh, J. 1971. Distribution on environmental gradients: theory and a preliminary interpretation of distributional patterns in the avifauna of the Cordillera Vilcabamba, Peru. Ecology 52: 23-40.

Terborgh, J. 1977. Bird species diversity on an Andean elevational gradient. Ecology 58: 1007-1019.

Terborgh, J. 1985. The role of ecotones in the distribution of Andean birds. Ecology 66: 1237-1246.

Terborgh, J., and J. S. Weske. 1975. The role of competition in the distribution of Andean birds. Ecology 56: 562-576.

CHAPTER 11

AMPHIBIANS AND REPTILES OF THE SOUTHERN VILCABAMBA REGION, PERU

Javier Icochea, Eliana Quispitupac, Alfredo Portilla, and Elias Ponce

INTRODUCTION

Knowledge of the herpetofauna of the cloud forests along the eastern slope of the Peruvian Andes is limited. Duellman (1988), using an index of species diversity for anurans in the South American tropics, determined that there were 23 species among nine genera in an area of the Coñispata River near Cusco (1700 meters in elevation). Duellman and Toft (1979) recorded 17 species of anurans in Peru's central forests of the Cordillera del Sira mountain range, Huánuco (690 m to 1280 m in elevation). Nine of those species were endemic to the area, and Archinger (1991) recorded an additional endemic anuran species from that mountain range. Reynolds and Icochea (1997) recorded 32 anuran species and 21 reptilian species along the Comainas River on the eastern slope of the Cóndor mountain range (665 m to 1750 m in elevation) in northeastern Peru.

Duellman and Wild (1993) recorded 21 anuran species for the Huancabamba mountain range in northern Peru, 10 of which were endemic to that range. Cadle and Patton (1988) established altitude ranges for 30 amphibian and 15 reptilian species for the Río Urubamba (Cusco) and Río Sandia (Puno) in the eastern Andean slopes. Those species were found in association with 5 vegetation types—lowland tropical forest, montane tropical forest, cloud forest, dwarf forest, and puna. Duellman (1979) determined that the southern part of Peru's Cordillera Oriental (Cusco and Puno) has the highest index of endemism of the central Andes (Peru and Bolivia). Leo (1993) lists 42 species of endemic anurans for the montane cloud forests of Peru. Péfaur and Duellman (1980) studied the community structure and Lynch (1986) studied the origins of the high Andean herpetological fauna.

The study area for this biodiversity assessment lies in the Vilcabamba mountain range between the Río Tambo, Río Apurímac, and Río Ene, along a proposed gas pipeline route extending from the Lower Urubamba region to the Peruvian Pacific coast. Rodríguez (1996) considers this geographically isolated area as a priority for investigation and conservation. Previous herpetological studies in the Apurímac Valley were limited to the slopes in the Department of Ayacucho, while very little information is available for the departments of Apurímac and Cusco. This is probably because an access road to the valley lies on the Ayacucho side, whereas there is no road on the Cusco side.

The species recorded for the Apurímac Valley in the literature include the salamander *Bolitoglossa digitigrada* from the Santa Rosa River between Pataccocha and San José at 1000 m, described by Wake et al. (1982). Several anurans have also been described: *Bufo veraguensis* from Tutumbaro along the Piene River at 1840 m in elevation (Duellman and Schulte 1992); *Truebella tothastes* from Carapa on the Tambo-Apurímac Valley road at 2438 m in elevation (Graybeal and Canatella 1995, original description); *Centrolenella bergeri* (=*Hyalinobatrachium bergeri*) from Tutumbaro along the Piene River at 1840 m in elevation (Canatella and Duellman 1982); *Centrolenella ocellata* (=*Cochranella ocellata*) from Huanhuachayocc on the Tambo- Apurímac Valley road at 1650 m in elevation (Canatella and Duellman 1982); *Centrolenella phenax* (=*Cochranella phenax*) from Tutumbaro along the Piene River at 1840 m in elevation (Canatella and Duellman 1982, original description); *Centrolenella spiculata* (=*Cochranella spiculata*) from San José along the Santa Rosa River at 1000 m in elevation (Canatella and Duellman 1982); *Gastrotheca pacchamama* from Abra Tapuna at 3710 m in elevation (Duellman 1987, original description); *Gastrotheca rebeccae* from Yuraccyacu on the Tambo- Apurímac Valley road at 2620 m in elevation (Duellman and Trueb 1988, original description); *Eleutherodactylus cruralis* from Huanhuachayocc on the Tambo- Apurímac Valley road at 1650 m in elevation (Lynch 1989); *Eleutherodactylus rhabdolaemus* from Huanhuachayocc at 1650 m in elevation, between Mitipucuro and Estero Ruana at 2400 m in elevation, and above Yuraccyacu at 2650 m in elevation (Duellman 1978a, original description); *Eleutherodactylus scitulus* from Yuraccyacu on the Tambo-Apurímac Valley road at 2620 m in elevation (Duellman 1978a, original description); *Leptodactylus griseigularis* from Sivia at 600 m in elevation (Heyer 1994); *Phrynopus pereger* from Ccarapa at 2460 m in elevation, Huanhuachayoc at 1650 m in elevation, Mitupucuru at 2425 m in elevation, and Yuracyacu at 2625 m in elevation (Lynch 1975); *Prionodactylus manicatus manicatus* from Luisiana along the Río Apurímac at 500 m in (Uzzell 1973); *Stenocercus apurimacus* from Cunyac bridge on the Río Apurímac at 1830 m in elevation and Apurímac and Limatambo at 2700 m in elevation (Fritts 1974); and *Stenocercus crassicaudatus* from Aina at 1400 m in elevation (Mertens 1952 in Fritts 1974).

METHODS

Study Sites

The study area lies primarily along the western slopes of the Vilcabamba mountain range,Apurímac River Valley, Echarate District, La Convención Province, Department of Cusco. We sampled at two locations along the proposed pipeline route, the Llactahuaman camp (12°51'55.5" S; 73°30' 46.0" W; elev. 1710 m) and the Wayrapata camp (12°50'10.1" S; 73°29'42.6" W; elev. 2445 m.). Both sites were on steep slopes.

Llactahuaman was dominated by *Chusquea* sp., a high-altitude bamboo found between 1600 m and 1900 m in elevation. A small area around the campsite was dominated by *Guadua* cf. *sarcocarpa* found at about 1600 m in elevation. Wayrapata was dominated by short, shrubby vegetation known as dwarf forest that is composed mostly of species in the Ericaceae family and found between 2300 m and 2500 m in elevation (see Comiskey et al., Chapter 4). The area around Wayrapata also contained concentrations of *Rhipidocladum* sp. on a ravine at 2200 m in elevation. Local inhabitants use this reed to construct musical instruments called Zampoñas.

Sampling

We conducted our sampling from July 13 to August 11, 1998, in teams of three people, two investigators and a local guide, on established trails using the proposed pipeline route as the reference point. All amphibian and reptile specimens found were sampled and preserved.

Transects

We established 17, 100-m-long transects along the proposed pipeline corridors and the trails. We sampled up to a height of 2 m on both sides of the pipeline route and trails during day and night.

Subplots

We sampled in 13 (20-m x 5- m) subplots located within the established botanical plots (Comiskey et al., Chapter 4), all of which were randomly chosen. During the day, ground litter at the plots was thoroughly and carefully searched, always moving from the middle of the plot to its edges. Day searches were complemented by night searches for activity over litter, branches, and the forest floor.

Sticky Traps

We used Victor sticky traps (28 cm x 13 cm) that we divided in half, each half counting as a unit. We attached traps to trees and branches at Llactahuaman and on the forest floor at Wayrapata.

Pan Traps

Pan traps are commonly used to sample terrestrial and aerial invertebrates. Deep plastic bowls were placed in holes along the sampling transects. Each bowl contained a mixture of water, detergent, and shampoo—a solution with little surface tension, making escape from the trap difficult.

Specimen Preservation

We preserved specimens according to established techniques (Pisani and Villa 1974, McDiarmid 1994). Each specimen was recorded in a field notebook by species (if known), sex (using secondary sexual characters), time and day of collection, mass, body length, tail length (when appropriate), and microhabitat where the specimen was found. Each adult was assigned a field number (larvae were assigned series numbers), fixed, and preserved in 10% formaldehyde.

Specimens identified in the field were later verified through the collection of the Museo de Historia Natural of the Universidad Nacional Mayor de San Marcos (MUSM). We used the following references: Avila-Pires (1995), Boulenger (1902, 1903) Campbell and Lamar (1989), Cannatella and Duellman (1982), Cope (1874), Dixon (1989), Dixon and Markezich (1979), Dixon and Soini (1975, 1977), Dixon et al. (1993), Donoso-Barros (1969), Duellman (1976; 1978a, b, c; 1979; 1987; 1988; 1991), Duellman and Ochoa (1991), Duellman and Schulte (1992), Fritts (1974), Kizirian (1996), Lynch (1980, 1989), Peters and Donoso-Barros (1970), Peters and Orejas-Miranda (1970), Reynolds and Foster (1992), Rodríguez (1994), Roze (1967), Savage (1960), Uzzell (1966, 1970, 1973), Vanzolini (1986), and Vitt and de la Torre (1996).

RESULTS AND DISCUSSION

At both sites, we recorded a total of 16 species of amphibians, all of them anurans (frogs and toads), and 13 reptile species—four lizards and nine snakes (Appendix 21). The numbers of species by family follow: Bufonidae (1), Centrolenidae (1), Hylidae (2), Leptodactylidae (12), Gymnophthalmidae (3), Tropiduridae (1), Colubridae (7), Elapidae (1), Viperidae (1). Eight species of amphibians and nine species of reptiles (four lizards and five snakes) were found at Llactahuaman, while 11 species of amphibians and seven species of reptiles (two lizards and five snakes) were recorded at Wayrapata (Appendix 21).

At Wayrapata, we recorded a new species of toad of the Bufonidae family belonging to the genus *Atelopus*. We also obtained the first record for Peru of the snake *Liophis andinus* (Colubridae). It was previously known only for the locality of Incachaca, Cochabamba, Bolivia (Dixon 1983, Carrillo and Icochea 1995). Individual numbers of specimens were 79 adult amphibians, seven larval series, as well as 43 reptiles (24 lizards and 19 snakes; Appendix 21). We recorded two species during diurnal sampling of transects: *Eleutherodactylus* sp. at Llactahuaman and *Euspondylus* cf. *rahmi* at Wayrapata. Nocturnal sampling of transects at Llactahuaman resulted in three species of amphibians (possibly four) including *Eleutherodactylus* sp., *E.* gr. *unistrigatus*, *E.* aff. *ventrimarmoratus*, and *Osteochepalus* sp. Nocturnal sampling of transects at Wayrapata yielded four species of amphibians (possibly five) including *Eleutherodactylus* sp., *E.* gr. *unistrigatus*, *E. cruralis*, *E. ockendeni*, and the snake *Oxyrhopus marcapatae*.

Diurnal sampling of the six subplots at Llactahuaman did not yield any specimens, while at Wayrapata, it produced one *Eleutherodactylus* gr. *unistrigatus*. Similarly, nocturnal sampling of the Llactahuaman subplots produced no captures, but at the Wayrapata subplots night sampling yielded four species of amphibians (perhaps five) including *Eleutherodactylus* gr. *unistrigatus*, *E. cruralis*, and *Eleutherodactylus* sp. 1 and 2.

We recorded three species of lizards from the sticky traps in Llactahuaman (*Euspondylus* cf. *rahmi*, *Stenocercus crassicaudatus*, *Proctoporus guentheri*), and two lizard species at Wayrapata (*E.* cf. *rahmi*, and *P. guentheri*). The sticky traps at Wayrapata produced a greater number (18) of individuals, however, than at Llactahuaman, where only four individuals of the three species were sampled. This difference may have been due to the location of the traps. At Wayrapata, we placed the traps on the forest floor where the greatest lizard activity appears to take place. At Llactahuaman, we placed the traps in trees, where lizard activity is apparently less common.

The pan traps at Wayrapata also yielded 18 individuals belonging to one lizard species (*Euspondylus* cf. *rahmi*,), and five amphibian species (*Eleutherodactylus cruralis*, *E. unistrigatus*, *Leptodactylus* sp., *Phrynopus*?, and *Phyllonastes* aff. *myrmecoides*) of which the last three were recorded only with this method.

Of all recorded anurans, the only species of Bufonidae found belonged to the genus *Atelopus*. We assumed that this species has not been previously described because none of its morphological characteristics matched species known for Peru (Boulenger 1902, 1903; Cope 1874; Gray and Canatella 1985; Peters 1973; Rodriguez et al. 1993) or Bolivia (De la Riva 1990, Donoso-Barros 1969, Reynolds and Foster 1992).

In the family Centrolenidae there was only one adult individual of *Cochranella* cf. *pluvialis* at Llactahuaman and only a tadpole belonging to this family at Wayrapata. Llactahuaman also produced the only specimen belonging to the family Hylidae, a tadpole sampled by a ravine. In addition, we heard the call of *Osteocephalus* sp. in the area's bamboo forest.

The greatest amphibian diversity was found in the family Leptodactylidae, which is typical of Andean montane forests. Within this family, the genus *Eleutherodactylus* with its reproductive mode of direct development (metamorphosis occurs within the egg, thus bypassing the need for standing bodies of water) is especially well adapted to the hills and mountains of the study area, and similar habitats (Lynch and McDiarmid 1987, De la Riva and Lynch 1997). There, the prevailing source of moisture is in the forms of mist and fog. We expected to find individuals of the genus *Telmatobius* based on the characteristics of the area, but we only found four tadpoles and no adults. The *Leptodactylus* found at Wayrapata was a juvenile and thus difficult to identify to species. The specimen of *Phyllonastes* aff. *myrmecoides* recorded at an elevation of 2450 m was very similar to the species found in low elevation forests. This was a surprising find because there are no previous records for this species at this elevation in the southeastern Andes. However, *Phyllonastes heyeri* and *P. lynchi* have been recorded in the northern Andes at elevations between 2500 m and 3100 m (Duellman 1991).

A specimen assigned to the genus *Phrynopus*? had characteristics of a Microhylid of the genus *Syncope*, but did not exhibit that genus's numerical reduction of the fingers. It also showed some similarities to *Phrynopus bracki*, a small leptodactilid only known to occur on the Yanachaga mountain range within Yanachaga-Chemillén National Park (Hedges 1990).

The species accumulation curve for amphibians at Llactahuaman levels off rather quickly (Fig. 11.1), possibly because when we conducted our study, the area was dry. Rain events were light and sporadic, therefore affecting amphibian activity. Also leaf litter on the forest floor

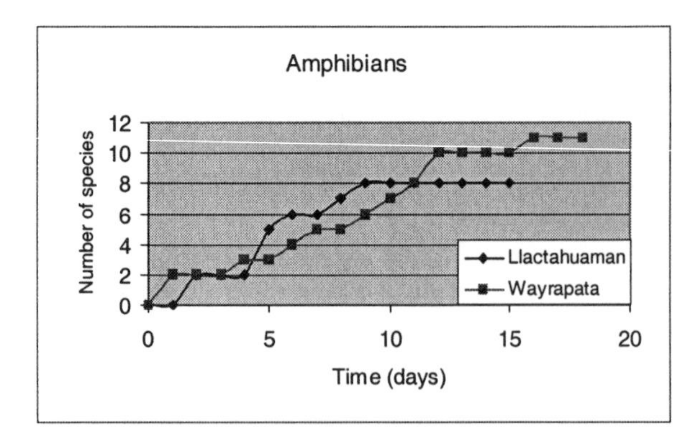

Figure 11.1. Species accumulation curves for amphibians at Llactahuaman (1710 m) and Wayrapata (2445 m) in the Southern Cordillera de Vilcabamba, Peru.

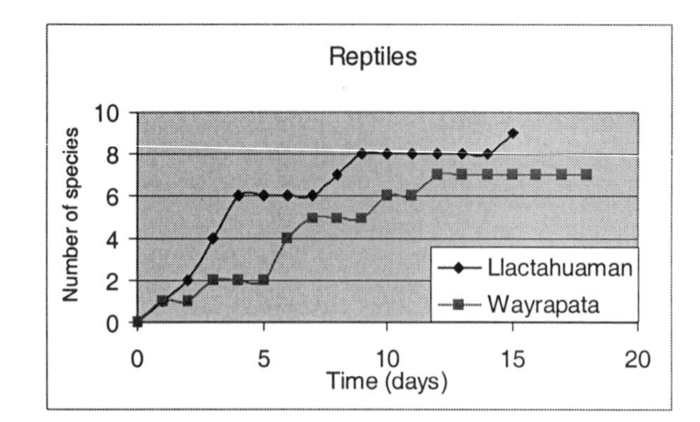

Figure 11.2. Species accumulation curves for reptiles at Llactahuaman (1710 m) and Wayrapata (2445 m) in the Southern Cordillera de Vilcabamba, Peru.

covered many cavities that provide shelter for amphibians during the dry season. In contrast, the amphibian species accumulation curve for Wayrapata continues to increase (Fig. 11.1). It was very wet from continuous fog on the hilltops when we sampled at Wayrapata. This condition likely favored the presence of amphibians.

These results suggest that the following species are relatively abundant in the study area: *Eleutherodactylus* gr. *unistrigatus* C and *Neusticurus ecpleopus* at Llactahuaman and *Eleutherodactylus cruralis, Euspondylus* cf. *rahmi*, and *Oxyrhopus marcapatae* at Wayrapata (see number of individuals in Appendix 21).

The species accumulation curves for reptiles were similar for the two sites, with both leveling off at 7 or 8 species by 9-12 days of sampling (Fig. 11.2).

The quantity of recorded species for both of these lowland montane forest locations was low when compared to other lowland montane forest sites in the Camisea River Basin. The only species recorded both at the Camisea Basin and the sites for this study was *Eleutherodactylus toftae*, although species similar to *Eleutherodactylus* aff. *ockendeni* and *Phyllonastes* aff. *myrmecoides*, which occur in the Camisea Basin, were also found at Llactahuaman and Wayrapata. Furthermore, the call we heard at Llactahuaman's bamboo forest was similar to *Osteocephalus leprieurii*, a species that is commonly heard in the bamboo forests of Camisea. Thus, we recorded it as an *Osteocephalus* sp. individual.

Because the vegetation type observed at Wayrapata is very similar to that of Machu Picchu (P. Núñez, pers. comm.), we compared findings related to herpetofaunal diversity from the two sites. The records for Machu Picchu were not abundant. *Bufo inca* was recorded by Duellman and Schulte (1992); *Stenocercus crassicaudatus* and *S. ochoai* by Fritts (1974) and Ochoa et al. (1996); *Pholidobolus anomalus* by Reeder (1996); *Liophis taeniurus* by Dixon and Markezich (1979); *Oxyrhopus doliatus* by Barbour (1913; see Peters and Orejas-Miranda 1970); *Oxyrhopus marcapatae* by Ruthven (1914; see Peters and Orejas-Miranda 1970); *Oxyrhopus petola* by Hurtado and Blanco (1994); *Micrurus* spp. by Cevallos (1992); and Bothrops andianus by Amaral (1923; see Peters and Orejas-Miranda 1970, Cevallos 1992, and Hurtado and Blanco 1994). The total includes one amphibian species, three lizard species, and at least seven snake species. One of these lizards, *Stenocercus crassicaudatus*, and three of these snakes (possibly four), *Liophis taeniurus, Oxyrhopus marcapate,* and *Bothrops andianus,* were also recorded at our study sites. It is possible that some of the *Micrurus* mentioned in Cevallos (1992) correspond to *Micrusurus annellatus* found at Llactahuaman and Wayrapata.

RESEARCH AND CONSERVATION RECOMMENDATIONS

- To obtain a better picture of the amphibian diversity in the Southern Cordillera de Vilcabamba, we recommend sampling during the rainy season.

- Of all sampling procedures employed in this study, the visual sampling technique yielded the most information. Thus, we recommend that this method be used in conjunction with subplots and transects in future studies.

- Pan traps are commonly used by entomologists, but they also yield good results for certain herpetological studies. We recommend that they be tested in other studies in different habitats to assess their effectiveness with amphibians and reptiles.

- Should construction of a gas pipeline occur where the study sites are located, we recommend the establishment of biological stations, with logistics overseen by the extracting company. This would allow for the continuous monitoring of species, particularly the endemic species that are likely to be the most sensitive to changes in their habitat.

LITERATURE CITED

Archinger, M. 1991. A new species of poison-dart frog (Anura: Dendrobatidae) from the Serranía de Sira, Peru. Herp. 47(1): 1-5.

Amaral, A.. 1923. New genera and species of snakes. Proc. New Eng. Zool. Club 8: 85-105.

Avila Pires, T. C. S. 1995. Lizards of Brazilian Amazonia (Reptilia: Squamata). Zool. Verhandel. 299: 1706.

Barbour, T. 1913. Reptiles collected by the Yale Peruvian expedition of 1912. Proc. Acad. Nat. Sci. Philadelphia 65: 505-507.

Boulenger, G. A. 1902. Descriptions of new batrachians and reptiles from the Andes of Peru and Bolivia. Ann. Mag. Nat. Hist. Ser. 7 X(59): 394-402.

Boulenger, G. A. 1903. Descriptions of new batrachians in the British Museum. Ann. Mag. Nat. Hist. Ser. 7 XII(71): 552-557.

Cadle, J. E., and J. L. Patton. 1988. Distributions patterns of some amphibians, reptiles, and mammals of the eastern Andean slope of southern Peru. Pages 225-244 In Proceedings of a Workshop on Neotropical Distribution Patterns (W. R. Heyer and P. E. Vanzolini, eds.) Academia Brasileira de Ciencias, Río de Janeiro.

Campbell, J. A., and W. W. Lamar. 1989. The Venemous Reptiles of Latin America. Cornell University Press, Ithaca, NY.

Cannatella, D. C., and W. E. Duellman. 1982. Two new species of *Centrolenella,* with a brief review of the genus in Peru and Bolivia. Herp. 38(3): 380-388.

Carrillo de Espinoza, N., and J. Icochea. 1995. Lista Taxonomica preliminar de los reptiles vivientes del Peru. Pub. del Mus. de Hist. Nat. UNMSM (A)49: 1-27.

Cevallos, B. I. 1992. Fauna del Santuario Histórico de Machu Picchu. Pages 79-89. In Machu Picchu (H. Chevarria, ed.) Devenir Histórico y Cultural. Editorial Universitaria, UNSAAC.

Cope, E. D. 1874. On some *Batrachia* and *Nematognathi* brought from the Upper Amazon by Prof. Orton. Proc. of the Acad. of Nat. Sci. of Philadelphia 26: 120-137.

De la Riva, Y. 1990. Lista preliminar comentada de los anfibios de Bolivia con datos sobre su distribución. Bol. Mus. Reg. Sci. Nat. Torino 8(1): 261-319

De la Riva, I., and J. D. Lynch. 1997. New species of *Eleutherodactylus* from Bolivia (Amphibia: Leptodactylidae). Copeia 1997 (1): 151-157.

Dixon, J. R. 1983. Systematics of *Liophis reginae* and *L. williamsi* (Serpentes, Colubridae), with a description of a new species. Ann. of Carnegie Mus. 52(6): 113-138.

Dixon, J. R. 1989. A key and checklist to the neotropical snake genus *Liophis* with country lists and maps. Smithsonian Herp. Info. Serv. 79: 1-30.

Dixon, J. R., and A. L. Markezich. 1979. Rediscovery of *Liophis taeniurus* Tschudi (Reptilia, Serpentes, Colubridae) and its relationship to other Andean Colubrid snakes. *J. Herp.* 13(3): 317-320.

Dixon, J. R., and P. Soini. 1975. The reptiles of the upper Amazon Basin, Iquitos region, Peru. I. lizards and amphisbaenians. *Milwaukee Pub. Mus.* 4:158.

Dixon, J. R., and P. Soini. 1977. The reptiles of the upper Amazon Basin, Iquitos region, Peru. II: crocodilians, turtles and snakes. *Milwaukee Pub. Mus.* 12:191.

Dixon, J. R., J. A. Wiest, Jr., and J. M. Cei. 1993. Revision of the neotropical snake genus *Chironius* Fitzingeri (Serpentes, Colubridae). Monogr. Mus. Regionale di Sci. Nat. No. XIII.

Donoso-Barros, R. 1969. Una nueva especie de atelopodidae de Bolivia. Physis. 28(77): 327-330.

Duellman, W. E. 1976. Centrolenid frogs from Peru. *Occas. Pap. Mus. Nat. Hist. Kansas Univ.* 52: 111.

Duellman, W. E. 1978a. Three new species of *Eleutherodactylus* from Amazonian Peru (Amphibia: Anura: Leptodactylidae). Herp. 34(3): 264-270.

Duellman, W. E. 1978b. Two new species of *Eleutherodactylus* (Anura: Leptodactylidae) from the Peruvian Andes. Trans. Kansas Acad. Sci. 81: 65-71

Duellman, W. E. 1978c. New species of Leptodactylid frogs of the genus Eleutherodactylus from the Cosñipata Valley, Peru. Proc. Biol. Soc. Wash. 91(2): 418-430.

Duellman, W. E. 1979. The herpetofauna of the Andes: patterns of distribution, origin, differentiation, and present communities. Pages 371-459 In The South American Herpeofauna: Its Origin, Evolution, and Dispersal (W. E. Duellman, ed.) Monogr. Mus. of Nat. Hist. Kansas Univ.

Duellman, W. E. 1987. Two new species of marsupial frogs (Anura: Hylidae) from Peru. *Copeia* 1987(4): 903-909.

Duellman, W. E. 1988. Patterns of species diversity in anuran amphibians in the American Tropics. Ann. Missouri Bot. Gard. 75: 79-104.

Duellman, W. E. 1991. A new species of Leptodactylid frog, genus *Phyllonastes* from Peru. Herp. 47(1): 9-13.

Duellman, W. E., and O. Ochoa. 1991. A new species of *Bufo* (Anura: Bufonidae) from the Andes of Southern Peru. Copeia 1991(1): 137-141.

Duellman, W. E., and R. Schulte. 1992. Description of a new species of Bufo from northern Peru with comments on phenetic groups of South American toads (Anura: Bufonidae). Copeia 1992(1): 162-172.

Duellman, W. E., and C. A. Toft. 1979. Anurans from Serranía de Sira, Amazonian Peru: Taxonomy and Biogeography. Herp. 35(1): 60-70.

Duellman, W. E., and L. Trueb. 1988. Cryptic species of Hylid marsupial frogs in Peru. J. Herp. 22(2): 159-179.

Duellman, W. E., and E. R. Wild. 1993. Anuran Amphibians from the Cordillera de Huancabamba, Northern Peru: Systematics, Ecology, and Biogeography. Occas. Pap. Mus. Nat. Hist. Kansas Univ. 157: 1-53.

Fritts, T. H. 1974. A multivariate evolutionary analysis of the Andean Iguanid lizards of the genus *Stenocercus*. Mem. San Diego Soc. of Nat. Hist. 7: 1-89.

Gray, P., and D. C. Cannatella. 1985. A new species of *Atelopus* (Anura, Bufonidae) from the Andes of northern Peru. Copeia 1985(4): 910-917.

Graybeal, A., and D. C. Cannatella. 1995. A new taxon of Bufonidae from Peru, with descriptions of two new species and a review of the phylogenetic status of supraspecific Bufonid taxa. Herp. 51(2): 105-131.

Hedges, S. B. 1990. A new species of *Phrynopus* (Anura: Leptodactylidae) from Peru. Copeia 1990(1): 108-112.

Heyer. W. R. 1994. Variation within the *Leptodactylus podicipinus-wagenri* complex of frogs (Amphibia: Leptodactylidae). Smithsonian Contr. to Zool. 546: 1-124.

Hurtado, J.L., and D. Blanco. 1994. Nuevo registro de ofidios del bosque nublado del Valle de Q'osñipata, Cusco. Bol. del Lima XVI(91-96): 49-52.

Kizirian, D.A. 1996. A review of Ecuadorian *Procotoporus* (Squamata: Gymnophthalmidae) with descriptions of nine new species. Herp. Monogr. 10: 85-155.

Leo, M. 1993. The importance of tropical montane cloud forest for preserving vertebrate endemism in Peru: The Río Abiseo National Park as a case study. Pages 126-133 In Proceedings of an International Symposium of Tropical Montane Cloud Forests (L. Hamilton, J. Juvik and F. Scatena, eds.) San Juan, Puerto Rico.

Lynch, J. D. 1975. A review of the Andean Leptodactylid frog genus *Phrynopus*. Occas. Pap. Mus. Nat. Hist. Kansas Univ. 35: 1-51.

Lynch, J. D. 1980. A taxonomic and distributional synopsis of the Amazonian frogs of the genus *Eleutherodactylus*. Am. Mus. Novitiates 2696: 1-24.

Lynch, J. D. 1986. Origins of the high Andean herpetological fauna. Pages 478-499 In High Altitude Tropical Biogeography (F. Vuilleumier and M. Monasterio, eds.) Oxford University Press, Oxford.

Lynch, J. D. 1989. Intrageneric relationships of mainland *Eleutherodactylus* (Leptodactylidae). I. A Review of the frogs assigned to the *Eleutherodactylus discoidalis* species group. Contrib. in Biol. & Geol. Milwaukee Pub. Mus. 79: 1-25

Lynch, J. D., and R. W. McDiarmid. 1987. Two new species of *Eleutherodactylus* (Amphibia: Anura: Leptodactylidae) from Bolivia. Proc. Biol. Soc. Wash. 100(2): 337-346.

McDiarmid, R. W. 1994. Preparing amphibians as scientific specimens. Pages 289-297 *In* Measuring and Monitoring Biological Diversity: Standard Methods for Amphibians (W. R. Heyer, M. A. Donnelly, R. W. McDiarmid, L.-A. Hayek, and M. S. Foster, eds.) Biological Diversity Handbook Series, Smithsonian Institution Press, Washington, DC.

Ochoa, C., J. A. Achcahuala, and J. C. Chaparro. 1996. Distribución del género *Stenocercus* (Tropiduridae) en el departamento del Cusco. K'ente 8: 7-8.

Péfaur, J. E., and W. E. Duellman. 1980. Community structure in high Andean herpetofaunas. Trans. Kansas Acad. Sci. 83(2): 45-65.

Peters, J. A. 1973. The frog genus *Atelopus* in Ecuador (Anura: Bufonidae). Smithsonian Contrib. to Zool. 145: 1-49.

Peters, J. A., and R. Donoso Barros. 1970. Catalogue of the neotropical squamata: Part II. lizards and amphisbaenians. Bull. U.S. Nat'l. Mus. 297: 1293.

Peters, J. A. , and Orejas-Miranda. 1970. Catalogue of the neotropical squamata: Part I. Snakes. Bull. U.S. Nat'l. Mus. 297: 1347.

Pisani, G. R., and J. Villa. 1974. Guía de técnicas de preservación de anfibios y reptiles. Herpeto.l Circ. Soc. for the Study of Amphib. and Rep. 2: 1-28.

Reeder, T. W. 1996. A new species of *Pholidobolus* (Squamata: Gymnophthalmidae) from the Huancabamba depression of Northern Peru. Herp. 52(2): 282-289.

Reynolds, R. P., and M. S. Foster. 1992. Four new species of frogs and one new species of snake from the Chapare region of Bolivia, with notes on other species. Herp. Monog. 6: 83-104.

Reynolds, R.. P., and J. Icochea. 1997. Amphibians and reptiles of the upper Río Comainas, Cordillera del Cóndor. Pages 82-90, 204-206. In The Cordillera del Cóndor Region of Ecuador and Peru: A Biological Assessment (L. O. Rodríguez, ed.) RAP Working Papers No.7.

Rodríguez, L. 1994. A new species of the *Eleutherodactylus conspicillatus* group (Leptodactylidae) from Peru, with comments on this call. Alytes 12(2): 49 63.

Rodríguez, L.O. 1996. Areas prioritarias para conservación de anfibios en el Peru. Pages 87-92. In Diversidad Biologica del Peru. Zonas prioritarias para su conservación (L.O. Rodríguez, ed.) FANPE, Proyecto de Cooperación Técnica Ayuda en la Planificación de una estrategia para el Sistema Nacional de Areas Naturales Protegidas GTZ-INRENA.

Rodríguez, L. O., J. H. Córdova, and J. Icochea. 1993. Lista preliminar de los anfibios del Peru. *Publ. Mus. Hist. Nat. UNMSM* (A) 45: 1-22.

Roze, J. A. 1967. A check list of the New World venomous coral snakes (Elapidae), with descriptions of new forms. Am. Mus. Novitates 2287: 1-60.

Savage, J. 1960. A revision of the Ecuadorian snakes of the Colubrid Genus Atractus. Misc. Pub. Mus. Zool. Univ. of Michigan 112: 186.

Uzzell, T. M., Jr. 1966. Teiid lizards of the genus *Neusticurus* (Reptilia, Sauria). Bull. Am. Mus. Nat. Hist. 132(5): 279-327.

Uzzell, T. 1970. Teiid lizards of the genus *Proctoporus* from Bolivia and Peru. Postilla. Peabody Museum, Yale University 142: 1-39.

Uzzell, T. 1973. A revision of lizards of the genus *Prionodactylus*, with a new genus for P. leucostictus and notes on the genus Euspondylus (Sauria, Teiidae). Postilla. Peabody Museum, Yale University 159: 1-67.

Vanzolini, P. E. 1986. Addenda and corrigenda to the catalogue of neotropical squamata. *Smithsonian Herp. Info. Serv.* 70: 125.

Vitt, L. J., and S. de la Torre. 1996. Guía para la investigación de las lagartijas de Cuyabeno. Museo de Zoología P.U.C.E., Monografía 1: 1165.

Wake, D. B., A. H. Brame, Jr., and R. Thomas. 1982. A remarkable new species of salamander allied to *Bolitoglossa altamazonica* (Plethodontidae) from southern Peru. Occas. Pap. Mus. Zool. Lousiana State Univ. 58: 1

CHAPTER 12

FISH FROM THE EASTERN SLOPE OF THE CORDILLERA DE VILCABAMBA, PERU

Fonchii Chang

INTRODUCTION

This biological inventory took place during a three week period in May 1997 along several tributaries of the Río Picha. The Río Picha is a tributary of the lower Río Urubamba, whose waters drain the eastern slopes of the Cordillera DE Vilcabamba. Surveys were completed along various tributaries of the Río Picha and the Río Paratori (a major tributary of the Río Picha). We visited three traditional communities that are situated along these streams: Camana, Mayapo, and Puerto Huallana, in that order. On this journey we collected at 19 stations. The first 10 collecting stations were on the Río Paratori and tributary streams (Manitiari, Maseriato, Jorrioni, Kamariampiveni and Serimpiai); five stations were sampled between the Río Picha and Río Parotori, and in the streams (Mayapo and Maenvari) near the community of Mayapo. Finally, four stations were in the vicinity of Puerto Huallana on the Río Picha (streams Mapichiriato and Camonashiari).

These whitewater rivers are characterized by gravel and sand bottoms with scattered boulders. The Río Picha flows through a large, sinuous canal with rapids and backwaters. River margins vary from banks of moderate slope to zones of abrupt, steep slopes, where erosion exposes sediment of fluvial soil, grainy and rocky in places and solid rock in others. Beach areas are large and contain grainy soils and boulders. The riverine vegetation is dominated by cetico *(Cecropia)*, bubinsana *(Calliandra)*, topa *(Ochroma)*, caña brava *(Gynerium)*, pajaro bobo *(Tessaria)* and grasses (Poaceae). A significant portion of the journey along these rivers presents large areas of bamboo *(Guadua)*, between upland forest and cultivated plots of yuca (manioc), plantain, and achiote close to the communities. The area in this zone that is subject to seasonal flooding is small and limited to the low areas close to the rivers.

Among the characteristics of the Río Picha and the Río Paratori that are typical of white waters are: high conductivity ranging from 74–210 moh/cm, temperature of 24–28 °C, total density between 70-148 ppm, calcium concentrations of 52-132 ppm, magnesium concentrations of 16-32 ppm and a pH of 7-7.2. These environments are very rich in species composition, as nutrients (related to the density of Ca and Mg) and conductivity are higher than in those environments with clear and dark waters. White water streams had lower conductivity than white water rivers, but mean temperature (24ºC) and pH (7.2) were comparable.

Blackwater streams exhibited conductivity of 16 –31 moh/cm, temperature between 23-26 °C, total density of 16-35 ppm, magnesium concentrations of 20-24 ppm, calcium concentrations of 12-16 ppm, and a pH of 6.3-6.6. These environments generally are limited in their number of species, but remain unique as various species of dark water ecosystems are adapted to these habitats.

METHODS

We used dragnets, canting nets, and hooks and lines to collect fish specimens. Some of the larger fish species were photographed in the field. Specimens are deposited in the Department of Ichthyology of the Museo de Historia Natural de Universidad Nacional Mayor de San Marcos, Lima. In addition, a fish survey was administered to local residents. Photographs of different fish species collected were shown to the residents to help determine whether or not these species were known locally. In addition, the following physical/chemical properties of the water were collected: conductivity, pH, temperature, total density, calcium density and magnesium density (see Introduction).

RESULTS

Eighty-six species of fish, representing 19 families, were identified (Appendix 22). The most common groups were Characiformes and Siluriformes, which numbered 48 and 31 species, respectively (Appendix 22). The ichthyofauna of these rivers and streams was diverse, comprised of upland forest species and some migratory species (e.g., *Prochilodus nigricans*, *Piaractus brachypomus*, *Zungaro zungaro*, *Pseudoplatysoma fasciatum*, and *P. tigrinum*). Although there is not a strict elevational stratification, several species that are characteristic of the Andean piedmont (*Creagrutus* sp., *Ceratobranchia delotaenia*, *Bryconacidnus ellisi*, *Hemibrycon jelskii*, *Prodontocharax melanotus* and a genus similar to *Rhinopetitia*) were common and the density of some of them is relatively high.

Tyttocharax sp. was one of the most common species found in dark water streams. This species seems to be related to *T. tambopatensis* of Madre de Dios but the latter is very small at 15.5 mm of Standard Length (SL), while the former is over 30 mm SL. These species also vary in other external physical characteristics. One species of a new genus, the description of which is currently in progress (R. Vari, pers. com.), is listed here as "aff. *Rhinopetitia* sp." (Appendix 22). The catfish *Corydoras* sp. (Callichthyidae) also may be a new species, sharing physical characteristic with *C. weitzmani* from Cusco and *C. panda* from Ucayali and Loreto.

USE OF ICHTHYOLOGICAL RESOURCES BY THE MATSIGENKA

Fish constitute an important daily source of protein for the Matsigenka people and fishing is extensive. All species, regardless of size, are consumed. The larger fish are eaten more frequently and typically are caught with dragnets, canting nets, bow/arrow, and hook/line. Among the larger species that are consumed regularly are boquichico (*Prochilodus nigricans*), tarpon (*Brycon erthropterus*), alpaca (*Piaractus brachypomus*), palometa (*Mylossoma duriventris*), doncella (*Pseudoplatystoma fasciatum*), pumazungaro (*Pseudoplatystoma tigrinum*), and achacubo (*Sorubimichthys planiceps*) (Appendix 23).

Children and adults of both sexes participated in fishing, although equipment such as dragnets (of 3-5 m diameter), canting nets, bow/arrow, and hook/line, is utilized exclusively by the male adults. A fish poison, *barbasco*, however, is used collectively by adult males and females of various families. This task involves damming or diverting the course of a stream or river. Rocks are used as an initial barrier which is further supplemented by filling it in with sand and covering the structure with big palm leaves or bijao. After the water flow has dropped, the barbasco is crushed and washed repeatedly into different sections of the water. Once the poison has taken effect, the fish are trapped and float to the surface, and all sizes of fish are collected, including boquichico, raya, sardinas, and carachamas. On other occasions the barbasco is applied in small backwaters or oxbows left by the river, allowing the immediate exploitation of all fish present.

CONSERVATION RECOMMENDATIONS

We found no indications of overfishing or of the misuse of the aquatic ecosystem. Nevertheless, the inhabitants of the region stated that fish were more abundant in earlier years, and some thought that the apparent decline might be due to the use of barbasco in fishing. However, typical use of this poison is after the rainy period, near the outlets of streams or in backwaters of the rivesr. Its effect is temporary and local, and affects the species less adapted to a lack of oxygen; those fish that survive an application of *barbasco* later may die as the backwaters in which they are trapped continue to shrink as the dry season advances. For these reasons, this does not seem to represent a true threat to the preservation of fish and their habitat in this region.

The principal threats to the conservation of the fish fauna, and aquatic ecosystems in general, are deforestation, chemical pollution, high discharges of organic wastes, and the introduction of exotic (non-native) fish species. Many species will disappear locally or may be displaced and replaced by species that are better adapted or are resistant to drastic changes in the ecosystem. Species such as *Tyttocharax* and *Scopaeocharax* are susceptible to disappearance if their habitat is modified.

CHAPTER 13

BIODIVERSITY ASSESSMENT OF THE AQUATIC SYSTEMS OF THE SOUTHERN VILCABAMBA REGION, PERU

Raúl Acosta, Max Hidalgo, Edgardo Castro, Norma Salcedo, and Daisy Reyes

INTRODUCTION

Cloud forests are found in large expanses in the mountains of Peru. Young and Valencia (1992) defined cloud forests as those between 1000 and 5000 meters (m) in elevation, including *Polylepis* forests, on the eastern and western slopes of the Andes. This forest type is common to the headwaters of the rivers Mayo, Huallaga, Pachitea, Perené, Ene, Urubamba and Madre de Dios in Peru (Brack 1992). Cloud forests are characterized by high biodiversity—for some groups, even higher than that of the Amazon Basin (Brack 1992). They are also very important for hydrological purposes, especially on the eastern slopes, and for the economic benefits derived from them. However, current rates of deforestation are causing major perturbations to these forests in many areas (Brack 1992, Young and Valencia 1992).

The present study was conducted in the Southern Cordillera de Vilcabamba, Peru along a proposed gas pipeline route. We studied the biology of several creeks near two base camps along the pipeline near vegetation plots established by the Smithsonian floristics team (Comiskey et al., Chapter 4). Our work included an assessment of water quality and macro-invertebrate and fish diversity in the creeks. These data can serve as baseline information for future monitoring programs.

METHODS

Study Sites

The study was conducted in a montane forest between 1480 m and 2500 m on the eastern side of the Apurimac River, District of Kimbiri, Province of La Convencion, Department of Cusco. We established our base at Llactahuaman and Wayrapata camps along the proposed pipeline, about 4 and 6 kilometers (km), respectively, from Pueblo Libre, a small town on the river bank of the Apurimac.

Llactahuaman

We sampled the aquatic systems near Llactahuaman camp (12°51'55.5" S , 73°30'46.0" W; 1710 m in elevation) that were accessible between 1480 m and 1675 m. This portion of the cloud forest is dominated by *Chusquea* sp., a bamboo from the highlands of the Andes. Because of the elevation and the nature of the terrain, the creeks at this camp flowed steeply down and were characterized by transparent, oxygenated waters with temperatures ranging from 14° to 16° C.

Four trails were cut leading away from the camp, two of which were used to reach the creeks. The Hananpata Trail led to the upper segment of Bagre Creek, while the Tawachaki Trail started from the helipad and led to the lower parts of Bagre and Puma creeks. Both creeks are tributaries of the Río Apurímac.

Three samples were taken along the upper portion of Bagre Creek—one from a high elevation (12°51'46.3" S, 73°30'46.4" W; 1675 m in elevation), one from a middle elevation (1565 m in elevation), and one from a lower elevation (12°51'55.8" S, 73°30'26.9" W; 1480 m in elevation) stations. The lower station was located 850 m from the middle station. We also took a sample from a station at Puma Creek (12°51'52.1" S, 73°30'23.8" W; 1490 m, in elevation) 500 m from the middle station at Bagre Creek.

We sampled during the dry season between July 13 and July 29, 1998.

Wayrapata

This camp (12°50'10.1" S, 73°29'42.6") was set in habitat dominated by plants of the Ericaceae, Saviaceae, Lauraceae, Clusiaceae and Bromeliaceae families, vegetation that is typical of elfin forests (Comiskey et al. this volume). The camp was established at the highest point along a mountainous chain at 2500 m. Creeks to the north flow into the Río Cumpirusiato, a tributary of the Río Urubamba. Creeks to the south flow into the Río Apurímac. Both the Río Urubamba and the Río Apurímac, located east and west of the Vilcabamba mountains, flow north and later join to form the Río Ucayali, which empties into the Río Amazonas.

As at the Llactahuaman camp, most creeks near the camp had little permanent water and relatively low volume (less than 0.05 m^3 per second), with a strong current, high oxygen content, and temperatures of 14° to 16° C. Three trails were cut leading away from the camp site. We used the east trail to reach three creeks. We sampled at two stations along Mamagwayjo Creek— the upper station (2230 m in elevation) and middle station (2190 m in elevation), which flows into the Río Cumpiriosato. Cascada Creek, a tributary of Mamagwayjo Creek, was sampled at a lower station (12°50'01.0" S, 73°29'13.1" W; 2215 m in elevation) located 300 m from the middle station on Mamagwayjo Creek. Afluente Creek, which also flows into the Mamagwayjo, was sampled at one lower station (2220 m in elevation) located 25 m from the middle station along the Mamagwayjo.

We conducted sampling at Wayrapata between August 1 and August 14, 1998.

Initial procedures

At each of the camps, we spent three to four days studying the characteristics of the areas and selecting creeks for sampling. Once a creek was chosen, we determined the average width and depth, water volume, substrate, and the presence of dead organic material (e.g., leaves and small branches) and vegetation (e.g. algae, mosses). To determine the volume of water (m^3 per second) of the creek, we used a floating devise to estimate the time (t) it took it to travel a given distance (L), given the average width (w) and depth (r) of the creek (volume = w*L*r*0.8/t). The type of substrate was determined based on the size classification of the rocks. We sorted all samples in the temporary laboratory at the camp, separating the macro-invertebrates (mostly aquatic flies, beetles, and immature stages of dragonflies).

Physical and chemical analysis of water

At each sampling station, we registered the water and ambient temperatures and the following chemical parameters, using a Hach Limonological Kit (Model AL-36B): pH, oxygen and CO_2 concentrations (milligrams per liter [mg/l]), alkalinity and total hardness (grains per gallon, gpg, of $CaCO_3$, grains=0.0648 grams; gallon = 3.785 liters), and nitrates (mg/l). We studied the physical and chemical characteristics at two points during quantitative sampling (see below), at the upper and lower portions of Bagre Creek because the water had a peculiar color at Llactahuaman, and on Afluente Creek at Wayrapata.

Biological Assessment

Quantitative sampling

The quantitative sampling protocol was designed to study macro-invertebrates associated with various substrates. Ten sites along each creek were sampled. The first site at each creek was selected randomly, and the rest were separated by a distance equal to the width of the creek at a reference point (e.g., at a trailhead when two creeks joined).

Samples were gathered moving up-stream from point to point using a Suber net (30 cm x 30 cm). One net was employed at each sampling station to agitate the substrate for one minute. We sorted the samples, removing large objects (e.g., big leaves and rocks). At the camp, we used a dissecting microscope to separate the invertebrates. Specimens were preserved in 70% alcohol and stored in glass vials for further identification in Lima.

Qualitative sampling

This sampling protocol was directed to study all types of substrates (e.g., leaf litter accumulation, mosses in waterfalls). At least three sites were sampled for each substrate using the Suber net. Specimens were also sampled haphazardly when found in the creeks and missed by the net. All samples were processed through screening and specimens were preserved with 70% alcohol.

We sampled specimens swimming on the water surface using a small fish net, and we sampled fishes with nets. These specimens were fixed with 10% formaldehyde for 24 hours and then preserved in 70% alcohol.

Specimens sampled at Llactahuaman were later identified at the Ichthyology laboratory at the Museo de Historia Natural de la Universidad Nacional Mayor de San Marcos. The macro-invertebrate samples from Wayrapata were sent to the Entomology museum of the Universidad Nacional Agraria La Molina. We used references by Ochs (1954), Benedetto (1974), Flint (1978), Correa et al. (1981), Flint and Angrisano (1985), Flint (1992), Dominguez et al. (1992), Spangler and Santiago –Fragoso (1992), and Merritt and Cummins (1996) to determine macro-invertebrate species. Several species could not be identified at levels lower than family because they were immature stages of species not well known. We used Fowler (1945a, 1945b) and Burgess (1987) to determine fish species.

Data analysis

For quantitative data, we used Margalef (1986) to obtain the species diversity index of Shannon- Wiener based on composition (H' = the number of taxa and their relative abundances), the species richness of Margalef (DMg =

number of species/area), and the evenness index of Shannon (E = uniformity of the distribution of the species). We also used "jack-knife" procedures proposed by Quenouille (1956—modified by Tukey (1958) and recommended by Magurran (1990)—to correct values of H' for taxa with clumped distributions (Zahl 1977).

We used the biotic indices of Ephemeroptera, Plecoptera and Trichoptera (EPT) and the Chironomids and Annelids (CA) to determine the biological status of the creeks. High values of EPT are usually associated with pristine conditions since members of these orders are very susceptible to organic pollution, while high values of CA are associated with disturbance. We compared similarities of the creeks at the two camps using the Morisita-Horn index (M-H), where one equals total similarity and zero total dissimilarity (Wilson and Mohler 1983). For the qualitative data, we also used the Brillouin Index (BI), the recommended index for comparing similarities between non-random samples (Magurran 1990).

RESULTS AND DISCUSSION

Physical and chemical analysis of water

Water sampled from the creeks had cool temperatures with a maximum of 15.0° C at the middle and lower station on Bagre Creek and the lower station on Puma Creek and a minimum of 13.5° C at the upper station on Mamagwayjo Creek (Table 13.1). The pH of the creeks at Llactahuaman was almost neutral (6.8 to 7.3), although the upper station at Bagre Creek was acidic (5.0), similar to the headwaters of other creeks (Jeffries and Mills 1990). Creeks at Wayrapata were more acidic (4.5 to 6.5). In general, pH in non-polluted waters ranges 6.0 to 9.0 (Cole 1983), although some creeks can have normal values as low as 5.0 (Jeffries and Mills 1990).

Table 13.1. Physical and chemical characteristics of the creeks at Llactahuaman (1710 m) and Wayrapata (2445 m).

	SAMPLING STATIONS			
Llactahuaman	**LLA(BAG)H**	**LLA(BAG)M**	**LLA(BAG)L**	**LLA(PUM)L**
Air Temp (°C)	15.5	15.0	16.0	16.3
Water Temp (°C)	14.0	15.0	15.0	15.0
pH	6.8	6.9	7.0	7.3
Dissolved oxygen (mg/l)	8.8	8.0	8.0	8.0
Dissolved CO^2 (mg/l)	2.0	2.0	1.5	2.0
Total hardness (gpg)	2.5	2.0	1.5	2.0
Alkalinity (gpg)	2.5	1.0	1.5	2.0
Nitrates (mg/l)	0.0	0.0	0.0	0.0
Volume (m^3/seg)	0.0144	0.0141	0.0095	0.0223
Wayrapata	**WAY(MAM)H**	**WAY(MAM)M**	**WAY(CAS)L**	**WAY(AFL)L**
Air Temp (°C)	14.0	13.8	16.3	14.0
Water Temp (°C)	13.5	14.0	14.5	14.0
pH	5.4	5.9	6.5	4.5
Dissolved oxygen (mg/l)	8.0	7.5	7.5	7.5
Dissolved CO^2 (mg/l)	1.5	1.5	1.0	2.0
Total hardness (gpg)	1.0	1.0	1.0	1.0
Alkalinity (gpg)	1.0	1.0	1.0	1.0
Nitrates (mg/l)	0.0	0.0	0.0	0.0
Volume (m^3/seg)	0.0004	0.0022	0.0077	-

Sampling Stations (see also text):
LLA (BAG) H = Bagre Creek, High elevation
LLA (BAG) M = Bagre Creek, Mid elevation
LLA (BAG) L = Bagre Creek, Low elevation
LLA (PUM) L = Puma Creek, Low elevation
WAY (MAM) H = Mamagwayjo Creek, High elevation
WAY (MAM) M = Mamagwayjo Creek, Mid elevation
WAY (CAS) L = Cascada Creek, Low elevation
WAY (AFLI) L = Afluente Creek, Low elevation

Water alkalinity and hardness are parameters that are closely linked and that generally neutralize acidity (Jeffries and Mills 1990). Thus, most soft waters have less alkalinity and low conductivity (Ward 1992). These two parameters were relatively low at both camps (Llactahuaman, lower than 2.5 gpg; Wayrapata, lower than 1.0 gpg, where the pH is more acidic). Oxygen concentrations of the creeks at both camps were within normal ranges (Table 13.1).

Physical description of the creeks

The average width of the creeks at Llactahuaman was 1.2 m. At Wayrapata, the average width was 0.7 m (Table 13.2). The average depth at Llactahuaman was 11.4 centimeters (cm), while at Wayrapata, it was 4.6 cm. This relatively shallow depth may be due to the fact that the Wayrapata camp was at a higher elevation than Llactahuaman in an area with very steep slopes. The dominant substrates at Llactahuaman were gravel (2 to16 millimeters [mm] in size) and pebbles (16 to 64 mm in size), with low quantities of accumulated leaf litter. At Wayrapata, all creeks had high quantities of leaf litter and mosses.

Biological Assessment

Quantitative sampling

The total number of macroinvertebrate families found was very similar in all creeks, but the number of species (average 47 vs. 37) and the total number of individuals (average 594 vs. 110) was higher in Wayrapata than at Llactahuaman (Table 13.3). Both sites exhibited a pattern typical of diversity—most species were represented by few individuals, while a few species had many individuals (Appendix 24). Evenness of distribution was higher in Llactahuaman. All of these differences may be the result of the abundant substrate (leaf litter and mosses) at Wayrapata (Table 13.2).

The EPT and CA biotic indices were high and low, respectively, at Llactahuaman (Table 13.3). This indicates that the community of macro-invertebrates was composed of organisms that live in aquatic systems with high concentrations of oxygen and low organic matter.

At Wayrapata, we found very few species of the order Ephemeroptera, corresponding to the relatively low EPT index (Table 13.3). Ephemeroptera does not do well in waters with acidic pH (Jewell 1922, Ward 1992). In laboratory experiments Sutcliffe and Carrick (1973) found that the female Ephemeroptera *Baetis* sp. does not deposit eggs when the pH is lower than 6.0. In addition to low pH, low water temperatures slow decomposition of leaf litter, thus favoring those species that can shred the material, eliminating organisms that scrape algae from rocks (Willoughby and Mappin 1988). Plecoptera, on the other hand, does very well in acidic environments (Minshall and Kuehne 1969, Sutcliffe and Carrick 1973) and was well represented by *Anacroneuria* (Plecoptera: Perlidae). Trichoptera was also frequent at Wayrapata and was represented by members of the Hydropsychidae, Leptoceridae, and Calamoceratidae families (Appendix 24).

The biotic index CA (chironomids and annelids), that is normally considered to be high in disturbed creeks, was very high at Wayrapata (Table 13.3). There, the Chironomids (immature flies) were represented by at least 10 species in high densities. However, the evenness of the distribution of these species is characteristic of pristine environments, since there are usually one or two species with many individuals in disturbed creeks (Reiss 1981).

The Morisita-Horn biotic index was used to compare similarity of the benthic macroinvertebrate fauna between Llactahuaman and Wayrapata. We found that the sites were not very similar (M-H = 0.3), with only 17 shared families from the 28 recorded at Llactahuaman. A major difference between sites was the greater number of individuals per creek at Wayrapata compared to Llactahuaman (Table 13.3).

Qualitative sampling

This approach was aimed at studying substrates that were not sampled by the random selection of sites for the quantitative effort. We used the Brilloullin index and found that Llactahuaman had a higher number of species than Wayrapata (Table 13.3). This may reflect the greater habitat heterogeneity at Llactahuaman.

Table 13.2. Physical characteristics of the creeks at Llactahuaman (LLA)and Wayrapata (WAY). See Table 13.1 for locations of sampling stations.

Sampling Stations	Average Width (m)	Average Depth (cm)	Dominant Substrate (particle size, mm)	OBSERVATIONS
LLA(BAG)H	1.5	14.8	Gravel (2-16)	Leaf litter
LLA(BAG)M	1.0	11.5	Gravel (2-16)	
LLA(BAG)L	1.3	8.7	Pebble (16-64)	Leaf litter
LLA(PUM)L	0.9	10.6	Pebble (16-64)	
WAY(MAM)H	0.5	7.5	Bedrock	Leaf litter, moss
WAY(MAM)M	1.0	3.3	Pebble (16-64)	Leaf litter, moss
WAY(CAS)L	0.6	3.1	Pebble (16-64)	Leaf litter, moss

The final list of orders, families, and genera for aquatic and semi-aquatic macroinvertebrates registered for both sites is presented in Appendix 24. A total of 191 morphospecies distributed among 13 orders and 69 families were recorded in the study areas.

Along the ravines of Llactahuaman, the more frequently encountered species were *Baetodes* sp. (Ephemeroptera: Baetidae), *Traverella* sp. (Ephemeroptera: Leptophlebiidae), *Leptohyphes* sp. (Ephemeroptera: Leptohyphidae); one species of dragonfly (Odonata: Polythoridae); *Philloicus*? sp. (Trichoptera: Calamoceratidae), *Leptonema* sp. (Trichoptera: Hydropsychidae), Leptoceridae (Trichoptera), *Marilia* sp. (Trichoptera: Odontoceridae), *Macrelmis* sp., *Neoelmis* sp., and *Phanoceroides*? sp. (Coleoptera: Elmidae); *Hexatoma* sp. (Diptera: Tipulidae); and various species of the family Chironomidae (Diptera).

Species within the Order Ephemeroptera showed a preference for a variety of diets. *Baetodes* sp. is reported as a scraper of periphyton associated with algae and other substrata (Merritt and Cummings 1996). *Lepthohyphes* sp. feeds on fine particles of organic matter found in many types of substrates. *Travellera* sp. is a filter feeder of fine particles of organic matter suspended in the water column.

Within the Trichoptera, *Philloicus*? sp. and *Marilia* sp. are detritivores that feed on decomposing plant tissue. The former genus builds cocoon-like shelters from dry leaves, while the latter uses small stones (gravel). *Leptonema* sp. is a filter feeder of decomposing organic material suspended in the water column. Species in the family Elamidae are feeders of fine detritus that settle at the bottom or on any other substratum. The family Tipulidae is a very diverse group in terms of species richness. Its species exhibit several different feeding habits. Some of them such as *Hexatoma* sp. are predators, but the majority of species in this family are omnivores.

Along the ravines of Wayrapata, the most frequently encountered benthic macro-invertebrates were *Aesnha* sp. (Odonata), *Anacroneuria* sp. (Plecoptera: Perlidae), *Ochrotrichia* sp. (Tricoptera: Hydroptilidae), *Smicridea* sp. (Trichoptera: Hydrposychidae), *Polycentropus* sp. (Trichoptera: Polycentropodidae), and *Banyallarga* sp. (Trichoptera: Calamoceratidae). There were several species of the family Leptoceridae (Trichoptera) —*Neoelmis* sp., *Heterelmis* sp., and *Phanoceroides*? sp. (Coleoptera Elmidae) as well as *Scirtes* sp. (Coleoptera: Scirtidae), Tipulidae (Diptera), *Atrichopogon* sp., *Forcipomya* sp., *Palpomyia* sp. (Diptera: Ceratopogonidae), and the Chironomidae (Diptera) complex (Appendix 24).

These groups share as common habitat all kinds of substrata that are characterized by the presence of allochtonous material. Only *Ochrotrichia* sp. and *Scitres* sp. are frequently found in moss growing along the walls of waterfalls and on rocks.

Aeshna sp. and *Anacroneuria* sp. are predatory species that feed on other macro-invertebrates. The tricopteran *Smicridea* sp. builds shelters shaped like nets from small sticks, leaves, or moss to collect detritus. *Ochrotrichia* sp. builds small cocoons with two entrances using sand and algae filaments. This species attaches itself to moss structures from which it feeds and shows some behavior as a collector of detritus in the sediment. *Banyallarga* sp. was one of the most frequently found species. It was sampled at all stations at Wayrapata. It builds cocoons with dry leaves, which are abundant in these ravines, and feeds on decomposing plant tissue.

The family Leptocedeira builds cocoons from many materials, including sand grains, small rocks, or small sticks. *Polycentropus* sp. and *Atopsyche* sp. were the main predatory species sampled during the survey. *Atopsyche* sp. does not build cocoons during the larval period. Instead it builds solid rock cocoons fixed to a substrate during the pupal stage. *Polycentropus* builds a cocoon with fine silk that it covers with sand while it awaits for prey to approach.

The majority of Ceratopogonidae (*Atrichopogon* sp., *Forcipomyia* sp.) and chironomids are collectors of very fine organic particles that are deposited over different substrates such as rocks, leaf litter, or twigs. Some species are

Table 13.3. Abundance, richness, and estimates of diversity for aquatic macro-invertebrates at Llactahuaman (LLA) and Wayrapata (WAY). See Table 13.1 for locations of sampling stations.

Sampling Station	# Families	# Species	Abundance (total # individuals)	Shannon Index	Jack-knife	Richness Index	Eveness Index	EPT (%)	CA (%)	Brillouin Index
LLA(BAG)H	26	32	86	2.2	2.5	3.4	0.8	68.2	10.6	1.3
LLA(BAG)L	29	38	132	2.2	2.3	3.5	0.8	44.2	23.3	2.1
LLA(PUM)L	28	40	112	2.5	2.6	4.0	0.8	63.4	5.4	2.2
WAY(MAM)H	28	47	517	2.3	2.3	4.3	0.7	31.5	41.2	<1.1
WAY(MAM)M	26	35	600	1.7	1.7	3.9	0.5	22.5	41.8	1.1
WAY(CAS)L	28	58	664	2.3	2.4	4.2	0.7	37.5	27.9	1.8

considered predators, including *Palpomya* sp. (Ceratopogonidae) and the subfamily Tanypodinane (Chironomidae), both of which were present during our surveys at Wayrapata. Finding predatory species is important because they are frequently the species most affected by changes in an ecosystem.

The macroinvertebrate fauna that require atmospheric air (i.e., neustonic) at both campsites were composed principally of *Rhagovelia* sp. (Hemiptera: Veliidae), Gerridae (Hemiptera), and Gyrinidae (Coleoptera). These were frequently found in pools along the ravines, where they were observed searching for prey while moving rapidly along the surface of the water.

The results of the surveys at Llactahuaman and Wayrapata led to the following conclusions.

- The presence of abundant allochtonous material (leaf litter, twigs, branches, logs) in the Wayrapata ravine was an important determinant in the distribution of benthic macro-invertebrates, not only because it represented a continuous source of food, but also because it provided refuge, shelter, and a variety of favorable microhabitats.
- The highest values for abundance and richness of macro-invertebrates were recorded at the Wayrapata sampling stations. The lowest values were recorded at Llactahuaman along the upper portion of the Bagre Creek ravine.
- Shannon's diversity index values were similar for both locations. That of Llactahuaman was slightly larger.
- The Wayrapata ravines presented larger index values for family richness, although they also showed a low uniformity value in relation to abundance distribution compared to the Llactahuaman ravines.
- The Morisita-Horn similarity index had a value of 0.3, indicating relatively low similarity between the community structures of benthic macro-invertebrates at the two sites.
- The EPT biotic index had lower values for the Wayrapata ravines (23% to 37%) than for those at Llactahuaman (44% to 68%). These values indicate that species in the order Ephemeroptera are poorly represented at Wayrapata. The acidic pH seems to be a limiting factor.
- The great abundance of diverse species of chirinomids at Wayrapata determined a high value for the CA index, but did not appear to correspond to any type of anthropogenic disturbance.
- Brillouin's diversity index for the qualitative evaluation confirmed the results from the quantitative information.
- We recorded only two fish species (*Trichomycterus* sp. and *Astroblepus* sp.) at the three Llactahuaman sampling stations. Both belonged to the suborder Suliformes.

LITERATURE CITED

Benedetto, L. 1974. Clave para la determinación de Plecópteros sudamericanos. Studies in Neotropical Fauna 9:141-170.

Brack, A. 1992. Estrategias nuevas para la conservación del bosque montano. Memorias del Museo de Historia Natural UNMSM 21: 223-27.

Burgess, W. 1987. An atlas of freshwater and marine catfishes. A preliminary survey of the Siluriformes. T. F. H. Publ. Inc., Neptune City.

Cole, G. A. 1983. Textbook of Limnology. Mosby, St Louis, MO.

Correa, M., T. Machado, and G. Roldan. 1981. Taxonomía y ecología del orden Trichoptera en el departamento de Antioquia en diferentes pisos altitudinales. *Act. Biol.* 10(36): 35-48.

Dallmeier, F., D. Lemarie, and A. Temple. 1998. Suggested Standardized Protocols for the Assessment and Monitoring of Aquatic Systems of the Lower Urubamba, Peru. In Fase IV: Taller para la Estandarización de Protocolos y Planificación de la investigación. Smithsonian Institution, Washington, DC.

Dominguez, E. M., M. Hubbard, and W. Peters. 1992. Clave para ninfas y adultos de las familias y géneros de Ephemeróptera (Insecta) sudamericanos. Biología Acuática 16: 40.

Flint, O. S. 1978. Studies of Neotropical Caddis-flies, XXII: Hydropsychidae of the Amazonan Basin (Trichoptera). Amazoniana 6: 373-421.

Flint, O. S., and A. B. Angrisano. 1985. Studies of Neotropical Caddisflies, XXXV: The immature stages of *Banyallarga argentinica* Flint (Trichoptera: Calamoceratidae). Proc. Biol. Soc. Wash. 98(3): 687-697.

Flint, O. S. 1992. Studies of Neotropical Caddisflies, XLV: The taxonomy, phenology and faunistics of the Trichoptera of Antioquia, Colombia. Smithsonian Contributions to Zoology. Vol 520. Smithsonian Institution Press, Washington, DC.

Fowler, H. W. 1945a. Los peces del Perú: catálogo sistemático de los peces que habitan en agua peruanas. Museo de Historia Natural Javier Prado, Lima, Peru.

Fowler, H. W. 1945b. Descriptions of seven new freshwater fishes from Peru. Notulae 159: 1-11.

Jeffries, M. and D. Mills. 1990. Freshwater Ecology Principles and Applications. Belhaven Press, New York, pp. 285.

Jewell, S. G. 1922. The fauna of an acid stream. Ecology 3: 22-28

Magurran, A. E. 1990. Ecological Diversity and Its Measurements. Princeton University Press, Princeton, pp. 179.

Margalef, R. 1986. Ecología. Ediciones Planeta. 256 pp.
Merrit, R. W., and K. W. Cummings. 1996. Aquatic insects of North América. Kendall/Hunt Publishing Company, Dubuque, Iowa.
Minshall, G. W., and R. A. Kuehne. 1969. An ecological study of invertebrates of the Duddon, an English mountain stream. Arch. Hydrobiol. 66: 169-191.
Ochs, G. H. 1954. Die Gyriniden Perus Undder Übrigen Südamerikanischen Kordilleren. Pages 116-155 *In* Fauna Perus. (E. Titschack, ed.).
Quenouille, M. H. 1956. Notes on bias in estimation. *Biometrika* 43: 353-60.
Reiss, F. 1981. Chironomidae. Pages 261-268 *In* Aquatic Biota of Tropical South America. (S. Hurlbert, G. Rodriguez, and N. D. Santos, eds.) University of San Diego, San Diego, CA.
Spangler, P. J., and S. Santiago-Fragoso. 1992. The aquatic beetle subfamily Larainae (Coleoptera: Elmidae) in Mexico, Central America and the West Indies. Smithsonian Contributions to Zoology. Vol 528. Smithsonian Institution Press, Washington, DC.
Sutcliffe, D. W., and T. R. Carrick. 1973. Studies on mountain streams in the English Lake District. I. PH, calcium and the distribution of invertebrates in the River Duddon. Freshwat. Biol. 3: 437-462.
Tukey, J. 1958. Bias and confidence in not quite large samples (abstract). Ann. Math. Stat. 29: 614.
Ward, J. V. 1992. Aquatic Insect Ecology. John Wiley & Sons, Inc., New York. Willoughby, L. G., and R. G. Mappin. 1988. The distribution of Ephemerella ignita (Ephemeroptera) in streams: the role of pH and food resources. Freshwat. Biol. 19: 145-155.
Wilson, M. V., and Mohler. 1983. Measuring compositional change along gradients. Vegetatio 54: 129-41.
Young, G., K. R., and N. Valencia. 1992. Los bosques montanos del Perú. Memorias del Museo de Historia Natural. (21): 5-9.
Zahl, S. 1977. Jack-knifing an index of diversity. Ecology 58: 907-13.

CHAPTER 14

LEPIDOPTERA OF TWO SITES IN THE NORTHERN CORDILLERA DE VILCABAMBA, PERU

Gerardo Lamas and Juan Grados

INTRODUCTION AND METHODS

The butterflies and larger moths of the Cordillera de Vilcabamba were surveyed during the RAP expeditons at two sites in June 1997. Camp One (3350 m elevation) was surveyed between 6-20 June 1997 by the authors. At Camp Two (2015-2050 m elevation), butterflies only were collected by Antonio Sánchez, to whom we express our heartfelt thanks for having added the burden of collecting those specimens to his numerous other support tasks. Butterflies were collected with entomological nets and baited traps, while moths were collected at a white sheet, having been attracted to a 250V mercury vapor light source.

RESULTS

At Camp One we recorded 29 species of butterflies and 29 species of larger moths. These 58 species are listed in Appendix 25. At the second site, 19 species of butterflies were obtained (Appendix 25).

An analysis of the lepidopteran fauna found at Camp One indicates a high level of endemism, represented by 11 butterfly species which have never been recorded from anywhere else. Thus, 37.9% of the butterfly species present at the time of the survey have been found exclusively at this site of the Cordillera de Vilcabamba. The remaining 18 taxa comprise a mixture of widespread, weedy species (e.g. *Dione glycera, Vanessa braziliensis, Tatochila xanthodice*) and species found mostly in the upper Andes of central and southern Peru, and Bolivia. Among the moths, most sphingids are widespread species, although both species of *Euryglottis* are restricted to areas above 1000 m. *Bathyphlebia aglia* is a rare high-Andean saturniid, known only from a handful of specimens collected along the Andes from western Venezuela to southern Peru. The cercophanid *Janiodes bethulia* is another rare species, previously known only from the type-locality in Pasco department, central Peru. Seven species of *Amastus* (Arctiidae) were taken at this site; this large and taxonomically difficult genus includes many high Andean species.

The butterflies collected at Camp Two represent only a small sample of the fauna that is expected to occur there. Most of the species collected are widespread in Andean montane forests, but a new species of *Pedaliodes*, otherwise known only from the Vilcanota and Santa María valleys in Cusco, was found here.

CONCLUSIONS

Biogeographically, the lepidopteran fauna of the two sites surveyed is, as expected, closely related to the communities found both in the mountains to the north (Chanchamayo valley, Junín) and to the south (Vilcanota and Santa María valleys, Cusco). In general, it can be stated that the butterflies found at Camp One appear to be more closely related to the fauna of Machu Picchu, while those from Camp Two appear to be closely related to the Chanchamayo fauna. Nine species recorded at Camp One and 10 at Camp Two have also been found at the Machu Picchu Historical Sanctuary in Cusco, while 14 from Camp One and 17 from Camp Two also are known from Chanchamayo. When subspecific differences can be recognized, the populations from Camp One are almost identical to those from Machu Picchu, and different to the Chanchamayo populations.

By contrast, populations from Camp Two usually belong to the same subspecies found in Chanchamayo but are different from the Machu Picchu subspecies. Interestingly, the subspecies of *Lymanopoda galactea* found in the Cordillera de Vilcabamba was *L. g. siviae*, previously known only from its type-locality, at Sivia, Ayacucho, across the Apurímac River from the Cordillera (and which should occur in Chanchamayo too), whereas the subspecies found at Machu Picchu, *L. g. shefteli* Dyar, would have been expected at this site instead. It is assumed that the two specimens of *L. g. siviae* recorded at Camp One were stragglers, having flown up from lower areas, as *L. g. shefteli* is much more common at elevations below 2000 m at Machu Picchu than above 2500 m.

Although the lepidopteran fauna inhabiting elfin forests in Peru still is poorly known, the extremely high proportion of new taxa found at Camp One exceeded our expectations. On the other hand, we were somewhat disappointed at the scarcity of butterflies in general, as we had predicted that some 50 species might be found at Camp One. One new species of *Pedaliodes*, (the most speciose butterfly genus in the world, with some 240 species currently recognized, of which more than half are undescribed), was relatively abundant and was observed virtually every day of our stay. It was most appropriate that this undescribed species of *Pedaliodes*, discovered during the Vilcabamba Expedition, is the most colorful and beautiful representative of the genus.

CHAPTER 15

BIODIVERSITY ASSESSMENT OF ARTHROPODS OF THE SOUTHERN VILCABAMBA REGION, PERU

Jose Santisteban, Roberto Polo, Gorky Valencia, Saida Córdova, Manuel Laime, and Alicia De La Cruz

INTRODUCTION

The Smithsonian Institution/Monitoring and Assessment of Biodiversity Program conducted an initial biodiversity assessment in the Southern Cordillera de Vilcabamba along a proposed natural gas pipeline route. With the aid of satellite images, 10 habitat types were identified along the first stretch of the pipeline. Here we report the findings of the arthropod diversity assessment carried out at two locations along the proposed pipeline route on the eastern slopes of the Río Apurímac valley. The arthropod sampling and processing protocols used during this study were developed during earlier phases of the project (see Alonso and Dallmeier 1998, 1999, and Dallmeier and Alonso 1997).

Five distinct arthropod taxa—Araneae, Orthoptera, Coleoptera, Coleoptera-Scarabeidae, and Hymenoptera-Aculeata—were assessed at each study site. Voucher material from this study will be deposited in the Coleccion La Artropodas, Departamento de Entomologia, Museo de Historia Natural, Universidad Nacional Mayor de San Marcos, Lima, Peru; Museo de Historia Natural, Universidad San Antonio Abad del Cusco, Cusco, Peru; and Museo de Entomologia, Universidad Nacional Agraria La Molina, Lima, Peru.

METHODS

Study Area

The valley of the Río Apurimac serves as the geographic boundary between the departments of Cusco and Ayacucho in southeastern Peru. The river runs from south to north, forming a deep valley with altitudes reaching 4000 m on the western (Ayacucho) side and 3200 m on the eastern (Cusco) side. East of the valley is the Cordillera de Vilcabamba, a region that has attracted considerable attention in relation to biodiversity (Terborgh 1971, Young and Valencia 1992).

A number of different habitats comprise the valley. Its lower portion and river banks have been heavily impacted by human activity, the mid-elevations (up to 1500 m) less so particularly on the western side of the valley. Mid-elevations on the eastern side exhibit the least amount of disturbance, most likely because of their steeper slopes and forest types. Above 1600 m, no human activity is evident.

Study Sites

Two field camps were established—Llactahuaman and Wayrapata. The two camps were about 40 km south of Kimbiri, Ayacucho, the logistics base for this field work. Access to the camps was by helicopter flights. Llactahuaman (12°51'55.5" S, 73°30'46.0" W) was established at 1710 m in elevation about 4 km by air southeast of the village of Pueblo Libre along the Río Apurimac. The forest types in the vicinity of were typical of low to mid-elevation forests (1500 to 1600 m). They consisted primarily of mountain forest and mixed bamboo *Chusquea* forest at lower elevations.

We selected three plots for sampling arthropods: the first, LL02 (corresponding to the botany study plot # 1; see Comiskey et al., Chapter 4) was located about 200 m northeast of the camp at an elevation of 1735 m on a 35° slope; the second, LL04, was situated east of the camp site along a small creek at an elevation of 1580 m; and the third, LL05, was located 350 m west of the camp in a patch of bamboo mixed forest (elevation 1650 m).

The Wayrapata camp (12°50'10.1" S, 73°29'42.6" W) was established 3km by air east of Llactahuaman at an elevation of 2445 m. The dominant forest type was the elfin forest formation, which consisted of very dense, low, shrubby vegetation restricted to the ridge tops and exposed peaks. Two of our sampling plots, WA04 and WA05, were placed in the elfin forest formation. The third, WA06, was located about 150 m below the camp in cloud forest.

Sampling Methods

The protocol we used in sampling arthropods was based on passive (traps) methods as described in detail in Santisteban et al. (1998). This protocol fosters repeatable and comparable sampling at different sites. It also provides the quantitative data need for further analysis. The sampling plots were designed as modified Whittaker vegetation plots corresponding to 0.1 hectares (ha; 20 x 50 m). See Comiskey et al. (Chapter 4) for a detailed description of the plots and the method for establishing them.

We targeted arthropods of three forest strata: the lower canopy level (8 to 10 m), the understory layer (up to 2.5 m), and the forest soil. We used canopy malaise traps, ground malaise traps, pan pitfall traps, and bait pitfall traps. To the extent possible, all traps were placed inside the plots. To obtain comparable results, we used the same number and types of traps and general trap layout at each of the sampling plots. All traps were installed and run at the same time except for the bait traps, which required a few days of bait conditioning for optimal results. We serviced the traps at 48-hour intervals. Potential interference with the traps was minimized by careful placement; we paid special attention to landscape, slope characteristics, and light among other factors. Because different traps are normally set for the different strata, we were able to obtain a complementary sampling of the arthropod community present during the sampling period.

At each location, at least three repetitions were conducted at each plot. Replicates allow more accurate sampling and help detect changes around the traps that are unrelated to disturbances caused when the traps are set. All specimens from each trap type were pooled for site analysis, but each specimen was recorded separately on the original data sheets in the field.

A brief description of each sampling method follows. For more detail, see Santisteban et al. 1998.

Ground malaise traps

The malaise trap takes advantage of the fact that when most flying insects meet an obstacle, they fly upward rather than around the obstacle. Basic malaise trap design consists of a tent made of a fine mesh with open sides and a partition in the middle. The tent sides and top are sloped, tapering to a collection chamber placed on top of the trap. When an insect flies into the trap, it reaches the partition and then flies or crawls up, eventually becoming trapped at the highest point. The collection chamber is a plastic cylinder with a jar attached to the bottom that contains a killing/preserving agent (70% ethanol). Five ground malaise traps were set at each plot. We used these traps in conjunction with malaise pan traps (see below).

Canopy malaise traps

Canopy malaise traps employ the same principle as ground malaise traps, but they are raised high up into the canopy. Their design differs in that they have a second collection chamber on the lower portion of the trap. We hung the aerial malaise traps from tree branches in suitable places in each plot. The traps were raised to the appropriate level using an attached string, which was lifted via a "sling shot" of fine fishing line and attached lead weights.

We installed two aerial malaise traps at each plot. During servicing, we lowered the traps. Samples from the upper and lower collection chambers were kept separate and placed in separate 18-ounce whirl-packs, then labeled and transferred to the field lab for processing and sorting.

Malaise pan traps

Some groups of insects, such as beetles and plants bugs, do not fly up when encountering an obstacle. They fall to the ground. For such insects, we used malaise pan traps, which consist of plastic recipients 30 centimeters (cm) in diameter, painted gray, and placed along the middle partition of the ground malaise traps. Pan traps contained a mixture of saturated salty water and a few drops of unscented baby shampoo to lower water surface tension. This liquid served as the killing/preserving agent. The edge of each trap was raised from the ground to prevent other soil-associated arthropods from falling into the trap and thus biasing the sample. The gray color was intended to prevent sampling of low flying, visually cued insects that are attracted by color. We placed three pan traps beneath three of the five ground malaise traps at each plot. Specimens were extracted from the traps with an aquarium dip net, washed, and placed in 18-ounce whirl-packs with 70% ethanol as preserving agent. The specimens from all three traps at each plot were pooled and placed into the same bag, labeled, and taken to the lab for further processing.

Pitfall pan traps

The pan pitfall traps we used were a combination of two collecting strategies. For those arthropods that live and move on the forest soil and fall into the trap upon encounter, and for those low flying, visually cued arthropods that are attracted to colors such as the bright yellow used in the traps. We placed eight pan pitfall traps at each sampling plot. The traps consisted of round, bright yellow plastic bowls 27 cm in diameter and 7.8 cm deep. The containers were buried up to their openings into the soil so as to produce the least disturbance around the trap edge. After installation, the bowls were filled with saturate saltwater to three fourths of their capacity. The salt was added to retard decomposition of specimens, particularly important for soft-bodied arthropods. We also added a few drops of liquid shampoo to lower water surface tension and sinking into the solution.

Equipment required for servicing included 18-ounce whirl packs, an aquarium dip net, a wash bottle, 70% ethanol, labels, alcohol-resistant ink pens, and a bucket for carrying water. In servicing the traps, we removed all arthropod specimens with the aquarium dip net. Leaves, twigs, and other items that may have fallen into the trap were removed prior to using the net. Specimens were then washed, transferred to 70% ethanol, labeled, and taken to the field lab. When servicing the traps, researchers ensured that the water was transparent enough to allow the yellow color of the trap bottom to be seen since dirt sometimes falls into the traps.

Bait traps

Bait traps are designed to attract arthropods that are specialized on decaying matter. A number of different substances in varying degrees of decomposition may be used, and each will attract different species. Arthropod specialization in such cases is widely diverse, from those with a broad variety of substrates such as fermenting fruit (saprophagous), to those specialized on rotting flesh (necrophagous) and those specialized on the droppings of vertebrate animals (coprophagous). Use of bait traps assures inclusion of specialized members of the arthropod community not readily sampled through more general methods.

Traps used consisted of 1-liter plastic containers with tight-fitting lids. Four symmetrically distributed 5 x 5 cm openings, or access "windows," were made on the sides of the container about five cm below the container lid. Inside each container, we suspended a smaller plastic container (a disposable plastic cup) by wire guides from the sides of the larger recipient and placed just below its lid. The smaller cup contained the bait, and the scent was released through small ventilation holes on the upper third of the container. About 100 grams of bait were used in each trap. The bottom of the smaller container never touched the preserving liquid, which assured that samples were usually clean and free of contaminants. About 100 milliliters of 70% ethanol were poured on the bottom of the large container to serve as the killing/preserving agent.

Traps were buried up to the lower level of the small side windows. In selection of sites for the traps, we chose level or somewhat raised places. If sloping, then a small drainage was dug in case of rain. The carrion traps (those containing chicken or fish meat) should be secured with small branches or stakes (about 40 to 50 cm long) inserted into the ground. This was necessary to prevent small carrion-feeding vertebrates from getting to the bait and damaging the trap. Finally, a large leaf such as "platanillo," *Heliconia* (Musaceae), or similar vegetation was fixed on top of the trap to provide further protection from rain.

Eight bait traps were placed at each of two sampling plots. Specimens were placed in separate 18-ounce whirl-packs and transported to the field lab for further processing and sorting.

The following baits were used:

- Saprophagous: consisting of a fermenting fruit "cocktail" (two-thirds smashed papaya and one-third ripe bananas, including peels) with a teaspoon of baking yeast dissolved in water to accelerate the rotting process. This bait was left to decay for 48 hours in a tightly closed container before use.
- Necrophagous: consisting of 100 grams of fish meat or 100 grams of chicken meat (eviscerated) and left to rot for 48 hours in a tightly closed container. We did not use fish in traps because of the difficulty in obtaining the bait.
- Coprophagous: human feces kept in a tightly closed container for 72 hours. We did not use this bait at Wayrapata because of logistical constraints.

Sample processing

As already noted, sample processing in the field consisted of labeling, washing, transferring, and sorting specimens to ordinal level. Samples were labeled according to a coding system that gave each sample a unique number (see data management section). Each sample was sorted to the following ordinal groups: Araneae (spiders), Orthoptera (crickets), Coleoptera (beetles), Hymenoptera (bees and wasps), and residue. The residue contained all other groups of arthropods not studied in this report. Lepidoptera (butterflies and moths) were also separated from other groups, dried, and stored in glassine envelopes. Specimens of the target groups were transferred to smaller vials with proper labeling. Samples were further sorted by locality and sampling repetition, and then arranged by taxonomic group. In the laboratory, specimens were identified to morphospecies. Data recorded included date of sampling, and sampling method as well as a sub-sample number. Codes for sampling techniques are "M" for ground malaise trap, "A" for canopy malaise trap, "P" for pan pitfall trap, and "C" for bait pitfall trap.

A typical sample code label looks like the following: LL01.0618P03. This label code means that the specimen was from Llactahuaman (LL01.0618P03) at a sampling plot number 1 (LL01.0618P03), was taken on 18 June (LL01.0618P03), and corresponded to pan pitfall trap (LL01.0618P03) number 03 (LL01.0618P03), the sub-sample number. All relevant data were stored in a custom-designed database using Microsoft Access, version 7.0, a commercial computer software program.

Target Groups

Our choice of taxonomic groups reflected the taxonomic expertise of the scientific personnel involved in the project and also corresponded to some of the most important and diverse arthropod groups. Identification was made to individual species or morpho-species. Each distinct or recognizable species was recorded, separated by family or subfamily, and assigned a consecutive number for that taxonomic grouping. The number of individuals for each particular species or morpho-species in each sample was recorded, providing information on relative abundance. Except for the Araneae, and because of storage constraints and the need to speed up the process, only a few specimens of each recognized species were mounted dry; all others were kept in alcohol. This reference collection of dried material was labeled and maintained using standard entomological techniques and materials. All Araneae specimens (reference and primary) were kept in alcohol.

Data analysis

We used EstimateS 5 version 5.0.1 (Colwell 1997) in analyzing our data to obtain randomized species accumulation curves and species richness estimators such as ICE and ACE (Colwell and Coddington 1994, Chazdon et. al 1998) that are based on sample coverage. The Incidence-based Coverage Estimator (ICE) incorporates species found in 10 or fewer sampling units (i.e., infrequent species), while the Abundance-based Coverage Estimator (ACE) incorporates species with 10 or fewer individuals in the sample (i.e., rare species). Abundance matrixes were used to compute the estimators and also to estimate the number of species shared among sites (similarity). The format chosen for data input into the program was species (rows) by samples (columns), which were directly extracted from the electronic data sheets. Parameters for diversity statistics were as follows: number of randomization=100, random seed number=17, with the upper limit for rare or infrequent species set at 10 (default). For comparative purposes, Fisher's and Shannon/Simpson's diversity indexes were computed, as well as the Morisita-Horn shared species index. Statistics and indexes were presented graphically for each group and sampling locality.

RESULTS AND DISCUSSION

The results of our work along the proposed pipeline route are given below by target group and separately for Llactahuaman and Wayrapata.

Araneae (Spiders)

A total of 184 individual spiders were recorded at Llactahuaman, belonging to 60 species and 16 different families (Table 15.1). The family Corinnidae was the most diverse with 12 species, followed by Salticidae with 10, Araneidae with nine, and Ctenidae with five species. Lycosidae was the most abundant with 88 individuals in just 4 species (Fig. 15.1). With the aid of the ICE and ACE coverage-based estimators, we estimate a total of 160 species for the area.

At Wayrapata, we recorded 121 individual spiders, belonging to 45 species and 18 different families (Table 15.2). The most diverse group was the family Salticidae with 12 species, followed by Anyphaenidae with six species and Araneidae with four species each. Again, Lycosidae were the most abundant with 45 individuals primarily belonging to Lycosidae species 01 (Fig. 15.2). These groups are mainly cursorial and active hunters. The estimated species accumulation curve for Wayrapata is shown in Figure 15.3. Forty-five species were found at the site and based on the abundance and incidence based estimators, the total number of species predicted to be found in the area reaches 100.

Orthoptera (Crickets)

At Llactahuaman, we registered 103 Orthoptera individuals (crickets that are primarily soil and litter inhabitants) from 22 species and 4 families (Table 15.3). Gryllidae was the most diverse group at this location with 12 species, followed by Tetrigidae with 6 species and Tridactylidae with 3 species. Gryllidae was also the most abundant with 58 individuals. Based on the ICE and ACE estimators, we estimate that 40 species could be present in the area.

We recorded 44 individuals of Orthoptera from 17 species and 4 families at Wayrapata (Table 15.4). Gryllidae was the most diverse group with 7 species, followed by Acrididae with 4 and Tetrigidae and Tettigonidae with 3 species each. Acrididae were the most abundant with 18 individuals. We estimate a total of 35 species for the area.

Coleoptera (Beetles)

Among the Coleoptera, we recorded 569 individuals at Llactahuaman from 166 species and 21 families, primarily from pan pitfall and malaise pan traps (Appendix 26). The most diverse group was the family Chrysomelidae (leaf beetles) with 52 species, followed by Staphylinidae (rove beetles) with 34 species and Scolytidae with 17 species. Chrysomelidae among the Cucujiformia and Staphylinidae among the Staphyliniformia were the most speciose for those taxa. The Staphylinidae was the most abundant with 201 individuals, followed by Chrysomelidae with 175 individuals. Based on the Coleoptera ICE and ACE coverage-based estimators, we calculate that more than 350 species could be found in the area (Fig. 15.4). The high incidence of Chrysomelidae may be explained by the effect of brightly colored pitfall pan traps that attract Alticinae, Galerucinae, and some Criocrinae. Usually, one can expect to find Chrysomelidae in malaise pitfall traps. The Coleoptera samples from Wayrapata had not been processed when this report was submitted.

Within the Coleoptera, the family Scarabeidae is one of the most important because of the high number of species worldwide and because of its ecological specialization. The specimens were found mainly in pitfall bait traps, pan pitfall traps, and malaise pitfall traps at both Llactahuaman and Wayrapata (Tables 15.5 and 15.6). At Llactahuaman we found a total of 24 species, while the ICE and ACE based estimators predicted 45 species. In contrast, at Wayrapata we found only three species with very low sampling success. he paucity of specimens from Wayrapata is likely related to the specialized habitat there and to the difficulty of sampling. It may also result from the low incidence of large mammals in the vicinity of Wayrapata (see Rodríguez and Amanzo Chapter 9). Among the sampled Scarabaeidae, species of the genus *Deltochilum* were the most diverse at Llactahuaman with 5 species, followed by *Dichotomius* and *Phyllophaga* with 3 species each. *Deltochium* was also the most abundant with 66 individuals, followed by *Dichotomius* with 45 and *Pleuroporus* with 31.

Hymenoptera (Bees and Wasps)

At Llactahuaman, we found 493 individuals of Hymenoptera-Aculeata (bees and wasps but no ants) in 102 species and 10 families (Appendix 27). The specimens were primarily from ground malaise traps, aerial malaise traps, and pan pitfall traps. The family Bethylidae was the most diverse with 34 species, followed by Apidae with 23 species and Halictidae with 14 species. Apidae were the most abundant with 244 individuals, followed by Bethylidae with 166 individuals. Sphecidae was represented by species of the genus *Liris*, a predatory specialist on crickets that are very diverse and abundant in the cloudforest. The ICE and ACE coverage-based estimators predicted more than 500 species for Llactahuaman.

At Wayrapata, we found 205 individuals of Hymenoptera-Aculeata (excluding the ants) in 67 species and 9 families (Appendix 28). The specimens were primarily from ground malaise traps and pan pitfall traps. The family Halictidae was the most diverse with 22 species, followed by Bethylidae with 11 species and Sphecidae with 9 species. Halictidae were the most abundant with 78 individuals, followed by Vespidae with 166 individuals and Apidae with 34 individuals. One hundred and fifty species were estimated based on ICE and ACE coverage-based estimators.

The exposed ridge tops, strong winds, stunted and densely compacted vegetation, and semi-permanent cloudiness may be favorable for nest-forming social species (*Trigona*, *Partamona*, *Nannotrigona*, *Melipona*) that appear to be better adapted to sudden environmental changes, which are likely common in these habitats.

Comparing Arthropod Diversity Patterns

Comparisons among the plots within the two study sites, based on taxonomic groups, are shown in Table 15.7. We computed shared species for the groups, whenever possible, using the EstimateS 5.0 and coverage-based estimators of shared species. Table 15.7 shows the number of species at each plot as well as the number of shared species. For comparative purposes, we included the Morisita-Horn Index of similarity, which clearly indicated that different groups presented different patterns of species similarity. In general, the sampling points at Wayrapata showed more similarity (values closer to one) while the sites at Llactahuaman did not have high similarity (values closer to zero).

We also compared Araneae, Orthoptera, and Hymenoptera-Aculeata results between the two sites (Table 15.8). The Araneae species composition was very similar between the two study sites while the Orthoptera, and Hymenoptera-Aculeata showed very low similarity.

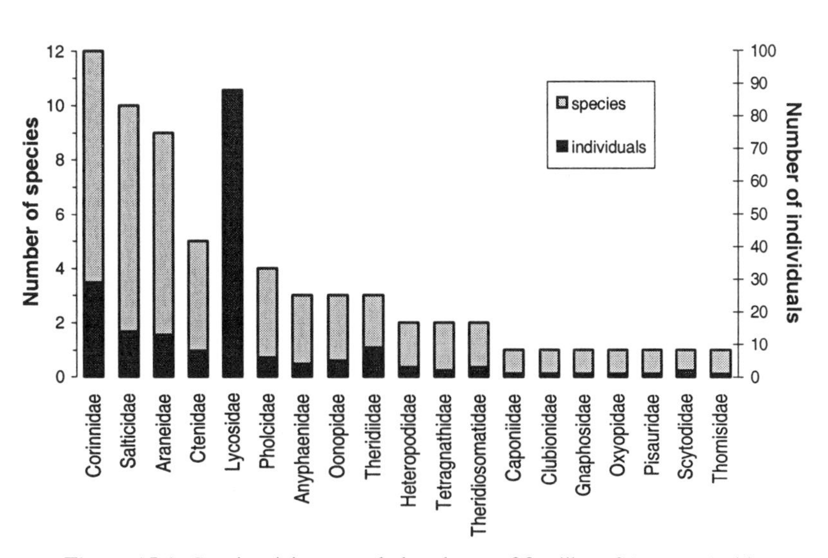

Figure 15.1. Species richness and abundance of families of Araneae (spiders) at Llactuahuaman (1710 m). Left scale indicates number of species (richness) and right scale indicates number of individuals (abundance).

Table 15.1. Abundance (numbers of individuals) of Araneae (spiders) sampled in three study plots at Llactahuaman (1710 m).

		Study Plot			
Family	**Species**	**LL02**	**LL04**	**LL05**	**Total**
Anyphaenidae	sp02			1	1
Anyphaenidae	sp04	2			2
Anyphaenidae	sp07		1		1
Araneidae	*Micratena* sp01		1		1
Araneidae	*Micratena* sp02		1		1
Araneidae	sp01			1	1
Araneidae	sp03	1			1
Araneidae	sp04	1			1
Araneidae	sp05			1	1
Araneidae	sp07	4			4
Araneidae	sp09		2		2
Araneidae	sp10	1			1
Caponidae	sp01		1		1
Clubionidae	sp01			1	1
Corinnidae	sp01	3		6	9
Corinnidae	sp02	1	1		2
Corinnidae	sp03			1	1
Corinnidae	sp04	4	1	1	6
Corinnidae	sp06	1			1
Corinnidae	sp07	2			2
Corinnidae	sp08	2			2
Corinnidae	sp09	1			1
Corinnidae	sp10	1			1
Corinnidae	sp11		1	1	2
Corinnidae	sp14		1		1
Corinnidae	sp12			1	1
Ctenidae	sp02	1			1
Ctenidae	sp03	1			1
Ctenidae	sp04		1		1
Ctenidae	sp06	3		1	4
Ctenidae	sp07			1	1
Gnaphosidae	sp01		1		1
Heteropodidae	sp01	1	1		2
Heteropodidae	sp02		1		1

		Study Plot			
Family	**Species**	**LL02**	**LL04**	**LL05**	**Total**
Lycosidae	sp01	41	7	28	76
Lycosidae	sp02	2		2	4
Lycosidae	sp03	2		5	7
Lycosidae	sp04	1			1
Oonopidae	sp01	1	2		3
Oonopidae	sp02	1			1
Oonopidae	sp04		1		1
Oxyopidae	sp01				
Pisauridae	sp01	1			1
Salticidae	sp02			1	1
Salticidae	sp04		1		1
Salticidae	sp05	2			2
Salticidae	sp06			1	1
Salticidae	sp08		1		1
Salticidae	sp09	2			2
Salticidae	sp10	2			2
Salticidae	sp11	1	1		2
Salticidae	sp13		1		1
Salticidae	sp17		1		1
Scytodidae	*Scytodes* sp01	1		1	2
Tetragnathidae	sp01		1		1
Tetragnathidae	sp02		1		1
Theridiidae	*Dipoena* sp02	4			4
Theridiidae	*Dipoena sp04*	4			4
Theridiidae	sp03	1			1
Theridiosomatidae	sp03			1	1
Theridiosomatidae	sp04	2			2
TOTAL		**98**	**31**	**55**	**184**

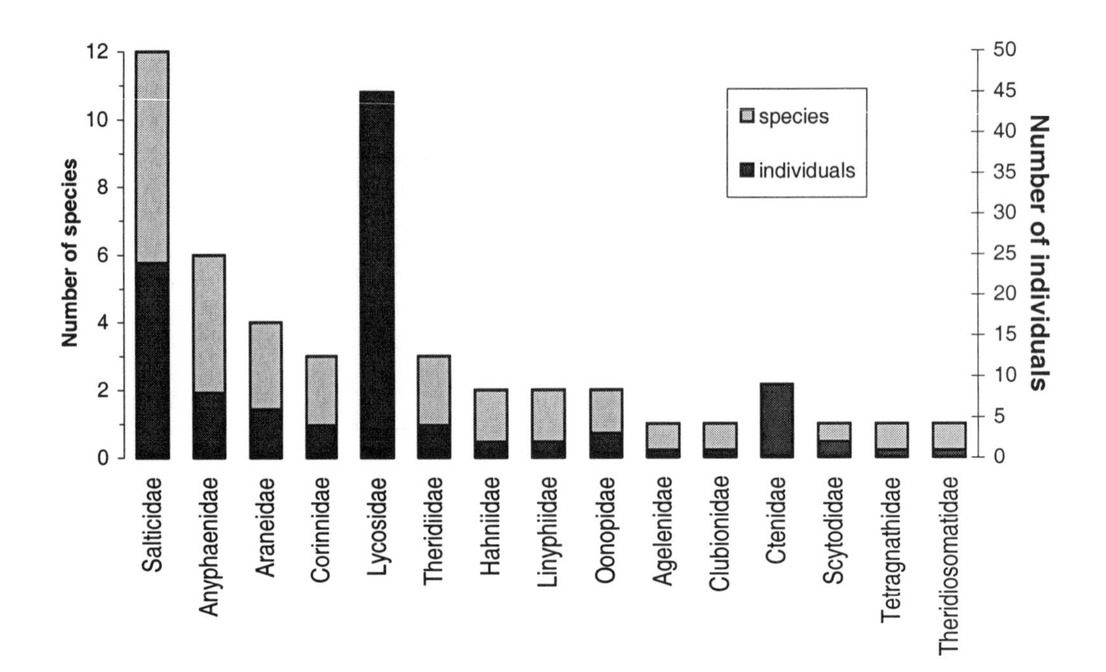

Figure 15.2. Species richness and abundance of families of Araneae (spiders) at Wayrapata (2445 m). Left scale indicates number of species (richness) and right scale indicates number of individuals (abundance).

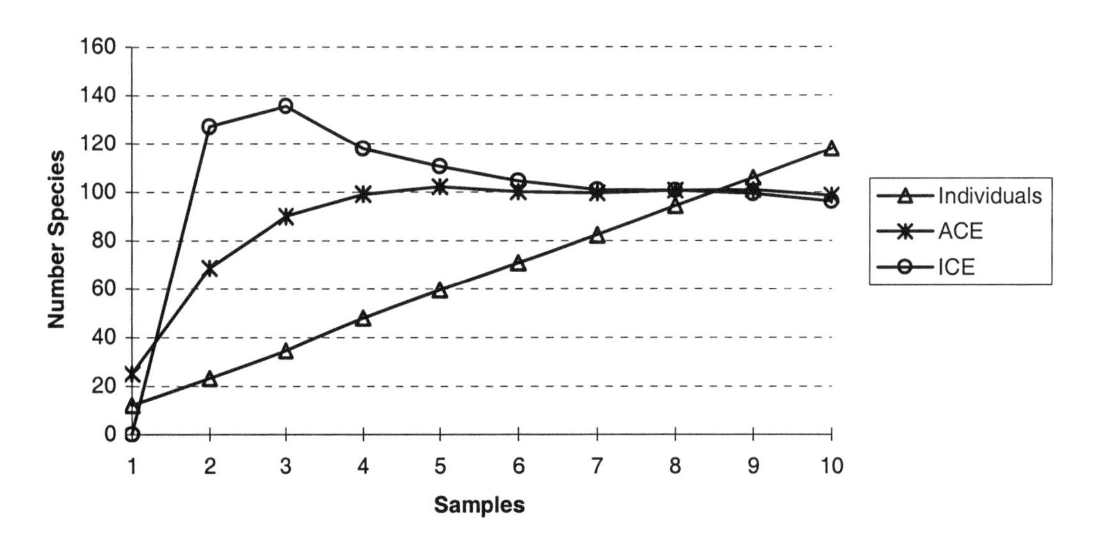

Figure 15.3. Abundance, ICE and ACE coverage-based estimators of species richness for Araneae (spiders) at Wayrapata.

Table 15.2. Abundance (number of individuals) of Araneae (spiders) sampled in three study plots at Wayrapata (2445 m).

		Study Plot			
Family	**Species**	**WA04**	**WA05**	**WA06**	**Total**
Agelenidae	sp01			1	1
Anyphaenidae	sp01	2			2
Anyphaenidae	sp02	1			1
Anyphaenidae	sp03	1			1
Anyphaenidae	sp05			1	1
Anyphaenidae	sp06			1	1
Anyphaenidae	sp08		1	1	2
Araneidae	sp02	1			1
Araneidae	sp06		1		1
Araneidae	sp08			2	2
Araneidae	sp11	1		1	2
Clubionidae	sp01			1	1
Coriniidae	sp04			1	1
Coriniidae	sp13			1	1
Corinnidae	sp05		1	1	2
Ctenidae	sp05	5		4	9
Hahnidae	sp01	2			2
Linyphidae	sp02			1	1
Linyphidae	sp03			1	1
Lycosidae	sp01	25	9	2	36
Lycosidae	sp02	6		1	7
Lycosidae	sp03	1	1		2
Oonopidae	sp02	1	1		2
Oonopidae	sp05			1	1
Oonopidae	sp02	1	1		2
Oxyopidae	sp01		1		1
Pholcidae	sp01	2			2
Pholcidae	sp02	1			1
Pholcidae	sp03			1	1
Pholcidae	und			2	2
Salticidae	sp01	1			1
Salticidae	sp02			2	2
Salticidae	sp03	1			1
Salticidae	sp04	1		1	2

		Study Plot			
Family	**Species**	**WA04**	**WA05**	**WA06**	**Total**
Salticidae	sp05	2	2	1	5
Salticidae	sp07	1			1
Salticidae	sp08	1	3	1	5
Salticidae	sp10		1		1
Salticidae	sp14			1	1
Salticidae	sp15		1	1	2
Salticidae	sp16		2		2
Salticidae	und			1	1
Scytodidae	*Scytodes* sp01	1		1	2
Tetragnathidae	sp01	1			1
Therididae	sp04	1			1
Theridiidae	*Dipoena* sp01	1		1	2
Theridiidae	*Twaithesia* sp01	1			1
Theridiosomatidae	sp02	1			1
Thomisidae	sp01	1			1
TOTAL		**63**	**24**	**34**	**121**

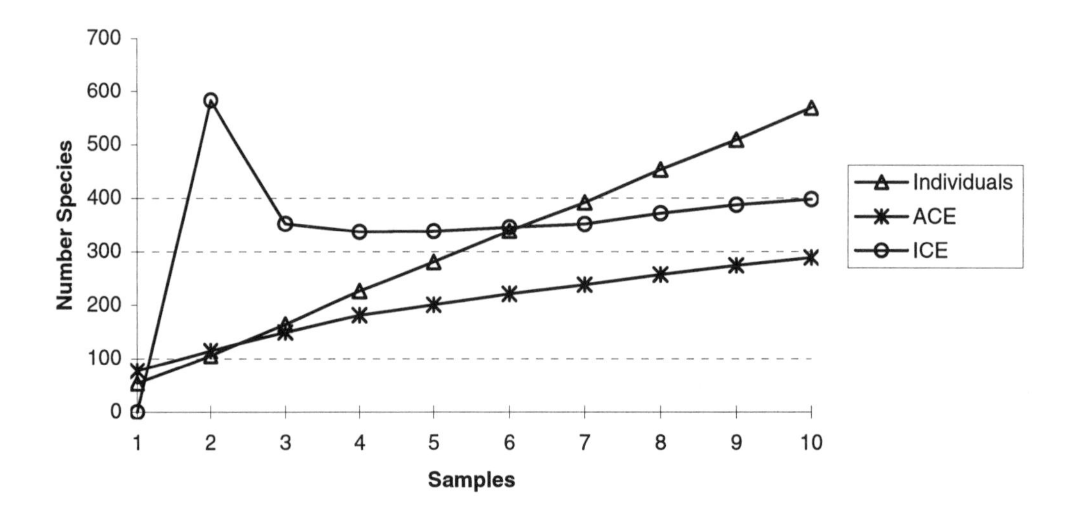

Figure 15.4. Abundance, ICE and ACE coverage-based species richness estimators for Coleoptera (beetles) from all samples at Llactahuaman.

Table 15.3. Abundance (number of individuals) of Orthoptera (crickets) sampled in three study plots at Llactahuaman

		Study Plot			
Family	**Species**	**LL02**	**LL04**	**LL05**	**Total**
Gryllidae	sp02	1			1
Gryllidae	sp03	1			1
Gryllidae Eneopterinae	sp04			2	2
Gryllidae Eneopterinae	sp05			3	3
Gryllidae Eneopterinae	sp10		3		3
Gryllidae Eneopterinae	sp11		6	2	8
Gryllidae Eneopterinae	sp12		4	19	23
Gryllidae Eneopterinae	sp13			1	1
Gryllidae Gryllinae	sp05	10	2	1	13
Gryllidae Gryllinae	sp06	1			1
Gryllidae Trigoniidae	sp09	1			1
Gryllidae Trigoniidae	sp14		1		1
Tetrigidae	sp01	2		3	5
Tetrigidae	sp02	1	1		2
Tetrigidae	sp03	2			2
Tetrigidae	sp04		1		1
Tetrigidae	sp05			2	2
Tetrigidae	sp07			2	2
Tettigonidae	sp01			1	1
Tridactylidae	sp01		12		12
Tridactylidae	sp02		17		17
Tridactylidae	sp03		1		1
TOTAL		**19**	**48**	**36**	**103**

Table 15.4. Abundance (number of individuals) of Orthoptera (crickets) sampled in three study plots at Wayrapata

		Study Plot			
Family	**Species**	**WA04**	**WA05**	**WA06**	**Total**
Acrididae	sp01	6	4	5	15
Acrididae	sp02	1			1
Acrididae	sp03			1	1
Acrididae	sp04		1		1
Gryllidae	sp01	2			2
Gryllidae	sp02	1	1		2
Gryllidae	sp05	1			1
Gryllidae Gryllinae	sp05	1			1
Gryllidae Gryllinae	sp07	2			2
Gryllidae Gryllinae	sp08	2			2
Gryllidae Trigoniidae	sp14	3			3
Tetrigidae	sp02	3	2		5
Tetrigidae	sp05	3	1		4
Tetrigidae	sp06	1			1
Tettigonidae	sp02	1			1
Tettigonidae	sp04		1		1
Tettigonidae	sp06	1			1
TOTAL		**28**	**10**	**6**	**44**

Table 15.5. Abundance (number of individuals) of Scarabaeidae sampled in three study plots at Llactahuaman.

	Study Plot			
Species	**LL02**	**LL04**	**LL05**	**Total**
Canthon sp. 10			1	1
Copris sp. 02	1			1
Copris sp. 03			1	1
Deltochilum sp. 08			10	10
Deltochilum sp. 11	8		13	21
Deltochilum sp. 12	6			6
Deltochilum sp. 12		3	24	27
Deltochilum sp. 13			2	2
Dichotomius sp. 06	1		3	4
Dichotomius sp. 07	2		26	28
Dichotomius sp. 08	1		12	13
Eurysternus sp. 04			5	5
Melolonthinae sp. 02	1			1
N.D. sp. 13	3	1	2	6
N.D. sp. 14			1	1
N.D. sp. 15			1	1
Phanaeus sp. 06	2			2
Phanaeus sp. 07			1	1
Phyllophaga sp. 01	1			1
Phyllophaga sp. 02	1			1
Phyllophaga sp. 03	1			1
Pleurophorus sp. 02			31	31
Plusiotis sp. 01		1		1
Scatimus sp. 02		1	3	4
TOTAL	**28**	**6**	**136**	**170**

Table 15.6. Abundance (number of individuals) of Scarabaeidae sampled in three study plots at Wayrapata.

	Study Plot			
Species	**WA04**	**WA05**	**WA06**	**Total**
Acanthocerus sp. 01	1		1	2
Anaides sp. 04	1	1		2
N.D. sp. 13	3			3
TOTAL	**5**	**1**	**1**	**7**

Table 15.7. Comparisons of arthropod species richness between sampling plots at Llactahuaman (LL) and at Wayrapata (WA), Southern Cordillera de Vilcabamba, Peru.

Sites Compared (Site 01- Site 02)	**Taxonomic Group**	**No. species Site 01**	**No. species Site 02**	**Shared Observed**	**Shared Estimated***	**Morisita-Horn Index**
Llactuahuaman						
LL02-LL04	Araneae	33	23	6		0.72
	Orthoptera	8	10	2	2.77	0.09
	Coleoptera	77	99	23	44.44	0.54
	Scarabeidae	12	4	1		0.07
	Hymenoptera	76	150	14	51.27	0.14
LL02-LL05	Araneae	33	18	7	10.64	0.93
	Orthoptera	8	10	2	5.33	0.08
	Coleoptera	77	42	12	28.71	0.34
	Scarabeidae	12	16	5	6.00	0.31
LL04-LL05	Araneae	23	18	3		0.64
	Orthoptera	10	10	3	3.33	0.20
	Coleoptera	99	42	20	31.14	0.25
	Scarabeidae	4	16	3		0.39
Wayrapata						
WA04-WA05	Araneae	27	12	5	11.93	0.85
	Orthoptera	14	6	4	5.68	0.70
	Hymenoptera	31	36	7	9.49	0.28
WA04-WA06	Araneae	27	27	9	52.75	0.34
	Orthoptera	14	2	1	1.00	0.43
	Hymenoptera	31	14	2	2.00	0.14
WA05-WA06	Araneae	12	27	6	17.25	0.28
	Orthoptera	6	2	1	1.00	0.69
	Hymenoptera	36	14	8	11.04	0.40

* Estimated using Estimate S 5.0 and coverage based estimators of shared species (Colwell 1999).

Table 15.8. Comparison of arthropod species richness between Llactahuaman (LL) and Wayrapata (WA), based on selected arthropod groups.

Arthropod Group	First Sample	Second Sample	# Species Obs LL	# Species Obs WA	Shared Observed	Shared Estimated*	Morisita-Horn Index
ARANEAE	LL	WA	60	45	13	26.03	0.88
ORTHOPTERA	LL	WA	22	17	5	6.11	0.06
HYMENOPTERA	LL	WA	102	67	17	40.13	0.11

* Estimated using Estimate S 5.0 and coverage based estimators of shared species (Colwell 1999).

CONCLUSIONS

- The results of our arthropod diversity assessment in cloud forest formations along the eastern slopes of the Apurimac River valley indicate a high diversity of arthropod species at both study locations.
- Among the spiders, the most frequent species sampled in the traps at Llactahuaman and Wayrapata were cursorial, free-living, prey-hunting species (Coriniidae, Salticidae, Lycosidae). Lycosidae was the most abundant group at both locations.
- Among the Orthoptera, the Gryllidae was the most diverse at Llactahuaman and Wayrapata. Gryllidae was the most abundant at Llactahuaman, while Acrididae was the most abundant in Wayrapata.
- Among the Coleoptera, the families Chrysomelidae and Staphylinidae were the most speciose and diverse at Llactahuaman. Staphylinidae was also the most abundant.
- Among the Scarabaeidae at Llactahuaman, the genera *Deltochilum*, *Dichotomus*, and *Phyllophaga* were the most diverse and *Deltochium* was the most abundant.
- Among the Hymenoptera, Bethylidae, and Apidae were the most diverse at Llactahuaman, while Apidae was the most abundant. At Wayrapata, Halictidae and Bethylidae were the most diverse, and Halictidae the most abundant.
- Comparisons among sampling plots based on general species frequency and abundance of several arthropod taxa showed different patterns. In general, the sampling plots at the higher elevation (Wayrapata) were more similar to each other than sampling plots at the lower elevation (Llactahuaman). Comparisons between the two sites for three taxa—Araneae, Orthoptera, and Hymenoptera—showed a similar species composition only for spiders.

LITERATURE CITED

Alonso, A., and F. Dallmeier (eds.). 1998. Biodiversity Assessment and Long-term Monitoring of the Lower Urubamba Region, Peru: Cashiriari-3 Well Site and the Camisea and Urubamba Rivers. SI/MAB Series # 2. Smithsonian Institution/MAB Biodiversity Program, Washington, DC.

Alonso, A., and F. Dallmeier (eds.). 1999. Biodiversity Assessment and Long-term Monitoring of the Lower Urubamba Region, Peru: Pagoreni Well Site Assessment and Training. SI/MAB Series # 3. Smithsonian Institution/MAB Biodiversity Program, Washington, DC.

Chazdon, R. L., R. K. Colwell, J. S. Denslow, and M. R. Guariguata. 1998. Statistical methods for estimating species richness of woody regeneration in primary and secondary rainforests of NE Costa Rico. In Forest Biodiversity Research, Monitoring, and Modeling: Conceptual Background and Old World Case Studies (F. Dallmeier and J. Comiskey, eds.) Parthenon Publishing, Paris.

Colwell, R. K. 1999. EstimateS 5 version 5.0.1. Statistical Estimation of species richness and shared species from samples. Web site: viceroy.eeb.uconn.edu/estimates.

Colwell, R. K., and J. A. Coddington. 1994. Estimating terrestrial biodiversity through extrapolation. *Philos. Trans. of the Royal Soc.* (Series B) 345: 101-118.

Dallmeier, F., and A. Alonso (eds.). 1997. Biodiversity Assessment and Long-term Monitoring of the Lower Urubamba Region, Peru: San Martin-3 and Cashiriari-2 Well Sites. SI/MAB Series # 1. Smithsonian Institution/ MAB Biodiversity Program, Washington, DC.

Santisteban, J., R. Polo, S. Córdova, G. Valencia, F. Gomez, A. De La Cruz, and P. Aibar. 1998. Arthropods: biodiversity assessment at the Pagoreni well site. *In*: Biodiversity Assessment and Long-term Monitoring of the Lower Urubamba Region, Peru: Pagoreni Well Site Assessment and Training. (A Alonso and F. Dallmeier, eds.) SI/MAB Series # 3. Smithsonian Institution/MAB Biodiversity Program, Washington, DC.

CHAPTER 16

RESOURCE USE AND ECOLOGY OF THE MATSIGENKA OF THE EASTERN SLOPES OF THE CORDILLERA DE VILCABAMBA, PERU

Glenn Shepard, Jr. and Avecita Chicchón

INTRODUCTION

The Vilcabamba mountain range, known as *Chovivanteni Otishi* to local Matsigenka, rises between the Urubamba and Tambo-Ene river systems, extending north from the Andean foothills near Quillabamba towards the Ucayali basin. The mountains are named after the ruined city of Vilcabamba in the southern part of the range, the final refuge and short lived capital-in-exile of the Inca State after the fall of Cusco in the early years of Spanish conquest. The mountain chain apparently served as a corridor of trade and communication between lowland Amazonia and the Inca state (Savoy 1970; Camino 1977; Lyon 1981). Despite its tremendous historical and scientific importance, the network of stone roads, highland ruins and lowland trading centers remains virtually unstudied (Lee 2000).

On the eastern slopes of the Vilcabamba range lies the Río Picha and tributaries, the beginning of the traditional territory of the Matsigenka people, which extends east and south towards the headwaters of the Manu, Urubamba and Madre de Dios rivers. On the west lies the Tambo-Ene river basin, marking the southeastern limit of the territory of the Asháninka people, linguistic and cultural cousins of the Matsigenka. Both groups were ancient trading partners of the Inca (see Varese 1968). Today, they live in numerous, dispersed communities where they combine traditional subsistence patterns of farming, hunting, fishing and gathering with varying degrees of participation in the market economy, selling such products as coffee, cacao, *annato* (*Bixa orellana*) and timber.

The abrupt and relatively recent geological processes that created this mountain range contribute to an apparent (though previously undocumented) richness of biological diversity and other natural resources, including gas and petroleum. In view of its importance as a haven of both biological and cultural diversity, three adjacent protected areas have been proposed for the region within the Apurímac Reserved Zone (see Proposed Protected Areas in the Cordillera de Vilcabamba). On the western side is a proposed Communal Reserve in Asháninka territory. In the middle and protecting the mountains above approximately 2000 meters elevation lies a proposed National Park. On the eastern side is a second proposed Communal Reserve in Matsigenka territory. These reserves are envisioned as areas of natural resource protection or management involving the participation of indigenous groups.

In order to provide scientific data in support of the creation of such protected areas, Conservation International carried out rapid assessments (RAP) of biological diversity of the Cordillera de Vilcabamba in 1997 and 1998. The assessment also included a rapid field study of ethnoecology and resource use among the lowland Matsigenka populations of the Río Picha, adjacent to the proposed Communal Reserve. The data presented here (condensed from Shepard 1997) were gathered by Glenn Shepard, Avecita Chicchón and Mateo Italiano as part of this study.

STUDY COMMUNITIES AND METHODS

The team spent four to six days each in the Matsigenka communities of Camaná, Mayapo and Puerto Huallana during May 1997. The communities of Nuevo Mundo, Nueva Luz and the Catholic Mission of Kirigueti (all along the Urubamba river) were visited in passing. The three principal study communities had populations of more than 300 people each. The total population of the native communities of the Río Picha and tributaries is close to 1500.

For at least a century, Asháninka migrants have crossed the Cordillera and settled or intermarried with Matsigenka families, especially in the communities of Puerto Huallana and Kotsiri. Political violence in the Tambo-Ene basin during the 1980s and 1990s brought hundreds of Asháninka refugees from the Kutivireni region across the Cordillera de Vilcabamba to the Río Kotsiri near the Picha region. Most of those who chose to remain in the Picha are now located in the large community of Tangoshiari.

A keystone in the rapid ethnoecological appraisal method was the preparation of community resource maps. Community members were asked to draw and label schematic maps detailing the names and relative locations of rivers, mountains, past and present human settlements, gardens, trails, animal and plant resources and other landmarks surrounding each community. By comparing this information with topographic maps and satellite images, it was possible to infer approximate locations of resource use areas and other landmarks. Figures 16.1 and 16.2 present some of the ethnoecological and ethnogeographical data at the regional and community-level scales. Gardens and other areas of resource extraction were visited, and interviews about resource use were conducted. A preliminary study of Matsigenka forest classification was carried out. Interviews were conducted with male and female informants of different ages and backgrounds about the oral history and folklore of the region.

RESULTS AND DISCUSSION

Resource Use: Centrifugal and Centripetal Forces

Though the populations of these communities (250-300 people) seem small by urban standards, they are in fact much denser than traditional Matsigenka settlements of 30-80 people. Until the past few decades, Matsigenka settlements consisted of extended families living in a matrilocal residence pattern (Snell 1964) widely dispersed throughout smaller tributaries and interfluvial areas. A typical settlement would include an older man and woman, the families of their married daughters and the houses and gardens constructed with the help of the sons-in-law. The nearest settlement might be several hours or days away by foot or river. Several settlements might join together for manioc beer festivals on the full moon, but otherwise extended family settlements were autonomous and staunchly independent, and political and economic integration was loose. The tradition of autonomy and independence among extended family units is still strong, and proves to be a constant source of negotiation and conflict in larger, modern communities where families live at closer quarters and are subject to communal obligations.

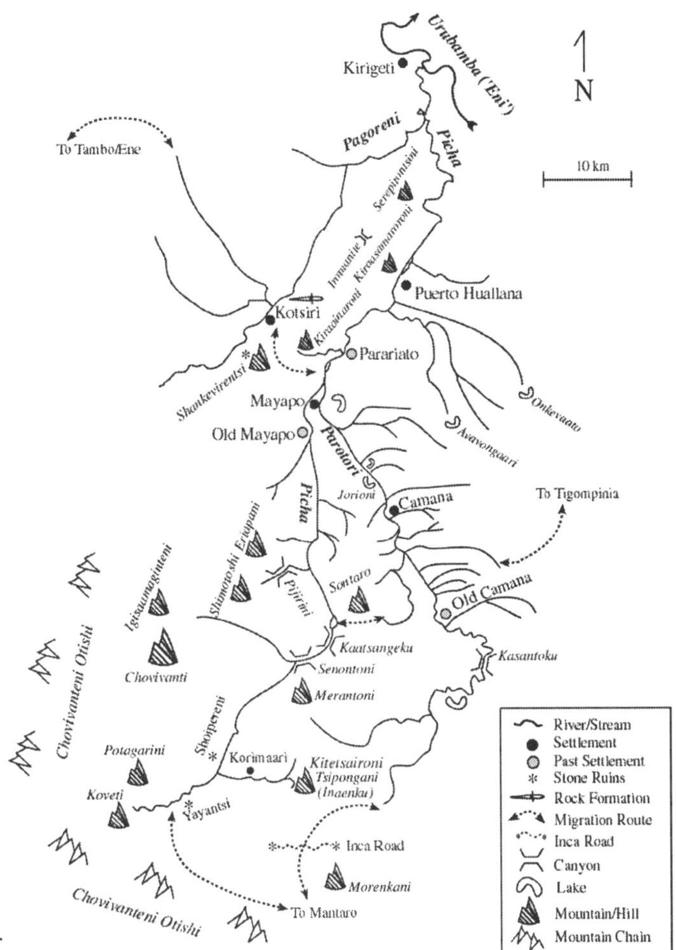

Figure 16.1. Picha River

In order take advantage of the diverse and scattered resources of the tropical forest and in order to circumvent local extinctions, the Matsigenka of the modern communities of the Río Picha engage in seasonal treks and occasional migrations. The joint centrifugal forces of resource depletion and community strife lead most Matsigenka families to maintain secondary homes, gardens and hunting camps at some distance from the main community. Nonetheless, the availability of health care, trade goods, formal literacy education and sporadic economic opportunities provides a strong centripetal force that maintains Matsigenka populations focused around semi-permanent communities along major rivers. Yet the denser the population in the central community, the more powerful the centrifugal forces of resource depletion and social strife become. For this reason, even the sedentary central communities are subject to periodic migrations. All three of the populations we visited had changed the location of their community en masse within the past twenty years. The tradition of trekking and migration contributes to the extent and detail of Matsigenka knowledge about the local environment, and provides a strong argument for the collective territorial rights of these people to the proposed Communal Reserve.

Agriculture

Matsigenka swidden agriculture, like that of other indigenous groups in the Amazon, is characterized by a relatively small area of forest disturbance, multi-cropping, great genetic diversity of crop cultivars and a rapid process of forest regeneration (Posey and Balée 1989, Boster 1984). The secondary vegetation in garden fallows is enriched with domesticated fruit trees that, in addition to providing food for humans, may also attract game animals (Hames 1980). As long as the fallow period is not reduced, such swidden agricultural practices exert a minimal impact on the health of tropical forests, both temporally and in terms of total area, when compared with the permanent disturbances caused by large-scale commercial monocropping or cattle ranching.

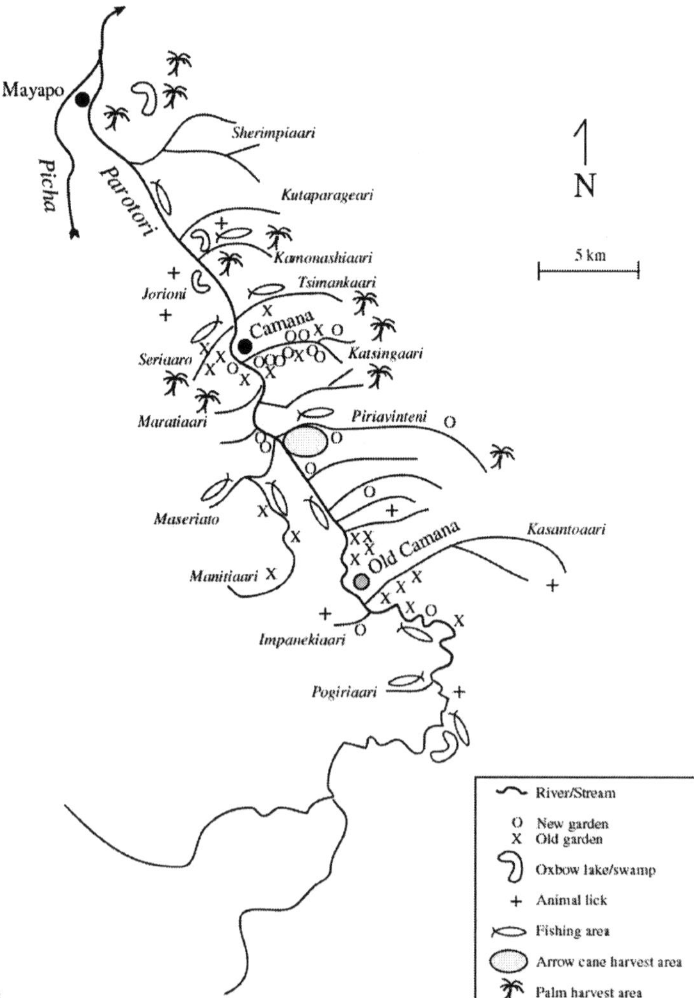

Figure 16.2. Camana Resource Use

During the short stays in each community, the authors visited several garden sites. The agricultural practices appear similar to those described by Johnson (1983).

Swiddens are cleared throughout the dry season (May-September), burned at the end of the dry season, and planted during the beginning of the rainy season in September or October. Selection of a garden site must take various factors into account, the most important being drainage and soil type. Tropical soils are notoriously low in organic material and nitrogen. The Matsigenka classify soils according to color, sand and clay content, presence of rocks or gravel, texture and drainage properties (Johnson 1983). The preferred soil for gardens is a dark sandy loam that drains well and is soft and easy to work. Land is classified between hillside (*otishi*) and river bank (*otapi*) locations. Nearness to a river or stream is an important consideration. Gardens located in the flood plains of streams or rivers may be subject to inundation, which can destroy the crops. However proximity to a stream is beneficial for two reasons: first, the garden site is often a house site as well, and houses always need water; second, the stream assures ease of transporting the produce. The ideal spot for a garden is on sandy loam soils along a gentle slope or plateau by a stream bend above the flood level. Though land is abundant in Matsigenka territory, preferred garden sites with good soils and convenient access are in fact somewhat scarce. In successive years of occupation in a single locality, prime garden sites near the settlement are converted to fallow, and new swiddens must be cleared at ever greater distances. Eventually, households or entire communities must relocate. Thus to maintain their agricultural practices over long periods of time, the Matsigenka require a large territory.

The principal staple crops are manioc, maize, plantains and bananas. Dozens of other crops are planted in smaller amounts throughout the garden including yam and cocoyam, pineapple, chili pepper, peanut, cotton, annato, papaya, sugar cane, squash, several bean species and a number of cultivated medicines. Maize appears to be a more sensitive

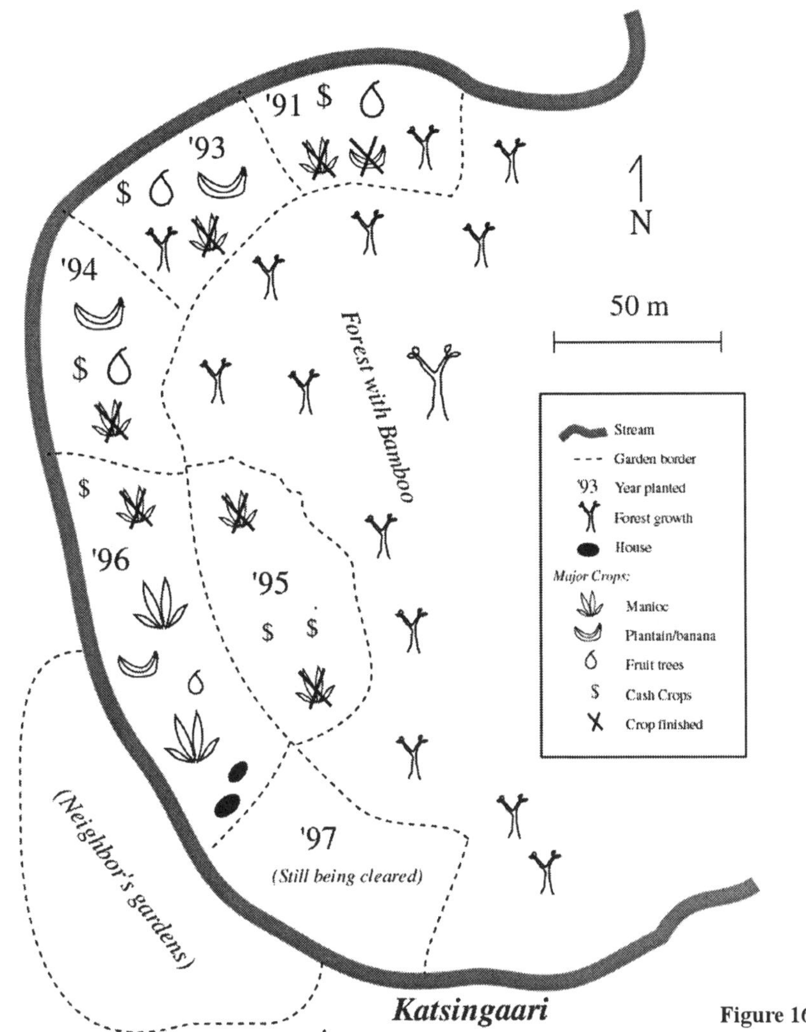

Figure 16.3. Daniel's Gardens

crop, subject to frequent failure. The Matsigenka may attribute the loss of a maize crop to ecological factors (poor soils, low rainfall, pests, etc.) as well as to certain taboo behaviors (eating certain foods, contamination by menstrual blood, consuming psychoactive substances during crucial moments of maize growth). Certain herbal medicines and other special practices are aimed at protecting the fragile maize crop.

Maize is planted only once in the life cycle of a garden, soon after the newly cleared swidden is burnt. The manioc crop takes three months to a year to mature, depending on the variety. Stems from harvested manioc are re-planted (*nokitatakero aikiro*) in the same spot to provide a second, and rarely, a third crop. Plantains and bananas begin to bear fruit in the second year, and continue to produce for several years after the manioc crop is finished. New swiddens must be cleared every one to two years to ensure an adequate supply of staple crops. Cash crops such as coffee, cacao and annato, fruits like guava, pineapple, avocado, peach palm, guaba (*pacae*) and zapote (*caimito*) as well as *barbasco* fish poison are planted alongside garden staples in preparation for a later stage of the garden's evolution. After three to five years, as the production of staples tapers off and the garden is left to fallow, fruit trees begin to mature and form part of natural secondary vegetation enriched with useful species. Swidden "fallows" continue to produce fruits and other useful species for ten years or more until secondary forest regeneration completes the agricultural cycle. Old fallows may be cleared and re-burned for planting, or simply allowed to regenerate fully to secondary forest.

The principal author visited a series of gardens cleared by "Daniel" (a pseudonym) along the stream Katsingaari (a tributary of the Parotori/Picha) about 50 minutes walk from his house in the community of Camaná. His gardens appear to be typical of the pattern described above, and serve as a microcosm for this brief study of Matsigenka agriculture (see Figure 16.3). Daniel began clearing gardens in this area in 1991, after having worked for several years along the Katsingaari somewhat closer to the community. Daniel observes that good agricultural land near the community already is growing scarce, though the community was established in its current location just ten years prior. The only good land left close to the community is in the river flood plain on the opposite side of the Parotori river from Camaná, and it can be risky to farm since floods can carry away an entire year's worth of crops (this happened to Daniel's neighbor last year).

In successive years, Daniel cleared a series of swiddens, working his way upriver along a stream bend. His 1991 garden is now in fallow, producing a number of fruit trees as well as small amounts of cash crops including coffee, cacao and annato. Continuing upriver are his next two swiddens, cleared in 1993 and 1994, still producing plantains alongside cacao and annato and maturing fruit trees. In 1995, he focused his efforts on planting a large crop of cacao for future commercialization, interspersed with manioc. Daniel's manioc crop from the 1995 garden ran out more quickly than expected because of family commitments involving the production of manioc beer. He was forced to depend on relatives for manioc during part of the following year. Because of this experience, he cleared a somewhat larger swidden in 1996, emphasizing traditional rather than cash crops. At the time of the interview, about one-third of the manioc from the 1996 garden had been harvested, and Daniel had begun clearing a new swidden to be burned and planted at the end of the 1997 dry season (approximately September-October).

Daniel's 1996 garden provides a good example of the diversity of Matsigenka crops. In this single small garden (less than 1/3 ha), Daniel maintains twelve different varieties of manioc, distinguished by various stem, leaf and tuber characteristics as well as maturation periods varying from three months to one year. At least five varieties of plantain and banana are present, and two varieties of corn were harvested last year. Sugar cane, two varieties of cotton and two species of fish poison (*Lonchocarpus, Tephrosia*) are also cultivated. In the flood plain of the Parotori, Daniel has a garden with two varieties of peanut. In addition to these crops, the garden also contains twelve other species of root and seed crops, eleven species of fruit and six species of medicinal plants. Daniel's garden contains seven cultivated varieties of sedge (*Cyperus* spp.), some of which he uses to improve his aim for hunting specific animals, and others of which belong to his wife Maria, and which she uses to treat various illnesses, especially among children. In all, Daniel cultivates 46 species of plants in a single garden (see Appendix 29). Johnson (1983) has collected a total of eighty species of plants cultivated by the Matsigenka. The diversity of crop species and varieties is crucial for the success of Matsigenka agriculture. Since two of the three staples (manioc and plantains) are reproduced clonally rather than from seed, genetic diversity is especially important to prevent pests from overtaking a clonal line and destroying the entire crop. This diversity is maintained by frequent trading of cultivars with extended family members and visitors from other communities. The ultimate source of new crop cultivars for the Matsigenka, however, is the spirit world, accessible only to shamans during trace (Shepard 1998, 1999).

Cash Economy

Though cash is not important in obtaining daily food staples, the Matsigenka of the region appear eager to have cash to pay for their children's schooling (notebooks, pencils), to pay off debts incurred at the local health post, and to obtain trade goods such as clothes, metal implements, flashlights, radios, etc. The availability of cash-earning opportunities has increased in recent years, and will probably continue to increase as hydrocarbon resources of the region are developed. People often complain, however about the shortage of cash in their communities. Commercial crops such as coffee, cacao and annato are produced on a limited scale and sold or bartered sporadically to traveling mestizo salesmen (*comerciantes*).

All three communities have communal rice gardens, the production of which is used primarily to feed school children at lunch time, or sold to pay for school-related or other community expenses. Certain individuals, notably male school teachers, have personal rice gardens they use both for consumption and commercialization. In general, however, rice farming and harvesting appears to be more labor intensive than traditional agriculture, and more risky. Some families have tried out rice farming, only to have "birds come eat it all."

Fishing

Most of the protein requirements of the Matsigenka come not from agriculture but from wild resources such as fish, game, fruits, nuts and palm hearts. Fishing in traditional Matsigenka communities occurs mostly in the dry season, since water levels are low and indigenous fish-catching technologies such as barbasco fish poison (*kogi*), weirs (*shirimentotsi*), fish traps (*shimperintsi*), woven nets (*shiriti*) and stream dams (*avitantarira*) are more effective. As Matsigenka communities along the Río Picha have become more permanent and densely populated, fish capture using various traditional and introduced technologies has become more important throughout the year (Baksh 1995). Fish species consumed and fishing techniques are discussed above by Chang.

Hunting

Though fishing has taken on a more significant role in the diet of the Río Picha communities in recent decades, hunting still is an important subsistence activity. Unlike fishing and gathering, hunting is an exclusively male activity. Since animal populations have been depleted in the immediate vicinity of the communities, hunters set out at a brisk walk or use motorized boats to get several kilometers away from the community before beginning to hunt. Men sometimes may walk a half day or more from the community to set up a hunting camp for several days before returning. Long-distance hunting trips may be combined with other resource-gathering activities such as gathering palm thatch, which also requires setting up a camp for a few days at a considerable distance from the community.

The Matsigenka hunt with bows made from the wood of the palm *kuri* (*Bactris gasipaes*) and cane arrows (*Gynerium*) tipped with bamboo or palm-wood points. There are several shotguns in each community, which are shared among male members of an extended family. Shotgun shells, however, are expensive and sometimes difficult to obtain because of military restrictions on arms and munitions sales. For this reason, Matsigenka hunters are frugal in their use of shotgun shells. Hunters will often travel in pairs (typically brothers-in-law) so that one can hunt with arrows for smaller game and birds while the other saves the precious shells for larger animals such as peccaries, tapir and pacas. Matsigenka men use a large number of wild plant species as well as cultivated sedges (*Cyperus* spp.) as hunting medicines (*kovintsari inchashi*) to 'improve their aim' (*ikovintsatake*) for hunting and fishing.

Notably absent in the communities and forests visited were spider monkeys (*osheto*) and woolly monkeys (*komaginaro*), both preferred edible species. Inhabitants stated that this situation is due to the vegetation of the region, dominated by bamboo, and not due to hunting pressure. They state that "spider monkeys never lived here as long as we know, they are not accustomed (*tera irametempa*) to living in bamboo stands." When explaining why other animals are abundant in the region, informants typically explain, "You see, squirrel monkeys, capuchin monkeys, brocket deer, they are all accustomed to (*irametempa*) bamboo stands."

In all three communities, resource-use maps indicate that woolly and spider monkeys only occur beyond a certain point along the upper courses of the major tributaries where the bamboo stands end and upland forest (*otishi*, *inchatoshi*) begins. This transition is clearly visible on satellite maps of the region where the yellow color of bamboo is replaced by dark brown to reddish brown of high-canopy forest. Community members sometimes trek several days by canoe in the rainy season to reach areas where spider and woolly monkeys are still abundant. It is unclear whether the presence of large stands of bamboo and the absence of these monkey species are natural phenomena, or whether they attest to a long history of intense human habitation in the region in prior centuries.

Apart from these two notably absent species, the forests in the vicinity of the visited communities appear relatively abundant in a number of important game species. The authors went on several all-day walks with male informants along hunting and resource acquisition trails in the three communities to observe resource-use areas. All encounters with game animals and animal signs were also recorded. Game is noticeably more abundant further from the communities. Appendix 30 lists Matsigenka and scientific names of important game animals, and the number and kind of animal signs encountered during these walks.

Other Animal Resources

Though mammals, birds and fish are the most important and consistent sources of meat, the Matsigenka gather and consume a number of other small animal species on a seasonal basis, including three species of turtles and their eggs, a dozen or more species of frogs, six species of snails and more than fifteen species of insect larvae. Turtle (*Podocnemis*) eggs are gathered during the dry season along beaches where the turtles once laid eggs prodigiously. Turtle populations near the communities have diminished drastically, but some streams with ample beaches still contain this highly desired food.

Photographs in Rodriguez and Duellman's (1994) field guide were shown to Matsigenka informants to obtain approximate scientific identifications of edible frog species, which included *tonoanto* (probably *Hyla* sp.) and *punto* and *pirinto* (probably *Leptodactylus* spp.). Small doses of the eggs and skin secretions of the frog *masoki* (probably *Bufo* sp.) were used in former times by shamans to attain trance states (Shepard 1998). Some edible frog species have toxic skin secretions that must removed with scalding water, while other species are poisonous no matter how they are prepared. A recent incident in the community of Nuevo Mundo, in which several people were intoxicated from eating poisonous frogs, underscores the sometimes drastic consequences of the erosion of traditional knowledge.

It was not possible to identify the numerous snail and insect larvae species consumed by the Matsigenka, however this would be an interesting area for further ethnobiological research. Named snail species include *mapoto*, *toturo*, *machoeri*, *koriti*, *sankero* and *turumviki*, some of which occur only at higher elevations. Common edible insect larvae mentioned during interviews on dietary diversity were the palm weevils *pagiri*, *pijiro* and *tuyoki*, the caterpillars *shigopa*, *tsiaro* and *kota*, and *pama*, an insect larva inhabiting *Guadua* bamboo internodes.

Wild Plant Resources

The Matsigenka gather a great diversity of wild plant resources including fruits, nuts, palm hearts, mushrooms, house construction materials (posts, beams, vines, walls, floors, roof thatch) and medicinal plants. Palms are among the most intensely and diversely used resources in the Matsigenka economy (Appendix 31). *Kamona*, *Iriartea deltoideae*, is a good example of the multiple use of a single resource: the heart is eaten, the leaves can be used for temporary shelter, the outer bark is used to make floors, the trunk is hollowed out to make kegs for manioc beer and the stumps attract edible beetle larvae which can be harvested several months after felling. Many other species of palm fruits are eaten at the appropriate season of the year, notably *Mauritia* and *Mauritiella*, which form dense stands in swampy areas. *Euterpe* and *Oenocarpus* also form noticeable stands in moist upland areas and are harvested intensively when their fruits become ripe. *Attalea phalerata* (formerly *Scheelea*) and *Astrocaryum* are common lowland palms whose thick, hard nuts are cracked to reveal a delicious, almond-like seed.

The leaves of several species of understory palms (*Geonoma* spp., *Chamaedora* spp. *Hyospathe* sp.) are used as thatch material for houses. The availability of thatch palm is one of the principal limiting factors (along with agricultural land and game availability) that leads to migrations. Thatch roofs must be replaced every five to ten years, depending on the species of thatch chosen and the quality of the thatching. Thatching an average house requires the harvest of hundreds of individual palm plants.

Larger communal buildings (schools, meeting areas) require much more thatch. If a community stays in the same area for many years, thatch species become locally unavailable, and constructing and maintaining a house becomes less and less convenient. The people of Puerto Huallana made an experimental plantation of thatch palms close to the community that proved successful. Unfortunately, Chevron workers inadvertently cleared the patch to make room for a helicopter landing pad during seismic survey activities carried out in 1995 and 1996. Further experiments in thatch palm management could prove important in conserving these important traditional resources.

Arboreal fruits are gathered and consumed mostly during the rainy season. These include Moraceae species *pochariki* (*Pseudolmedia laevis*), *meronki* (probably *Brosimum lactescens*), *etsiki* (*Helicostylis*) and other species in the family; *peroki* (probably *Ocotea*) and other Lauraceae; Leguminosae including *intsipa* (*Inga* spp.), *koveni* (*Hymenea oblongifolia*) and *katsiriki* (*Bocoa alterna*); wild cacao species *sarigemini* (*Theobroma cacao*) and *tangaroki* (*T. speciosum*); wild *Annona* species; several Sapotaceae species including *segorikashi* (*Pouteria* sp.); and other plants from diverse botanical families such as *tsigeroki* (*Calatola venezuelana*), *tsigoroki* (*Spondias mombin*), *pojaroki* (*Clavija weberbaueri*) and *kachopitoki* (*Chrysochlamys ulei*).

Resource Use: Spatial and Temporal Distribution

Effective hunting, gathering and fishing require knowledge of the spatial and seasonal distribution of resources through a large territory. The Matsigenka possess a wide knowledge of geographical landmarks and ecological zones throughout their territory. Old and new garden sites are found far beyond the immediate vicinity of the community center. Hunters make long treks to take advantage of macaw clay licks (*irapitari kimaro*) and salt licks (*itsimini*) frequented by specific animals: monkeys, tapir, peccary or birds. Various streams are known to be abundant in certain species of fish at specific times of the year. The location of palm stands is another important aspect of the Matsigenka's cognitive maps of their territory. Many palm species are prone to patchy distribution, and are sensitive indicator species for different soil and terrain types. As discussed below, palm species figure prominently in the Matsigenka classification of forest types.

Matsigenka Forest Classification

In their daily interactions with their environment, and in the accumulation of such observations over generations, the Matsigenka have amassed a rich body of knowledge about the diversity of plant species and habitats in their territory. Satellite images provide a useful and easy tool for viewing patterns of habitat diversity over large areas (Salo *et al.* 1986, Tuomisto *et al.* 1994, 1995). Yet little is still known about the actual vegetational characteristics on the ground of different "colors" seen in satellite maps. "Ethnobotanical ground-truthing" (Shepard *et al.* in press) could prove to be a powerful method for locating and characterizing habitat diversity in areas that are new to Western scientists. Matsigenka forest classification depends on a complex set of variables which fall into three broad categories:

1) Topography: The lay of the land is of primary importance in Matsigenka forest classification. Matsigenka forest terminology includes words for ridges (*otishi*), flat terraces (*pampa*), gradual slopes (*agiringira*), rocky outcroppings (*imperita*), eroded cliffs (*otarangira*) and mountainous foothills (*otishipaketira*). Elevation, soil types and temperature are also important in the classification of forest habitats.

2) Hydrology: Matsigenka terminology for forest types includes words for seasonally inundated flood plains (*ovogea*), areas that were inundated at some time in the past (*nigankivoge*), swamps or lakes (*inkaare*), river and stream banks (*otapi*), springs and waterfalls (*okonteaatira*) and the swampy headwaters of small streams (*osateni*).

Topographic and hydrologic features are used by the Matsigenka to distinguish three broad categories of forest, corresponding to current scientific classifications: lowland or floodplain forest (*ovogeshi*), upland or interfluvial forest (*nigankipatsa*) and montane forest (*otishpaketira*). Vegetation types are used to refer to a finer level of differentiation within these broad types.

3) Vegetation: The general word for forest is *inchatoshi*, 'tree leaves, or *inkenishiku*, 'entering among the leaves.' Most names for specific forest types are formed by adding the affix *-shi*, 'leaf' to the name of the dominant or indicator species. For example *kapiroshi* is a common forest type dominated by the spiny bamboo *kapiro*, *Guadua weberbaueri*; *kepitoshi* is an important upland forest type characterized by dense concentrations of the palm *kepito, Wettinia augusta*, in the understory. Other names refer to general characteristics of the forest, for example, high canopy forest (*kurayongashi*, *karororoempeshi*), low-canopy forest (*oshavishitira*), dwarf forest (*otyomiani inchato*) and liana forest (*shivitsasemai*).

During the brief study period, Matsigenka informants in the three study communities named 21 distinct vegetation types found in floodplain and interfluvial forest, 11 types of disturbed forest vegetation, and 13 types of montane vegetation. Including data gathered in a more extensive study carried out in Manu National Park (Shepard *et al.* in press), the Matsigenka were found to distinguish a total of some 68 vegetationally-defined habitats, 26 abiotically-defined habitats, and at least seven habitats associated with specific faunal indicators. Appendix 32 summarizes the forest types encountered and/or described in the Picha, organized according to major categories: floodplain forest, interfluvial forest, disturbed or inundated forest, and montane forest.

Lowland floodplain forest types on the Picha named by the Matsigenka consist mostly of successional zones in the immediate vicinity of the river (Appendix 32). Mature lowland forest (*nigankivoge*) appears to be uncommon along the Picha due to the ubiquitous presence of bamboo. Only one small patch of intact, mature lowland forest was seen during the study, located on the east bank of the Parotori river about 8 km south of its confluence with the Picha. This forest patch is visible as a small island of light brown within the broad yellow expanse of bamboo. Most of the river margin along the Picha and affluents is dominated by a wide swath of dense *Guadua* bamboo stands (*kapiroshi*) extending from several meters to several kilometers on either side of the river basin, even along the smallest streams. This vegetation is clearly visible on satellite maps we have examined as a light yellow coloration extending in a dendritic pattern from the mouth of the Picha and tributaries up most of the length of the river. It is clear from the satellite images as well as the observations of local people that human settlements tend to increase the size of *Guadua* stands. However the dendritic pattern of bamboo stands along stream tributaries suggests a natural spread of bamboo throughout the territory (Nelson 1994). Both spiny *Guadua* and non-spiny *Aulonemia* bamboo species grow in disturbed upland forest areas, especially on steep hillsides. Other disturbed forest types include swamps, lakes, liana forests and agricultural clearings (Appendix 32). These are visible as tiny blue-purple patches or reddish-brown streaks on the satellite images.

The light yellow of bamboo contrasts sharply with the rich brown or reddish brown of upland forest on the available satellite images. The greatest diversity of habitat types recognized by the Matsigenka occurs in upland forest (Appendix 32), an observation corroborated by ecologists working in Manu and elsewhere in the Peruvian Amazon. Palms and especially understory palms are important as indicator species in Matsigenka forest classification, especially in upland forest. This fact has been confirmed empirically by ecologists, who have found palms to be highly sensitive indicators of a variety of localized edaphic factors. Many of the palm species used as indicators by the Matsigenka are understory palms, not typically included in the standard forest plots used by ecologists which count only trees greater than 10 cm in diameter. More detailed work on indigenous forest classification could provide valuable insights in the study of forest heterogeneity in Amazonia, an area of ecology that is still in its infancy.

Perhaps the most surprising finding of the study was to learn that some individuals living in Río Picha communities maintained detailed ecological knowledge about high-elevation plant communities including cloud forest, pajonal and high-Andean vegetation (Appendix 32), despite the great distance and remoteness of these regions from current Matsigenka settlements. Some informants were able to describe details of the colors, forms, odors and other characters of montane vegetation sufficient to allow an educated guess as to the scientific identification of certain indicator species. Oral histories reveal that the Matsigenka migrated back and forth across the Vilcabamba mountain range into the adjacent Tambo and Mantalo river systems throughout the past century. Apparently, environmental knowledge about high-elevation habitats has been remembered and transmitted across generations.

Folklore, Biodiversity, and Concepts of Ecological Balance

The Matsigenka's knowledge and appreciation of biological diversity goes beyond utilitarian facts. Myths concerning the origins of biodiversity are a source of entertainment and philosophical speculation. Animal species are anthropomorphized in folk tales, spiritual beliefs, and personal anecdotes. Metaphors drawn from ecological processes are used to understand the human condition, while human intentions and emotions are projected into the ecological sphere. At the core of the Matsigenka world view, biological diversity is an expression of the prowess and virtuosity of shamans. The first shaman and god of creation *Tasorintsi*, 'Blowing Spirit,' used the transformative powers of tobacco to breathe diversity into the animal and plant kingdoms. Legendary human shamans ascended into the spirit world in hallucinogenic trance to obtain cultivated plants and other technological innovations for Matsigenka society. Modern shamans are responsible for obtaining new varieties of crops and medicines, for calling game animals out of their enchanted hiding places, and for fighting off illness and other demonic forces (Shepard 1999). According to Matsigenka beliefs, without the shaman's transformative powers, there can be no adaptation, innovation, or evolution — cultural or biological.

Matsigenka food taboos are intimately associated with cosmological and medical beliefs. The Matsigenka believe that deer (*maniro*) can harbor a demonic spirit (see also Milton 1991, 1997). For these reasons, some members of the study communities maintain a loosely-enforced taboo against eating deer meat. Other carnivorous or omnivorous animal species, for example capybara (*iveto*), armadillos (*kinteroni, etini*), jaguars and ocelots (*matsontsori*), caiman (*saniri*) and a number of carnivorous fish ('fish with teeth') are also subject to dietary taboos by some individuals or families. Taboos against eating certain meats and other foods are also practiced by sick people, girls during their puberty initiation, pregnant women, and couples with newborn children.

Certain game animals are believed to take 'vengeance' (*pugasetagantsi*) on vulnerable people, especially newborns and infants. Vengeful animals are said to 'steal the soul' (*yagasuretakeri*) of young children, causing a wide range of illnesses including sleeplessness, skin rashes, nosebleed, fever, gastrointestinal distress, and even death. Women give newborn infants daily baths of *Cyperus* and other fragrant and succulent plants to protect them from vengeful animals and other illness-causing spirits. Gender specialization is apparent in Matsigenka ethnobotanical knowledge: men specialize in plant medicines used to improve their aim for hunting, while women have specialized knowledge of child-care medicines and fragrant plants to protect newborn infants from game animal spirits (Shepard in press). On the one hand, these practices reveal the complementary gender roles in Matsigenka subsistence and child-rearing. At another level, these beliefs imply a system of ecological checks and balances in which the role of predator and prey become reversed as Nature settles its scores.

Segamai: Survival of the Pleistocene ground sloth?

Matsigenka folk tales describe a number of dangerous creatures inhabiting the mountains. Fierce animals such as jaguars and spectacled bears as well as enchanted animals like deer and owls abound in these remote territories. Folk tales describe gigantic monkeys, malevolent dead spirits and humanoid demons with enormous penises who wander through the hills in search of human victims. One tale in particular describes a creature so vividly and convincingly that it suggests a basis in fact. *Segamai* is described as being an animal about the size of a cow, that can walk on all fours or erect. It is described as having matted fur and a snout similar to that of the giant anteater. It is said to reside in caves and remote forests of the foothills and cloud forest, where it eats the pith of palms and Cyclanthaceae. The name *segamai* means '*Oenocarpus* fibers,' referring to the dark, matted hair of the creature which is said to be like the fibers surrounding the leaf stems of the palm *Oenocarpus*. As recently as twenty-five years ago, a man claims to have seen the creature from a distance. The Matsigenka say *segamai* exists to this day in some locations. The Matsigenka are deathly afraid of *segamai*. It is said to be highly aggressive, emits a bloodcurdling roar, is impervious to weapons, and emanates a strong odor or supernatural force that dazes, stupefies or renders unconscious those who come near it.

These descriptions are remarkably like those of the *Mapinguari*, a folkloric creature from the Brazilian Amazon. Oren (1993) suggests the *Mapinguari* and similar folk tales in Amazonia may represent evidence for a quite recent extinction — or perhaps even continued existence — of Pleistocene ground sloth species, thought to be extinct in the Americas for a few thousand years. (In an intriguing coincidence, a Matsigenka student once saw a life-size model of an extinct ground sloth at the natural history museum in Lima and returned to his community exclaiming, "Remember that story about the Segamai? Well it's not just a story! I saw one at the Museum!"). Descriptions of the creature's size, armored skin, coloration, locomotion, vocalizations and other behavior are similar throughout Brazil, and are consistent with the paleontological evidence. Oren believes that some Mapinguari sightings as recently as the 1970s represent possible evidence of extant ground sloths. Oren's views are admittedly speculative and controversial.

However a careful study of *segamai* folklore in the Vilcabamba range could yield important data for this intriguing debate, and possibly the most spectacular zoological discovery of the century.

CONCLUSION: THE SPIRITUAL VALUES OF BIODIVERSITY AND THE CONCEPT OF THE MATSIGENKA HOMELAND

Increasingly, tropical countries and international organizations have come to understand the practical value of protecting biodiversity. The values of biodiversity include global climatic maintenance, bioprospecting, and ecotourism. Many have come to appreciate the contributions of indigenous peoples to this body of practical knowledge, for example in traditional plant medicines, agricultural varieties and systems of agroforestry management. However important scientists and writers, from John Muir and Henry David Thoreau to John Terborgh and Darrell Posey, contend that nature has an inherent beauty, a spiritual content that transcends practical applications and material value. Indigenous cultures incorporate such views into their basic philosophy of being (Posey 1999). If biodiversity is viewed sheerly in practical terms, then tropical nature and the planet itself are perhaps doomed.

The mountains, rivers and forests of the Vilcabamba range have been inhabited, explored, used, studied and pondered by the Matsigenka people for centuries. While ever practical in their knowledge and use of natural resources for sustenance in daily life, the Matsigenka have also breathed their history into a landscape which echoes with the triumphs and tragedies of the past. It is in this landscape of rugged peaks and lush valleys that the Matsigenka equivalents of the Homeric epics and the Biblical tales were composed and are told to this day. In these Andean foothills, the Matsigenka experienced a cultural Renaissance under the aegis of the Inca State. Stone buildings, roads, coca plantations, and commerce in metal tools were among the benefits that the Inca introduced during this time. Matsigenka story tellers reflect bitterly and ironically on the collapse of this Golden Era of their civilization. More recently, the Vilcabamba's mountain passes were used as escape routes by entire villages seeking refuge from various Holocausts over the past one hundred years, from the "rubber fever" at the turn of the century to the ongoing political violence in Peru.

The Vilcabamba range is also a place of great spiritual power and danger. Fierce animals, seductive spirits, and bloodthirsty demons wander remote forests and rocky outcroppings. Contagious vapors and infernal flames rage up from the bowels of the earth bringing sickness and death. Some Matsigenka fear that recent oil-drilling activities in their territory may unleash evil vapors from deep in the ground, releasing new illnesses into the atmosphere. Their fears are probably justified. However the eternal souls of Matsigenka shamans and their luminous spirit mentors, the *Sangariite*, still return through portals hidden in the foothills to fight off these destructive forces and thus ensure the survival of the Matsigenka people and the whole of humanity.

The eastern Vilcabamba basin is the Promised Land of the Matsigenka people. The rivers and mountains are alive with their collective hopes and fears. Here is written the Matsigenka's past and also their future. The importance of the region for biodiversity conservation is complemented by the historical depth and complex patterns of human occupation. A largely undocumented wealth of archeological and cultural history awaits discovery. As steps are taken in ensuing years to protect the region's biodiversity and develop its petroleum and gas resources, we hope that the indigenous inhabitants will be involved in the management and control of their territory, and that the eventual protected areas will prove to be bastions not only of biodiversity but also of indigenous culture, history and spirit.

LITERATURE CITED

Baksh, M. S. 1995 Change in resource use patterns in a Machiguenga community. *In:* Sponsel, L. (Ed.) Indigenous Peoples and the Future of Amazonia: An Ecological Anthropology of an Endangered World. Tucson: University of Arizona Press.

Boster, J. S. 1984. Classification, cultivation, and selection of Aguaruna cultivars of *Manihot esculenta* (Euphorbiaceae). Advances in Economic Botany 1: 34-47.

Camino, A. 1977. Trueque, correrías e Intercambio entre los Quechas Andinos y los Piros y Machiguenga de la montaña peruana. Amazonia Peruana 1(2): 123-140.

Hames, R. B. 1980. Resource depletion and hunting zone rotation among the Ye'kwana and Yanomamo of Amazonas, Venezuela. *In* Hames, R. B. (Ed.). Working papers on South American Indians Number 2: Studies in hunting and fishing in the Neotropics. Bennington, Vermont: Bennington College Press.

Johnson, A. 1983. Machiguenga gardens. *In* Hames, R. and W. Vickers (Eds.) Adaptive responses of Native Amazonians. New York: Academic Press.

Lee, V. R. 2000 Forgotten Vilcabamba: Final Stronghold of the Incas. n.p.: Empire Publishing.

Lyon, P. J. 1981. An imaginary frontier: Prehistoric highland-lowland interchange in the southern Peruvian Andes. *In*: Francis, P.D., F.J. Kense and P.G. Duke (Eds.) Networks of the Past: Regional Interaction in Archeology. Calgary: University of Calgary Archeological Association.

Milton, K.. 1997. Real men don't eat deer. Discover (June 1997), 46-53.

Milton, K.. 1991. Comparative aspects of diet in Amazonian forest-dwellers. Philosophical Transactions of the Royal Society of London (Series B) 334: 253-263.

Nelson, B. 1994. Natural forest disturbance and change in the Brazilian Amazon. Remote Sensing Review 10: 105-125.

Oren, D. C. 1993. Did ground sloths survive to recent times in the Amazon region? Goeldiana Zoologia 19: 1-11.

Posey, D.A. (Ed.). 1999. Cultural and Spiritual Values of Biodiversity. London: United Nations Environmental Programme and Intermediate Technology Publications.

Posey, D. A. and W. Balée (Eds). 1989. Resource Management in Amazonia: Indigenous and Folk Strategies. Advances in Economic Botany, Vol. 7. New York: New York Botanical Gardens.

Rodriguez, L.O. and W.E. Duellman. 1994. Guide to the Frogs of the Iquitos Region, Amazonian Peru. Lawrence, Kansas: Asociación de Ecología y Conservación and the Natural History Museum, University of Kansas.

Salo, J., R. Kalliola, I. Häkkinen, Y. Mäkinen, P. Niemelä, M. Puhakka, and P. D. Coley. 1986. River dynamics and the diversity of Amazon lowland forest. Nature 322:254-258.

Savoy, G. 1970. Antisuyo. New York: Simon and Schuster.

Shepard, G. H. Jr. 1997. History and ecology of the eastern slopes of the Vilcabamba Cordillera. Unpublished report to Conservation International (Peru). To be published elsewhere.

Shepard, G. H. Jr. 1998. Psychoactive plants and ethnopsychiatric medicines of the Matsigenka. Journal of Psychoactive Drugs 30(4): 321-332.

Shepard, G. H. Jr. 1999. Shamanism and diversity: A Matsigenka perspective. *In*: Posey, D.A. (Ed.) Cultural and Spiritual Values of Biodiversity. London: United Nations Environmental Programme and Intermediate Technology Publications, 93-95.

Shepard, G. H. Jr. *in press*. Primates in Matsigenka subsistence and worldview. *In*: Fuentes, A. and L. Wolfe (Eds.) The Conservation Implications of Human and Nonhuman Primate Interconnections. Cambridge, U.K.: Cambridge University Press.

Shepard, G. H. Jr., D. W. Yu, B. W. Nelson and M. Italiano. *in press* Ethnobotanical ground-truthing and forest diversity in the Western Amazon. Advances in Economic Botany.

Snell, W. W. 1964. Kinship relations in Machiguenga. M.A. Thesis. The Hartford Seminary Foundation.

Tuomisto, H., A. Linna, and R. Kalliola. 1994. Use of digitally processed satellite images in studies of tropical rain forest vegetation. International Journal of Remote Sensing 15: 1595-1610.

Tuomisto, H., K. Ruokolainen, R. Kalliola, A. Linna, W. Danjoy and Z. Rodriguez. 1995. Dissecting Amazonian biodiversity. Science 269: 63-66.

Varese, S. 1968. La sal de los cerros: Notas etnográficas e históricas sobre los Campa de la selva del Peru. Lima: Universidad Peruana de Ciencias y Tecnología.

GAZETTEER

Coordinates were taken with Trimble® hand-held GPS receivers, map datum WPS-84.

CI-RAP EXPEDITIONS

Peru: Departamento de Junín, Provincia de Satipo

Camp One
11° 39' 36" S, 73° 40' 02" W, 3350 m
Located in the headwaters of the Río Pomureni, just to the west of the crest of the divide that separates the Río Apurímac/Ene drainages from those of the Poyeni and Urubamba. Most work at this camp was at close to the same elevation and below, to the N and NE of the campsite.

Open pampas (*Sphagnum* bogs) provided easy access to many kilometers in various directions. There were no flowing streams of surface water, but many water-filled large potholes and sinkholes. The water source for our camp was about 400 m above the campsite, on a flat limestone crest. Access was by helicopter, although landing was difficult due to softness of the water-logged pajonal. 6-21 June 1997.

Camp Two
11° 33' 35" S, 73° 38' 28" W, 2050 m
In contrast to the relatively gently sloping Río Ene side of the Cordillera de Vilcabamba, the Amazonian side drops off precipitously from its summit in a series of spectacular cliffs, waterfalls and steep-walled canyons. Among the few helicopter landing spots in this otherwise high-angled terrain were a series of boggy meadows, or pajonales perched on a small plateau above the canyon of the Río Poyeni, surrounded by cloud forest and *Chusquea* grass thickets. Camp Two was situated within patchy forest at the edge of one of these meadows.

In the headwaters of the Río Poyeni, on a small flat bench. We worked 2-3 km in all directions except north from our campsite near a *Sphagnum* bog. Streams largely flowed underground, below rocks and the rootmat, surfacing occasionally. Our camp was in a bamboo thicket over a groundwater sheet and next to a section of exposed stream. Access was by helicopter, which landed in a large open *Sphagnum* bog. 21 June-4 July 1997.

Peru: Departamento de Cusco, Provincia de La Convención

Tangoshiari
11° 46' 00" S, 73° 19' 33" W, ca. 500 m
On the upper Río Pagoreni. Has a small unpaved airstrip. The forest behind the community is tall, but is increasingly fragmented due to advancing cultivation. C. Rivera collected along a small stream, and M. Arakaki, H. Beltrán and B. Holst collected plants in the fragmented forests, clearings, and along streams 1-2 km WSW of Tangoshiari at 550-650 m elevation.

Camp Three
11° 46' 46.8" S, 73° 20' 26.7" W, ca 1000 m
The lower eastern slopes of the Cordillera de Vilcabamba present a formidable region for exploration—access is difficult and topography is extreme. During May 1998, the CI-RAP team studied a 1450 m high limestone hill, which is part of a chain of hills that separates the main Vilcabamba massif from the lowland Amazon region. With its base at an elevation of 500 m, near the Asháninka community of Tangoshiari, the hill that we explored provided a wide range of habitat types, from humid, deep shaded canyons with high herbaceous plant diversity, to sharp limestone ridges dominated by palms, to mid-montane cloud forests with high tree species diversity.

Located on a small flat saddle along the crest of a ridge ascending the slopes of a hill 2 km (straight line distance) west of the community of Tangoshiari. A large thicket of bamboo just above camp, along the trail and extending from there to the crest of the hill at 1450 m. Very little surface water; drinking water was collected off of tarps over the tent, or from a very small stream a half hour walk below camp. 29 April-23 May 1998.

SI/MAB EXPEDITION

Peru: Departamento de Cusco, Provincia de La Convención

Llactahuaman
12° 51' 55" S, 73° 30' 46.0" W, ca 1710 m
The "Llactahuaman" camp was located on the western slopes of the mountain range about 4 km NE of Pueblo Libre. The elevation of the site ranged from 1675 to 1735 m, with slopes between 35 to 70°, and the primary aspect is south. Soils were composed mainly of clay and organic matter. Llactahuaman can be characterized as upper montane forest zone, with high epiphyte loading on the trees. The camp was set in an exposed ridge where the common bamboo of the highlands, *Chusquea*, was less dominant than on the surrounding slopes. Two streams, Bagre Creek and Puma Creek, passed near the camp. The water was transparent and oxygenated, and temperatures ranged from 14 to 16 °C. Both streams had permanent water and belong to the watershed that empties into the Río Apurímac. 11-29 July 1998.

Wayrapata
12° 50' 10.1" S, 73° 29' 42.6" W, ca 2445 m
The "Wayrapata" camp was located on the top of a ridge about 10 km NE of Pueblo Libre. The elevation of plots ranged between 2285 and 2465 m, and slopes ranged from 30 to 65°. Soils were composed almost entirely of organic matter. The floor structure is composed of a complex vertical web of root systems. The site can be described as an Elfin forest or subalpine tropical forest, and was also located within the cloud forest range. Average temperatures during the day range between 24 and 26°C, falling noticeably during the night to about 8 to 10°C. Fog and wind are characteristic of the site.

At the site, streams to the north flow into the Río Cumpirusiato and eventually into the Río Urubamba, while streams to the south flow into the Río Apurímac. Most streams near the camp had little permanent water and relatively low volume, with temperatures ranging from 14 to 16°C. 29 July – 11 August 1998.

ITINERARY

CI-RAP EXPEDITIONS

1997

6 June: Arakaki, Beltrán, Boyle, Emmons, Evans, Grados, Lamas, Rodríguez, Romo, Schulenberg, and Servat from Ayacucho to Camp One.

21 June: Arakaki, Beltrán, Boyle, Emmons, Evans, Rodríguez, Romo, and Schulenberg from Camp One to Camp Two. López and Luna from Pichari to Camp Two. Grados, Lamas, and Servat from Camp One to Ayacucho.

4 July: Arakaki, Beltrán, Boyle, Emmons, Evans, López, Luna, Rodríguez, Romo, and Schulenberg from Camp Two to Ayacucho.

1998

28 April: Emmons, Evans, Romo and Schulenberg from Satipo to Tangoshiari.

29 April: Emmons, Evans, Romo and Schulenberg from Tangoshiari to Ridge Camp.

2 May: Beltrán, Holst, and Rivera to Tangoshiari and on to Ridge Camp.

6 May: Valdes to Tangoshiari and on to Ridge Camp.

7 May: Arakaki and Luna to Tangoshiari and on to Ridge Camp.

10 May: Schulenberg and Valdes to Tangoshiari.

11 May: Schulenberg and Valdes to Lima via Kochiri and Satipo.

21 May: Arakaki, Beltrán, Holst, Rivera to Tangoshiari.

23 May: Emmons, Evans, Luna, and Romo to Tangoshiari.

25 May: Arakaki, Beltrán, Emmons, Evans, Holst, Luna, Rivera, and Romo to Satipo.

SI/MAB EXPEDITION

1998

11 July: Baldeon, Martinez, Rodas, Santisteban, Udvardy, Yucra C., Yucra G., and the field support from Concessions and Catering, SA, formed the advanced team, from Pueblo Libre to Llactahuaman camp.

13 July: Acosta, Amanzo, De la Colina, Hidalgo, Icochea, Nauray, Pequeño, Polo, Ponce, Rodriguez J. J., Salazar, Valencia, and Velazco from Pueblo Libre to Llactahuaman camp.

25 July: Baldeon, Gutierrez C., Gutierrez H., Lopez, Rodriguez R., Santisteban, Udvardy, and the field support from Concessions and Catering, SA, formed the second advanced team, from Llactahuaman to Wayrapata camp.

29 July: Acosta, Amanzo, De la Colina, Hidalgo, Icochea, Nauray, Nuñez, Pequeño, Polo, Ponce, Rodriguez J. J., Salazar, Valencia, and Velazco from Llactahuaman to Wayrapata camp. Baldeon, Martinez, Rodas, Yucra C., Yucra G., from Wayrapata camp to Pueblo Libre.

11 Aug.: Acosta, Amanzo, De la Colina, Gutierrez C., Gutierrez H, Hidalgo, Icochea, Lopez, Nauray, Nuñez, Pequeño, Polo, Ponce, Rodriguez J. J., Rodriguez R., Salazar, Valencia, and Velazco from Wayrapata camp to Pueblo Libre.

APPENDICES

APPENDIX 1 Proposed Long-term Biodiversity Monitoring of the Cordillera de Vilcabamba, Peru

Francisco Dallmeier and Alfonso Alonso

A long-term monitoring program should be established within the Cordillera de Vilcabamba to track trends within the biological groups that have been studied during the RAP and SI/MAB surveys. A monitoring program will help determine changes in forest structure and ecosystem functions that may result from any development activities in the region.

The following discussion presents the rationale for monitoring of six biological groups, along with recommendations for the structure and operation of the monitoring program.

Vegetation

Biological systems, including tropical forests, are extremely complex networks of biotic and abiotic interactions. While all components of a complex system are essential, plants comprise perhaps the most important structural and functional component of the forest ecosystem. They play many roles, and they are crucial in multi-taxa monitoring programs.

In every forest ecosystem, certain plants are considered "keystone" species in that they provide food and shelter for numerous species during periods when fruits are not readily available. Without keystone species (for example, figs), many bird and bat population levels would decline. In turn, plants that depend on these animals for seed dispersal or pollination would experience declines in their populations.

There are several other aspects of forest ecosystems that should not be overlooked. Forest plants have tremendous economic, medical, and agricultural benefits for humans. As well, they are important in many cultural rituals, as building materials, and in crafts. Knowledge of their potential in these areas requires more research.

Primary objectives of the vegetative monitoring program include the following:

- Continue with baseline data gathering. This includes completion of the assessments already initiated and establishment of new assessments at new sites. The results will be increased information through the incorporation of more sites and the collection of data on species dynamics as different locations are sampled at different times.
- Increase the focus on habitat types, vegetation types, and spatial distributions of species. This will assist in determining the impacts of human activity on vegetation patterns and in studying forest regeneration. Human-caused changes in forest structure will promote the establishment of second-growth pioneer plant species, which form a forest quite different from the surrounding undisturbed forest. Such a successional sequence will have a significant impact on fauna at the site.
- Vigilantly register quantitative information concerning how representative permanent forest monitoring plots are compared to the multi-scale vegetation plots and the larger unsampled landscape. Because this forest is very large, it is possible to sample only a few hectares. Thus, it is important that the monitoring sites be representative of the larger area.

Birds

Many tropical birds are rare because of their preference for very specific habitats. While this factor contributes to high diversity, it makes such animals vulnerable to changes in the environment.

In monitoring birds in the Cordillera de Vilcabamba region, it is therefore important to pay close attention to population and distribution changes that may be caused by disturbances of even small patches of forest. Priorities for monitoring include:

- Conduct monitoring at each site using mist-net and point-count surveys.
- Establish transect routes and conduct transect surveys to monitor game birds and other exploited species.
- Teach two- to three-week orientation and training courses to ensure a reliable cadre of field workers for bird monitoring over the long term.

Mammals

Bats

The monitoring program for bats continues to feature a variety of observation and sampling techniques that are determined best for different species. Included are direct roost counts, disturbance counts at roosts, nightly dispersal counts, direct visual counts, counts using motion detectors, ultrasonic bat detection, and direct capture. Monitoring should take place in undisturbed habitats and along the proposed pipeline corridor.

Small non-flying mammals

The major goals of monitoring for small non-flying mammals are to: 1) ascertain population distributions and abundances, 2) take note of trends over the long term, and 3) analyze the data collected to understand any changes that may occur. An adequate number of representative habitats must be chosen for long-term monitoring. Methodologies include use of appropriate traps.

Medium and large mammals

The monitoring program for medium and large mammals contains elements that, as in monitoring plans for the other biological groups selected for this study, focus on a variety of habitat types near and far from human activities. In addition to conducting monitoring at these locations in accordance with standardized protocols, we recommend that training manuals and support materials be developed to assist future researchers in their environmental assessment efforts. The publications should include a compilation of track, scat, and sign keys and biomonitoring technique manuals for South American mammals.

Amphibians and Reptiles

Amphibians reach their greatest abundance in tropical regions. They are diverse predatory animals with complex life cycles. Amphibians possess aquatic larval stages and terrestrial adult stages unique to vertebrates. This complicated life history means that they are sensitive to changes in aquatic, terrestrial, and atmospheric perturbations and are ideal candidates in studies of environmental changes.

Monitoring amphibians in the Cordillera de Vilcabamba will help determine whether changes in species distribution are taking place because of alterations/management or habitat or natural processes. Over several years, it will be possible to ascertain which, if any, species have been affected. Changes in the fauna because of climate shifts or atmospheric perturbations are not expected in the near future. Still, analysis of changes in amphibian abundance and distribution must take into account possible changes in the abiotic environment.

Major steps in the amphibian/reptile monitoring program, carried out in accordance with standardized monitoring protocols, should include:

- Evaluating the effects on amphibian/reptile communities from the operations associated with the proposed pipeline and other human activities.
- Document long-term trends in amphibian/reptile community structure and species distributions in relation to changes in habitats and climate.
- Enhance the Cordillera de Vilcabamba faunal inventories and knowledge of the natural history of each species at each site using data from the monitoring project.

Aquatic Systems

The Cordillera de Vilcabamba region has a high annual precipitation and is considered a tropical rainforest. Many creeks and streams within the Cordillera are headwaters of Amazonian tributaries. The quality of the water is thus of great importance to the indigenous communities as well as to plants and animals in the region. Priorities for monitoring the aquatic systems include:

- Continue assessing water quality in streams, particularly along the proposed gas pipeline route and areas susceptible to siltation. The data should be compared to that from streams that are not affected by the pipeline corridor.
- Monitor taxonomic groups of aquatic invertebrates and plankton that are good indicators of water quality.
- Continue studies of fish diversity and use by local communities of the fish species in streams, creeks, and rivers.

Invertebrates

Because arthropods (insects, spiders, mites, and their relatives) are the most biologically diverse group of organisms in terrestrial ecosystems and because of their numerous essential roles in ecosystem functioning, these creatures can provide information on virtually all macrohabitats and microhabitats within an ecosystem. Monitoring the status of arthropods is therefore important in biodiversity studies. In the Cordillera de Vilcabamba, researchers recommend the following components for the arthropod monitoring program:

- Restrict monitoring to groups that are taxonomically known (that is, specimens can be determined). This is necessary because the study area contains a great number of species, not all of which can be identified and monitored. Choosing those groups that are better known ensures that many forms of arthropods are taken into account.
- Monitor at least two plots at each site, including at least one undisturbed (control) plot. These plots should be used for each successive census.
- Based on these two recommendations, the following objectives apply:
- Quantify the rate of change in arthropod species assemblages that can be associated with any disturbances or natural change.
- Assess the impact of changes in arthropod species assemblages in the study area.
- Develop an arthropod fauna inventory of the region using data obtained from the baselines and monitoring program.

  We note that with each additional phase of this project, we expect to generate information on most levels of the ecosystems surrounding the gas well sites through the arthropod monitoring effort. That information will include data about arthropod soil, litter, vegetation, and canopy communities. We continue to estimate that about 15,000 species of arthropods will ultimately be included in this study.

APPENDIX 2

Vegetation transects from Camp One (3300-3400 m) in the Northern Cordillera de Vilcabamba, Departments of Junín and Cusco, Peru

Brad Boyle, Mónica Arakaki, and Hamilton Beltrán

For all transects, subsamples are based on size, as follows: Herbs, < 0.5 m tall; Shrubs, > 0.5 m tall and < 1.0 cm dbh. Abundance for Herbs subsample is the number of subunits in which species was recorded. Abundance for Shrubs subsample is number of individuals. "Habit" is the normal growth form of an adult plant: H = terrestrial herb, E = epiphyte, Hem = hemiepiphyte, L = climber (includes herbaceous vines), S = shrub, T = tree, "nc" = specimen not collected.

A. Transect 1. Dwarf *pajonal*. Abundance of species in all subsamples. Herbs subsample 0.5 x 2 m (x 10; = 10 m^2), shrubs subsample 2 m x 50 m (= 100 m^2).

			ABUNDANCE IN SIZE CLASS	
FAMILY	GENUS, SPECIES	HABIT	HERBS	SHRUBS
POACEAE	*Calamagrostis* sp. 1	H	10	.
GENTIANACEAE	*Gentiana sedifolia* Kunth	H	10	.
ASTERACEAE	*Hieracium* sp. 1	H	9	.
POACEAE	*Chusquea* sp. 1	S	9	.
ASTERACEAE	*Senecio hygrophyllus*	H	8	.
APIACEAE	Indet. sp. 1	H	8	.
GENTIANACEAE	*Gentianella* sp. 2	H	8	.
CYPERACEAE	*Rhynchospora* sp. 2	H	7	.
ASTERACEAE	Indet. sp. 1	H	7	.
CYPERACEAE	*Carex* sp. 2	H	6	.
CYPERACEAE	*Carex* sp. 4	H	6	.
ASTERACEAE	*Diplostephium* sp. 2	S	4	3
ERICACEAE	*Pernettya ?/Disterigma ?* sp. 1	S	4	.
GENTIANACEAE	*Gentianella* sp. 1	H	3	.
GERANIACEAE	*Geranium* sp. 1	H	3	.
SCROPHULARIACEAE	*Bartzia pedicularoides* Benth.	H	3	.
ASTERACEAE	*Laestadia muscicola* Wedd.	H	3	.
XYRIDACEAE	*Xyris subulata* R. & P. var *breviscapa* Idobro & L.B. Smith	H	1	.
ORCHIDACEAE	*Aa ?/Myrosmodes ?* sp. 1	H	1	.
ASTERACEAE	*Oritrophium* sp. 1	H	1	.
VALERIANACEAE	*Valeriana* sp. 2 (sp. nov. probably)	H	1	.
XYRIDACEAE	*Xyris* sp. 1	H	1	.
	Totals		**113**	**3**

B. Transect 2. Boggy dwarf *pajonal*. Abundance of species in all subsamples. Herbs subsample 0.5 x 2 m (x 10; = 10 m^2), shrubs subsample 2 m x 50 m (= 100 m^2).

			ABUNDANCE IN SIZE CLASS	
FAMILY	**GENUS, SPECIES**	**HABIT**	**HERBS**	**SHRUBS**
POACEAE	*Calamagrostis* sp. 1	H	10	.
ASTERACEAE	*Hieracium* sp. 1	H	10	.
GENTIANACEAE	*Gentianella* sp. 2	H	9	.
GENTIANACEAE	*Gentiana sedifolia* Kunth	H	9	.
VALERIANACEAE	*Valeriana* sp. 2 (sp. nov. probably)	H	9	.
ASTERACEAE	*Diplostephium* sp. 2	S	7	2
PTERIDOPHYTA	*Grammitis moniliformis* (Lag. ex Sw.) Proctor	H	6	.
ASTERACEAE	*Oritrophium* sp. 1	H	6	.
ERICACEAE	*Gaultheria vaccinioides* cf. Wedd.	S	5	.
CYPERACEAE	*Rhynchospora* sp. 2	H	4	.
ERIOCAULACEAE	*Paepalanthus pilosus* aff. probably undescribed	H	3	.
GERANIACEAE	*Geranium* sp. 1	H	3	.
JUNCACEAE	*Luzula* sp. 1	H	3	.
ERICACEAE	*Gaultheria glomerata* (Cav.) Sleumer	S	3	.
ERICACEAE	*Pernettya prostrata* (Cav.) DC.	S	2	.
APIACEAE	Indet. sp. 1	H	2	.
ASTERACEAE	Indet. sp. 1	H	2	.
ASTERACEAE	*Senecio hygrophyllus*	H	1	.
PTERIDOPHYTA	*Huperzia eversa* (Poiret) B. Ollgaard	H	1	.
ASTERACEAE	*Chaptalia* sp. 1	H	1	.
	Totals		**96**	**2**

C. Transect 3. Tall *pajonal*. Abundance of species in all subsamples. Herbs subsample 0.5 x 2 m (x 10; = 10 m^2), shrubs subsample 2 m x 50 m (= 100 m^2).

			ABUNDANCE IN SIZE CLASS	
FAMILY	**GENUS, SPECIES**	**HABIT**	**HERBS**	**SHRUBS**
POACEAE	*Calamagrostis* sp. 2	H	10	.
GENTIANACEAE	*Gentiana sedifolia* Kunth	H	10	.
POACEAE	*Calamagrostis* sp. 1	H	8	.
GENTIANACEAE	*Gentianella* sp. 2	H	8	.
CYPERACEAE	*Carex* sp. 2	H	7	.
ERICACEAE	*Pernettya prostrata* (Cav.) DC.	S	6	.
ASTERACEAE	Indet. sp. 1	H	6	.
ASTERACEAE	*Oritrophium* sp. 1	H	6	.
RUBIACEAE	*Arcytophyllum* sp. 1	H	5	.
GERANIACEAE	*Geranium* sp. 1	H	5	.
CYPERACEAE	*Rhynchospora* sp. 1	H	4	.
CYPERACEAE	*Rhynchospora* sp. 2	H	4	.
ERIOCAULACEAE	*Paepalanthus pilosus* aff. probably undescribed	H	4	.
PTERIDOPHYTA	*Grammitis moniliformis* (Lag. ex Sw.) Proctor	H	4	.
POACEAE	*Chusquea* sp. 1	S	3	.
ERICACEAE	*Gaultheria vaccinioides* cf. Wedd.	S	3	.
ASTERACEAE	*Hieracium* sp. 1	H	3	.
ASTERACEAE	*Diplostephium* sp. 2	S	3	6
XYRIDACEAE	*Xyris subulata* R. & P. var *breviscapa* Idobro & L.B. Smith	H	2	.

			ABUNDANCE IN SIZE CLASS	
FAMILY	**GENUS, SPECIES**	**HABIT**	**HERBS**	**SHRUBS**
PTERIDOPHYTA	*Jamesonia scalaris* Kunze	H	2	.
ASTERACEAE	*Laestadia muscicola* Wedd.	H	2	.
XYRIDACEAE	*Xyris* sp. 1	H	2	.
APIACEAE	Indet. sp. 1	H	1	.
CYPERACEAE	*Carex* sp. 1	H	1	.
SCROPHULARIACEAE	*Bartzia pedicularoides* Benth.	H	1	.
GENTIANACEAE	*Gentiana* sp. 1	H	1	.
ROSACEAE	*Hesperomeles weberbaueri* C. Schneider	S	.	1
	Total		**112**	**7**

D. Transect 4. Dwarf *pajonal*, with *Puya*. Abundance of species in all subsamples. Herbs subsample 0.5 x 2 m (x 10; = 10 m^2), shrubs subsample 2 m x 50 m (= 100 m^2).

			ABUNDANCE IN SIZE CLASS	
FAMILY	**GENUS, SPECIES**	**HABIT**	**HERBS**	**SHRUBS**
POACEAE	*Calamagrostis* sp. 1	H	10	.
VIOLACEAE	*Viola pygmaea* Juss. ex Poir.	H	10	.
BROMELIACEAE	*Puya* sp. 1	H	10	.
XYRIDACEAE	*Xyris* sp. 1	H	8	.
ASTERACEAE	*Oritrophium* sp. 1	H	8	.
XYRIDACEAE	*Xyris subulata* R. & P. var *breviscapa* Idobro & L.B. Smith	H	8	.
ASTERACEAE	*Diplostephium* sp. 2	S	7	2
ERICACEAE	*Gaultheria glomerata* (Cav.) Sleumer	S	7	.
ERICACEAE	*Gaultheria vaccinioides* cf. Wedd.	S	6	.
GENTIANACEAE	*Gentiana sedifolia* Kunth	H	6	.
ASTERACEAE	*Hieracium* sp. 1	H	6	.
ERIOCAULACEAE	*Paepalanthus pilosus* aff. probably undescribed	H	5	.
GENTIANACEAE	*Gentianella* sp. 1	H	4	.
HYPERICACEAE	*Hypericum struthiolifolium* Juss.	S	3	4
ASTERACEAE	*Chaptalia* sp. 1	H	2	.
MYRTACEAE	*Myrteola nummularia* (Poir.) Berg.	S	2	.
ERICACEAE	*Gaultheria* sp. 1	S	2	4
CYPERACEAE	*Rhynchospora* sp. 2	H	1	.
PTERIDOPHYTA	*Jamesonia scalaris* Kunze	H	1	.
PTERIDOPHYTA	*Grammitis moniliformis* (Lag. ex Sw.) Proctor	H	1	.
ASTERACEAE	*Senecio hygrophyllus*	H	1	.
CAMPANULACEAE	*Lysipomia brachysiphonia* var. *brevifolia* (A. Zahlbr.) E. Wimm.	H	1	.
PTERIDOPHYTA	*Blechnum auratum* (Fee) Tryon & Stolze	T	1	1
POACEAE	*Chusquea* sp. 1	S	1	.
ERICACEAE	*Pernettya prostrata* (Cav.) DC	S	1	.
APIACEAE	Indet. sp. 1	H	1	.
PTERIDOPHYTA	*Elaphoglossum* sp. 1	H	1	.
ERICACEAE	*Vaccinium floribundum* H.B.K.	S	.	5
ROSACEAE	*Hesperomeles weberbaueri* C. Schneider	S	.	1
	Totals:		**114**	**17**

E. Transect 5. Tall *pajonal*. Abundance of species in all subsamples. Herbs subsample 0.5 x 2 m (x 10; = 10 m^2), shrubs subsample 2 m x 50 m (= 100 m^2).

			ABUNDANCE IN SIZE CLASS	
FAMILY	**GENUS, SPECIES**	**HABIT**	**HERBS**	**SHRUBS**
POACEAE	*Chusquea* sp. 1	S	10	.
ASTERACEAE	*Diplostephium* sp. 2	S	9	2
GENTIANACEAE	*Gentianella* sp. 1	H	9	.
ASTERACEAE	*Hieracium* sp. 1	H	9	.
PTERIDOPHYTA	*Grammitis moniliformis* (Lag. ex Sw.) Proctor	H	8	.
VALERIANACEAE	*Valeriana* sp. 2 (sp. nov. probably)	H	8	.
POACEAE	*Calamagrostis* sp. 1	H	7	.
GENTIANACEAE	*Gentiana sedifolia* Kunth	H	7	.
GERANIACEAE	*Geranium* sp. 1	H	5	.
ASTERACEAE	*Senecio hygrophyllus*	H	4	.
ERICACEAE	*Pernettya prostrata* (Cav.) DC	S	2	.
ERICACEAE	*Gaultheria glomerata* (Cav.) Sleumer	S	2	.
ERICACEAE	*Gaultheria vaccinioides* cf. Wedd.	S	2	.
APIACEAE	Indet. sp. 1	H	1	.
JUNCACEAE	*Luzula* sp. 1	H	1	.
ASTERACEAE	*Chaptalia* sp. 1	H	1	.
ERIOCAULACEAE	*Paepalanthus pilosus* aff. probably undescribed	H	1	.
	Totals:		**86**	**2**

F. Transect 6. Tall *pajonal* mixed with rosette plant association. Abundance of species in all subsamples. Herbs subsample 0.5 x 2 m (x 10; = 10 m^2), shrubs subsample 2 m x 50 m (= 100 m^2).

			ABUNDANCE IN SIZE CLASS	
FAMILY	**GENUS, SPECIES**	**HABIT**	**HERBS**	**SHRUBS**
PTERIDOPHYTA	*Grammitis moniliformis* (Lag. ex Sw.) Proctor	H	9	.
GERANIACEAE	*Geranium* sp. 1	H	9	.
POACEAE	*Calamagrostis* sp. 2	H	8	.
PLANTAGINACEAE	*Plantago tubulosa* Decne.	H	7	.
POACEAE	*Chusquea* sp. 1	S	6	.
GENTIANACEAE	*Gentianella* sp. 1	H	5	.
ASTERACEAE	*Hieracium* sp. 1	H	5	.
CYPERACEAE	*Carex* sp. 1	H	4	.
GENTIANACEAE	*Gentiana sedifolia* Kunth	H	4	.
POACEAE	*Festuca rigidifolia* Tovar	H	4	.
RUBIACEAE	*Nertera granadensis* (L.f.) Druce	H	3	.
ASTERACEAE	*Diplostephium* sp. 2	S	3	7
GENTIANACEAE	*Halenia weddelliana* cf. Gilg.	H	3	.
POACEAE	*Calamagrostis* sp. 1	H	2	.
VALERIANACEAE	*Valeriana* sp. 2 (sp. nov. probably)	H	2	.
CYPERACEAE	*Carex* sp. 3	H	2	.
SCROPHULARIACEAE	*Bartzia pedicularoides* Benth.	H	2	.
ASTERACEAE	*Senecio hygrophyllus*	H	2	.
SCROPHULARIACEAE	*Castilleja nubigena* H.B.K.	H	2	.
JUNCACEAE	*Luzula* sp. 1	H	1	.
ERICACEAE	*Pernettya prostrata* (Cav.) DC.	S	1	.
ERICACEAE	*Gaultheria vaccinioides* cf. Wedd.	S	1	.
ERICACEAE	*Gaultheria glomerata* (Cav.) Sleumer	S	1	.
ASTERACEAE	*Oritrophium peruvianum* (Lam.) Cuatrec.	H	1	.
	Totals:		**87**	**7**

G. Transect 7. Dwarf *pajonal*, with *Puya*. Abundance of species in all subsamples. Herbs subsample 0.5 x 2 m (x 10; = 10 m^2), shrubs subsample 2 m x 50 m (= 100 m^2).

			ABUNDANCE IN SIZE CLASS	
FAMILY	**GENUS, SPECIES**	**HABIT**	**HERBS**	**SHRUBS**
ERICACEAE	*Gaultheria glomerata* (Cav.) Sleumer	S	10	.
XYRIDACEAE	*Xyris* sp. 1	H	10	.
BROMELIACEAE	*Puya* sp. 1	H	10	.
POACEAE	*Calamagrostis* sp. 1	H	10	.
XYRIDACEAE	*Xyris subulata* R. & P. var *breviscapa* Idobro & L.B. Smith	H	10	.
ASTERACEAE	*Oritrophium* sp. 1	H	9	.
ASTERACEAE	*Diplostephium* sp. 2	S	8	.
HYPERICACEAE	*Hypericum struthiolifolium* Juss.	S	8	2
ERICACEAE	*Gaultheria vaccinioides* cf. Wedd.	S	7	.
GENTIANACEAE	*Gentiana sedifolia* Kunth	H	6	.
GENTIANACEAE	*Gentianella* sp. 1	H	6	.
MYRTACEAE	*Myrteola nummularia* (Poir.) Berg.	S	4	.
ASTERACEAE	*Hieracium* sp. 1	H	4	.
VIOLACEAE	*Viola pygmaea* Juss. ex Poir.	H	3	.
ERICACEAE	*Gaultheria* sp. 1	S	3	9
ERIOCAULACEAE	*Paepalanthus pilosus* aff. probably undescribed	H	2	.
APIACEAE	Indet. sp. 1	H	2	.
CAMPANULACEAE	*Lysipomia brachysiphonia* var. *brevifolia* (A. Zahlbr.) E. Wimm.	H	1	.
PTERIDOPHYTA	*Blechnum auratum* (Fee) Tryon & Stolze	T	1	.
ASTERACEAE	*Oritrophium* sp. 2	H	1	.
ROSACEAE	*Hesperomeles cuneata* Lindl.	S	1	.
ASTERACEAE	*Chaptalia* sp. 1	H	1	.
	Totals:		**117**	**11**

H. Transect 8. Shrubby *pajonal*. Abundance of species in all subsamples. Herbs subsample 0.5 x 2 m (x 10; = 10 m^2), shrubs subsample 2 m x 50 m (= 100 m^2).

			ABUNDANCE IN SIZE CLASS		
FAMILY	**GENUS, SPECIES**	**HABIT**	**HERBS**	**SHRUBS**	**1-10 cm DBH**
ERICACEAE	*Pernettya prostrata* (Cav.) DC.	S	10	.	.
POACEAE	*Calamagrostis* sp. 1	H	10	.	.
ERICACEAE	*Gaultheria glomerata* (Cav.) Sleumer	S	9	.	.
GENTIANACEAE	*Gentiana sedifolia* Kunth	H	7	.	.
ASTERACEAE	*Hieracium* sp. 1	H	7	.	.
ROSACEAE	*Hesperomeles cuneata* Lindl.	S	7	.	.
PTERIDOPHYTA	*Grammitis moniliformis* (Lag. ex Sw.) Proctor	H	7	.	.
POACEAE	*Chusquea* sp. 1	S	7	.	.
POACEAE	*Calamagrostis* sp. 2	H	6	.	.
CYPERACEAE	*Rhynchospora* sp. 2	H	4	.	.
ERICACEAE	*Gaultheria vaccinioides* cf. Wedd.	S	4	.	.
GENTIANACEAE	*Gentianella* sp. 1	H	4	.	.
PTERIDOPHYTA	*Lycopodium clavatum* L.	H	4	.	.
CYPERACEAE	*Carex* sp. 3	H	3	.	.
ASTERACEAE	*Diplostephium* sp. 2	S	2	1	.
ASTERACEAE	*Loricaria lucida*	S	2	37	3
ASTERACEAE	*Senecio argutidentatus* Cuatrec.	S	2	.	.
VALERIANACEAE	*Valeriana* sp. 2 (sp. nov. probably)	H	1	.	.
CUNONIACEAE	*Weinmannia fagaroides/microphylla* H.B.K.	T	1	1	.
SCROPHULARIACEAE	*Bartzia pedicularoides* Benth.	H	1	.	.
PTERIDOPHYTA	*Blechnum auratum* (Fee) Tryon & Stolze	T	1	.	.
PTERIDOPHYTA	*Jamesonia scalaris* Kunze	H	1	.	.
ERICACEAE	*Vaccinium floribundum* H.B.K.	S	1	4	1
GENTIANACEAE	*Gentiana* sp. 1	H	1	.	.
PTERIDOPHYTA	*Hymenophyllum ruizianum* cf. (Klotzsch) Kunze	E	1	.	.
XYRIDACEAE	*Xyris subulata* R. & P. var *breviscapa* Ido	H	1	.	.
MYRTACEAE	*Myrteola phylicoides var. glabrata* (O. Berg) Landrum	S	1	.	.
PLANTAGINACEAE	*Plantago tubulosa* Decne.	H	1	.	.
MELASTOMATACEAE	*Tibouchina* sp. 1	S	.	1	.
ASTERACEAE	*Diplostephium* sp. 1	S	.	1	.
MYRTACEAE	*Myrteola nummularia* (Poir.) Berg.	S	.	1	.
ROSACEAE	*Hesperomeles weberbaueri* C. Schneider	S	.	1	.
	Totals:		**106**	**47**	**4**

I. Transect 9. Mixed *Polylepis*-elfin forest. Abundance of species in all subsamples (Epiphytes subsample not collected). Herbs subsample 0.5 x 2 m (x 10; = 10 m²); shrubs subsample 1 m x 2 m (x 10; = 20 m²); 1-10 cm dbh 2 m x 100 m (200 m²); 10-30 cm dbh 4 m x 100 m (= 400 m²); > 30 cm dbh 8 m x 100 m (= 800 m²).

			ABUNDANCE IN SIZE CLASS				
FAMILY	**GENUS, SPECIES**	**HABIT**	**HERBS**	**SHRUBS**	**1-10 cm DBH**	**10-30 cm DBH**	**>30 cm DBH**
POACEAE	*Chusquea* sp. 1	S	7	3	.	.	.
ERICACEAE	*Pernettya prostrata* (Cav.) DC.	S	5	.	.	.	.
ERICACEAE	*Gaultheria vaccinioides* cf. Wedd.	S	4	.	.	.	.
PTERIDOPHYTA	*Grammitis moniliformis* (Lag. ex Sw.) Proctor	H	4	.	.	.	.
SCROPHULARIACE	*Bartzia pedicularoides* Benth.	H	4	.	.	.	.
POACEAE	*Chusquea* sp. 2	S	3	6	.	.	.
ASTERACEAE	*Senecio argutidentatus* Cuatrec.	S	3	2	.	.	.
APIACEAE	*Hydrocotile* sp. 1	H	3	.	.	.	.
MELASTOMATACE	*Miconia* sp. 2	T	3	.	.	.	.
ASTERACEAE	*Diplostephium* sp. 1	S	2	3	3	.	.
MELASTOMATACE	*Miconia* sp. 1	T	2	1	.	.	.
MELASTOMATACE	*Tibouchina* sp. 1	S	2	1	.	.	.
APIACEAE	Indet. sp. 1	H	2	.	.	.	.
CYPERACEAE	*Rhynchospora* sp. 2	H	2	.	.	.	.
GENTIANACEAE	*Gentiana sedifolia* Kunth	H	2	.	.	.	.
OXALIDACEAE	*Oxalis melilotoides* Zucc.	H	2	.	.	.	.
PTERIDOPHYTA	*Elaphoglossum* sp. 1	H	2	.	.	.	.
PTERIDOPHYTA	*Elaphoglossum* sp. 2	H	2	.	.	.	.
PTERIDOPHYTA	*Grammitis athyrioides* (Hk.) Morton	E	2	.	.	.	.
RUBIACEAE	*Nertera granadensis* (L.f.) Druce	H	2	.	.	.	.
CUNONIACEAE	*Weinmannia fagaroides/microphylla* H.B.K.	T	1	2	3	3	2
ASTERACEAE	*Diplostephium* sp. 2	S	1	1	.	.	.
ROSACEAE	*Polylepis sericea* cf. Wedd.	T	1	.	2	10	7
BROMELIACEAE	*Tillandsia* sp. 1	H	1	.	.	.	.
CYPERACEAE	*Carex* sp. 3	H	1	.	.	.	.
GENTIANACEAE	*Gentianella* sp. 1	H	1	.	.	.	.
ORCHIDACEAE	*Pachyphyllum* sp. 1	E	1	.	.	.	.
PTERIDOPHYTA	*Grammitis variabilis* (Mett.) Morton	E	1	.	.	.	.
MYRSINACEAE	*Myrsine dependens* (R. & P.) Spreng.	T	.	3	8	1	.
ROSACEAE	*Hesperomeles heterophylla* (R.& P.) Hook	S	.	3	.	.	.
MYRTACEAE	*Myrteola phylicoides var. glabrata* (O. Berg) Landrum	S	.	2	.	.	.
POACEAE	*Chusquea* sp. 3	S	.	2	.	.	.
ASTERACEAE	*Gynoxys* sp. 1	T	.	1	2	.	.
ERICACEAE	*Gaultheria buxifolia* Willd.	S	.	1	1	.	.
SYMPLOCACEAE	*Symplocos nana* Brand	S	.	1	1	.	.
ASTERACEAE	*Senecio* sp. 2	H	.	1	.	.	.
BERBERIDACEAE	*Berberis saxicola* Lechl.	S	.	1	.	.	.
LOGANIACEAE	*Desfontainea* sp. 1	S	.	1	.	.	.
PTERIDOPHYTA	*Blechnum loxence*	S	.	1	.	.	.
VALERIANACEAE	*Valeriana* sp. 1 (sp. nov. probably)	H	.	1	.	.	.
ASTERACEAE	*Gynoxys* sp. 2	T	.	.	8	.	.
POLYGALACEAE	*Monnina marginata* Presl.	T	.	.	3	.	.
AQUIFOLIACEAE	*Ilex ovalis* (R. & P.) Loesener	T	.	.	1	1	.
AQUIFOLIACEAE	*Ilex* sp. 1	T	.	.	1	.	.
SYMPLOCACEAE	*Symplocos psiloclada* Stahl	T	.	.	1	.	.
SYMPLOCACEAE	*Symplocos quitensis* Brand	T	.	.	1	.	.
ARALIACEAE	*Oreopanax* sp. 2	T	.	.	.	1	.
	Totals:		**66**	**37**	**35**	**16**	**9**

J. Transect 10. Boggy dwarf *pajonal*. Abundance of species in all subsamples. Herbs subsample 0.5 x 2 m (x 10; = 10 m^2), shrubs subsample 2 m x 50 m (= 100 m^2).

			ABUNDANCE IN SIZE CLASS	
FAMILY	**GENUS, SPECIES**	**HABIT**	**HERBS**	**SHRUBS**
POACEAE	*Calamagrostis* sp. 1	H	10	.
CYPERACEAE	*Carex* sp. 2	H	8	.
ERICACEAE	*Pernettya prostrata* (Cav.) DC.	S	7	.
ASTERACEAE	*Diplostephium* sp. 2	S	6	.
ASTERACEAE	*Hieracium* sp. 1	H	5	.
APIACEAE	Indet. sp. 2	H	5	.
GENTIANACEAE	*Gentiana sedifolia* Kunth	H	5	.
CYPERACEAE	*Carex* sp. 3	H	4	.
ERIOCAULACEAE	*Paepalanthus pilosus* aff. probably undescribed	H	4	.
XYRIDACEAE	*Xyris subulata* R. & P. var *breviscapa* Idobro & Smith	H	4	.
GENTIANACEAE	*Gentianella* sp. 2	H	4	.
PTERIDOPHYTA	*Grammitis moniliformis* (Lag. ex Sw.) Proctor	H	4	.
ERICACEAE	*Gaultheria vaccinioides* cf. Wedd.	S	4	.
BROMELIACEAE	*Puya* sp. 1	H	3	.
MYRTACEAE	*Myrteola nummularia* (Poir.) Berg.	S	3	.
XYRIDACEAE	*Xyris* sp. 1	H	3	.
SCROPHULARIACEAE	*Bartzia pedicularoides* Benth.	H	3	.
POACEAE	*Chusquea* sp. 1	S	2	1
VALERIANACEAE	*Stangea wandae* cf. Graebn	H	2	.
CYPERACEAE	*Rhynchospora* sp. 2	H	2	.
ASTERACEAE	*Oritrophium* sp. 1	H	2	.
HYPERICACEAE	*Hypericum struthiolifolium* Juss.	S	1	.
ERICACEAE	*Gaultheria glomerata* (Cav.) Sleumer	S	1	.
VIOLACEAE	*Viola pygmaea* Juss. ex Poir.	H	1	.
POACEAE	*Calamagrostis* sp. 2	H	1	.
ASTERACEAE	*Chaptalia* sp. 1	H	1	.
ASTERACEAE	Indet. sp. 1	H	1	.
	Totals:		**96**	**1**

K. Transect 11. *Polylepis* forest. Species encountered in all subsamples except Epiphytes (see Appendix 1L for Epiphytes subsample from this transect). Herbs subsample 0.5 x 2 m (x 15; = 15 m^2); shrubs subsample 1 m x 2 m (x 15; = 15 m^2); 1-10 cm dbh subsample 2 m x 150 m (= 300 m^2); 10-30 cm dbh subsample 4 m x 150 m (= 600 m^2); > 30 cm dbh subsample 8 m x 150 m (= 1200 m^2); epiphytes (6 host trees).

			ABUNDANCE IN SIZE CLASS				
FAMILY	**GENUS, SPECIES**	**HABIT**	**HERBS**	**SHRUBS**	**1-10 cm DBH**	**10-30 cm DBH**	**>30 cm DBH**
APIACEAE	*Hydrocotile* sp. 1	H	14	.	.	.	.
POACEAE	*Chusquea* sp. 1	S	11	8	.	.	.
PTERIDOPHYTA	*Grammitis moniliformis* (Lag. ex Sw.) Proctor	H	11	.	.	.	.
SCROPHULARIACEAE	*Bartzia pedicularoides* Benth.	H	9	.	.	.	.
POACEAE	*Calamagrostis* sp. 2	H	9	.	.	.	.
VALERIANACEAE	*Valeriana jasminoides* Briquet	H	8	1	.	.	.
GERANIACEAE	*Geranium* sp. 1	H	8	.	.	.	.
PTERIDOPHYTA	*Elaphoglossum* sp. 3	H	8	.	.	.	.
RUBIACEAE	*Nertera granadensis* (L.f.) Druce	H	6	.	.	.	.
ASTERACEAE	*Senecio* sp. 1	S	5	.	.	.	.
GENTIANACEAE	*Halenia weddelliana* cf. Gilg.	H	5	.	.	.	.
BERBERIDACEAE	*Berberis saxicola* Lechl.	S	4	1	.	.	.
CARYOPHYLLACEAE	*Arenaria lanuginosa* (Michx.) Rohrb.	H	4	.	.	.	.
PTERIDOPHYTA	*Jamesonia alstonii* A.F.Tryon	H	4	.	.	.	.
PLANTAGINACEAE	*Plantago tubulosa* Decne.	H	4	.	.	.	.
OXALIDACEAE	*Oxalis phaeotricha* cf. Diels	H	4	.	.	.	.
JUNCACEAE	*Luzula* sp. 2	H	4	.	.	.	.
ERICACEAE	*Gaultheria vaccinioides* cf. Wedd.	S	4	.	.	.	.
ASTERACEAE	*Diplostephium* sp. 2	S	3	2	.	.	.
ROSACEAE	*Polylepis* sp. 1	T	3	1	18	16	2
ROSACEAE	*Polylepis sericea* cf. Wedd.	T	3	1	12	4	1
MELASTOMATACEAE	*Miconia* sp. 4	T	3	.	.	.	.
GENTIANACEAE	*Gentiana sedifolia* Kunth	H	3	.	.	.	.
RUBIACEAE	*Arcytophyllum* sp. 1	H	3	.	.	.	.
GROSSULARIACEAE	*Ribes incarnatum* Weddell	H	3	.	.	.	.
ASTERACEAE	*Baccharis* sp. 1	S	2	4	.	.	.
ASTERACEAE	*Hieracium* sp. 1	H	2	.	.	.	.
ERICACEAE	*Pernettya prostrata* (Cav.) DC.	S	2	.	.	.	.
ERICACEAE	*Sphyrospermum cordifolium* Benth.	Hem	2	.	.	.	.
POACEAE	*Festuca rigidifolia* Tovar	H	2	.	.	.	.
CYPERACEAE	*Rhynchospora* sp. 2	H	2	.	.	.	.
ROSACEAE	*Hesperomeles weberbaueri* C. Schneider	S	1	.	3	.	.
GENTIANACEAE	*Gentianella* sp. 2	H	1	.	.	.	.
CYPERACEAE	*Carex* sp. 3	H	1	.	.	.	.
BROMELIACEAE	*Puya* sp. 2	H	1	.	.	.	.
PTERIDOPHYTA	*Blechnum auratum* (Fee) Tryon & Stolze	T	1	.	.	.	.
ASTERACEAE	*Senecio argutidentatus* Cuatrec.	S	1	.	.	.	.
APIACEAE	*Azorella crenata* ?	H	1	.	.	.	.
PTERIDOPHYTA	*Huperzia eversa* (Poiret) B. Ollgaard	H	1	.	.	.	.
ARALIACEAE	*Oreopanax* sp. 2	T	.	1	.	.	.
	Totals:		**163**	**19**	**33**	**20**	**3**

L. Transect 11. *Polylepis* forest. Abundance of species in Epiphytes subsample only (see Appendix 1K for remaining subsamples from this transect). Counts are total number of host trees (out of six) in which species was present.

FAMILY	GENUS, SPECIES	HABIT	TOTAL HOST TREES WITH SPECIES (OUT OF SIX)
PTERIDOPHYTA	*Grammitis athyrioides* (Hk.) Morton	E	6
PTERIDOPHYTA	*Elaphoglossum* sp. 3	H	6
POACEAE	*Chusquea* sp. 1	S	5
PTERIDOPHYTA	*Hymenophyllum ruizianum* cf. (Klotzsch) Kunze	E	5
PTERIDOPHYTA	*Grammitis moniliformis* (Lag. ex Sw.) Proctor	H	5
PTERIDOPHYTA	*Grammitis variabilis* (Mett.) Morton	E	4
ERICACEAE	*Pernettya prostrata* (Cav.) DC.	S	2
JUNCACEAE	*Luzula* sp. 2	H	2
PTERIDOPHYTA	*Elaphoglossum* sp. 1	H	2
PTERIDOPHYTA	*Jamesonia scalaris* Kunze	H	1
PTERIDOPHYTA	*Elaphoglossum paleaceum* cf. (Hook. & Grev.) Sledge	E	1
ASTERACEAE	*Senecio* sp. 1	S	1
MYRTACEAE	*Myrteola nummularia* (Poir.) Berg.	S	1
PTERIDOPHYTA	*Hymenophyllum* sp. 1	H	1
ERICACEAE	*Sphyrospermum cordifolium* Benth.	Hem	1
PTERIDOPHYTA	*Jamesonia alstonii* A.F.Tryon	H	1
ERICACEAE	Indet. sp. 1	E	1
BERBERIDACEAE	*Berberis saxicola* Lechl.	S	1
OXALIDACEAE	*Oxalis phaeotricha* cf. Diels	H	1
	Total:		**47**

M. Transect 12. Elfin forest. Abundance of species in Herbs and Shrubs subsamples (see Appendices 1N-O for remaining subsamples from this transect). Herbs subsample 0.5 x 2 m (x 13; = 13 m^2), shrubs subsample 1 m x 2 m (= 30 m^2).

			ABUNDANCE IN SIZE CLASS	
FAMILY	**GENUS, SPECIES**	**HABIT**	**HERBS**	**SHRUBS**
PTERIDOPHYTA	*Asplenium delicatulum* cf. Presl.	H	13	.
URTICACEAE	*Pilea* sp. 3	H	8	.
APIACEAE	*Hydrocotile* sp. 1	H	8	.
URTICACEAE	*Pilea* sp. 2	H	7	.
MELASTOMATACEAE	*Miconia* sp. 1	T	6	.
PTERIDOPHYTA	*Campyloneurum amphostenon* (Klotzsch) Fee	H	6	.
RUBIACEAE	*Manettia* sp. 1	L	5	4
CAMPANULACEAE	*Centropogon peruvianus* (E. Wimm.) Mc Vaugh	S	4	1
BROMELIACEAE	*Greigia* sp. 1	H	4	.
PTERIDOPHYTA	*Trichomanes diaphanum* H.B.K	H	4	.
URTICACEAE	*Pilea diversifolia* Wedd.	S	3	7
ROSACEAE	*Rubus* sp. 1	L	3	3
PIPERACEAE	*Peperomia* sp. 2	H	3	.
ALSTROEMERIACEAE	*Bomarea coccinea* (R. & P.) Baker	L	3	.
OXALIDACEAE	*Oxalis melilotoides* Zucc.	H	3	.
POACEAE	*Chusquea* sp. 3	S	2	9
ASTERACEAE	*Munnozia* sp. 1	L	2	3
GESNERIACEAE	*Drymonia* sp. 1	E	2	2
POACEAE	*Arthrostylidium* sp. 1	S	2	1
VALERIANACEAE	*Valeriana clematitis* Kunth	L	2	1
CARYOPHYLLACEAE	*Stellaria serpyllifolia* Willd. ex Schldl.	H	2	.
PIPERACEAE	*Peperomia galioides* Kunth	H	2	.
RUBIACEAE	*Psychotria/Palicourea* sp. 1	S	1	4
ASTERACEAE	*Pentacalia* sp. 1	L	1	2

			ABUNDANCE IN SIZE CLASS	
FAMILY	**GENUS, SPECIES**	**HABIT**	**HERBS**	**SHRUBS**
SYMPLOCACEAE	*Symplocos reflexa* A. DC.	S	1	2
MELASTOMATACEAE	*Miconia* sp. 6	T	1	1
MYRSINACEAE	*Myrsine dependens* (R. & P.) Spreng.	T	1	1
PTERIDOPHYTA	*Elaphoglossum* sp. 1	H	1	.
CYPERACEAE	*Carex* sp. 5	H	1	.
PTERIDOPHYTA	*Blechnum* sp. 1	T	1	.
ASTERACEAE	*Gynoxys* sp. 3	T	1	.
PTERIDOPHYTA	*Grammitis athyrioides* (Hk.) Morton	E	1	.
CHLORANTHACEAE	*Hedyosmum* sp. 1	T	.	3
ALSTROEMERIACEAE	*Bomarea* sp. 1	L	.	2
THEACEAE	*Freziera revoluta* cf. A. Weitzman	T	.	2
ASTERACEAE	*Mikania* sp. 1	L	.	1
ARALIACEAE	*Oreopanax* sp. 2	T	.	1
	Totals:		**104**	**50**

N. Transect 12. Elfin forest. Abundance of species in size classes > 1 cm dbh (see also Appendices 1M and 1O). Subsamples 1-2.5 cm dbh 2 m x 150 m (300 m^2); 2.5-10 cm dbh 2 m x 150 m (300 m^2); 10-30 cm dbh 4 m x 330 m (1320 m^2); > 30 cm dbh 8 m x 290 m (2320 m^2).

			INDIVIDUALS IN SIZE CLASS (DBH)			
FAMILY	**GENUS, SPECIES**	**HABIT**	**1-2.5 cm**	**2.5-10 cm**	**10-30 cm**	**>30 cm**
POACEAE	*Chusquea* sp. 3	S	11	.	.	.
CHLORANTHACEA	*Hedyosmum* sp. 1	T	9	14	5	.
ASTERACEAE	*Munnozia* sp. 1	L	7	.	.	.
SYMPLOCACEAE	*Symplocos reflexa* A. DC.	S	5	6	5	.
RUBIACEAE	*Psychotria/Palicourea* sp. 1	S	5	2	.	.
THEACEAE	*Freziera revoluta* cf. A. Weitzman	T	3	7	1	1
ARALIACEAE	*Oreopanax* sp. 3	T	3	5	2	.
MELASTOMATACEAE	*Miconia* sp. 2	T	3	3	6	4
ASTERACEAE	*Gynoxys* sp. 3	T	2	5	11	1
CUNONIACEAE	*Weinmannia fagaroides/microphylla* H.B.K.	T	1	1	2	27
ASTERACEAE	*Gynoxys* sp. 2	T	1	.	2	2
ASTERACEAE	*Pentacalia* sp. 1	L	1	.	.	.
CAPRIFOLIACEAE	*Viburnum ayavasense* H.B.K.	T	.	1	.	1
CAMPANULACEAE	*Centropogon peruvianus* (E. Wimm.) McVaugh	S	.	1	.	.
APIACEAE	*Azorella crenata* ?	L	.	1	.	.
ELAEOCARPACEAE	*Vallea stipularis* L. f.	T	.	1	.	.
VALERIANACEAE	*Valeriana clematitis* Kunth	L	.	1	.	.
MONIMIACEAE	*Siparuna thecaphora* cf. (Poeppig & Endl.) A. DC.	T	.	1	.	.
MYRSINACEAE	*Myrsine dependens* (R. & P.) Spreng.	T	.	.	3	2
ROSACEAE	*Polylepis pauta* ?	T	.	.	1	10
SOLANACEAE	*Saracha punctata* R. & P.	T	.	.	1	.
MELASTOMATACEAE	*Miconia* sp. 1	T	.	.	1	.
MELASTOMATACEAE	*Miconia* sp. 5	T	.	.	.	1
	Total individuals:		**51**	**49**	**40**	**49**

O. Transect 12. Elfin forest. Abundance of species in Epiphytes subsample only (see also Appendices 1M-N). Counts are total number of host trees (out of six) in which species was present.

FAMILY	GENUS, SPECIES	HABIT	TOTAL HOST TREES WITH SPECIES (OUT OF SIX)
ERICACEAE	Indet. sp. 1	E	6
PTERIDOPHYTA	*Asplenium cuspidatum* Lam.	E	5
ORCHIDACEAE	*Epidendrum* sp. 1	E	5
PTERIDOPHYTA	*Elaphoglossum paleaceum* cf. (Hook. & Grev.) Sledge	E	5
GESNERIACEAE	*Drymonia* sp. 1	E	5
ORCHIDACEAE	*Epidendrum* sp. 3	E	4
ERICACEAE	Indet. sp. 2	E	4
ERICACEAE	*Demosthenesia spectabilis* (Rusby) A.C. Sm.	E	4
PTERIDOPHYTA	*Elaphoglossum* sp. 1	H	4
ORCHIDACEAE	*Stelis* sp. 2	E	4
URTICACEAE	*Pilea* sp. 2	H	3
PTERIDOPHYTA	*Elaphoglossum* sp. 2	H	3
BROMELIACEAE	Indet. sp. 1	E	2
PTERIDOPHYTA	*Campyloneurum amphostenon* (Klotzsch) Fee	H	2
PTERIDOPHYTA	*Grammitis athyrioides* (Hk.) Morton	E	2
PTERIDOPHYTA	*Grammitis lanigera* (Desv.) C.V. Morton	E	2
PTERIDOPHYTA	*Grammitis moniliformis* (Lag. ex Sw.) Proctor	H	2
ASTERACEAE	*Pentacalia* sp. 1	L	2
URTICACEAE	*Pilea* sp. 3	H	2
ORCHIDACEAE	*Sobralia/Elleanthus* sp. 1 (nc)	E	1
ERICACEAE	Indet. sp. 3 (nc)	Hem	1
ASCLEPIADACEAE	*Marsdenia* sp. 1	L	1
DIOSCORIACEAE	*Dioscorea piperifolia* Humb. & Bonpl.	L	1
ONAGRACEAE	*Fuchsia apetala* R. & P.	E	1
ORCHIDACEAE	*Epidendrum* sp. 2	E	1
ALSTROEMERIACEAE	*Bomarea coccinea* (R. & P.) Baker	L	1
ORCHIDACEAE	*Pachyphyllum* sp. 1	E	1
VALERIANACEAE	*Valeriana clematitis* Kunth	L	1
ORCHIDACEAE	*Stelis* sp. 1	E	1
ORCHIDACEAE	Subtribe Cranichidinae sp. 1	E	1
OXALIDACEAE	*Oxalis melilotoides* Zucc.	H	1
PTERIDOPHYTA	*Vittaria graminifolia* Kaulf.	E	1
RUBIACEAE	*Emmeorrhiza umbellata* Spreng.	L	1
ORCHIDACEAE	*Oncidium* sp. 1 (nc)	E	1
	Total:		**81**

P. Transect 13. Tall hilltop forest. Abundance of species in all subsamples (> 1 cm dbh only; other size classes not sampled). Subsamples 1-2.5 cm dbh 1 m x 70 m (70 m^2); 2.5-10 cm dbh 2 m x 60 m (120 m^2); 10-30 cm dbh 4 m x 60 m (240 m^2); > 30 cm dbh 8 m x 60 m (480 m^2).

			INDIVIDUALS IN SIZE CLASS (DBH)			
FAMILY	**GENUS, SPECIES**	**HABIT**	**1-2.5 cm**	**2.5-10 cm**	**10-30 cm**	**>30 cm**
ASTERACEAE	*Pentacalia* sp. 1	L	5	.	.	.
ASTERACEAE	*Gynoxys* sp. 1	T	4	4	.	.
ERICACEAE	*Gaultheria erecta* cf. Vent.	Hem	4	2	.	.
POLYGALACEAE	*Monnina marginata* Presl.	T	4	.	.	.
AQUIFOLIACEAE	*Ilex* sp. 2	T	3	3	2	.
AQUIFOLIACEAE	*Ilex ovalis* (R. & P.) Loesener	T	3	1	.	.
ERICACEAE	*Gaultheria buxifolia* Willd.	S	3	.	.	.
ASTERACEAE	*Diplostephium* sp. 1	S	3	.	.	.
POACEAE	*Chusquea* sp. 3	S	3	.	.	.
SYMPLOCACEAE	*Symplocos psiloclada* Stahl	T	2	8	6	.
SYMPLOCACEAE	*Symplocos quitensis* Brand	T	2	3	.	.
ARALIACEAE	*Schefflera inambarica* Harms	T	2	2	.	.
MYRSINACEAE	*Myrsine dependens* (R. & P.) Spreng.	T	2	1	7	.
ERICACEAE	Indet. sp. 1	E	2	1	.	.
ASTERACEAE	*Senecio* sp. 2	H	2	.	.	.
ERICACEAE	*Siphonandra elliptica* (R. & P.) Klotzsch	T	2	.	.	.
MELASTOMATACEAE	*Miconia* sp. 1	T	1	5	.	.
ASTERACEAE	*Gynoxys* sp. 5	T	1	2	.	.
CUNONIACEAE	*Weinmannia fagaroides/microphylla* H.B.K.	T	1	1	2	1
ARALIACEAE	*Oreopanax* sp. 1	T	1	1	1	1
AQUIFOLIACEAE	*Ilex* sp. 3	T	1	.	3	.
LOGANIACEAE	*Desfontainea* sp. 1	S	1	.	.	.
LORANTHACEAE	*Gaiadendron punctatum* (R. & P.) G. Don	S	1	.	.	.
ASTERACEAE	*Gynoxys* sp. 2	T	.	10	.	.
MELASTOMATACEAE	*Miconia* sp. 2	T	.	2	1	.
MELASTOMATACEAE	*Miconia* sp. 3	S	.	2	.	.
CLETHRACEAE	*Clethra cuneata* Rusby	T	.	2	.	.
ASTERACEAE	*Gynoxys* sp. 4	T	.	1	.	.
SYMPLOCACEAE	*Symplocos apiciflora* cf. Stahl	T	.	1	.	.
ROSACEAE	*Polylepis sericea* cf. Wedd.	T	.	.	3	1
AQUIFOLIACEAE	*Ilex* sp. 1	T	.	.	2	1
	Total individuals		**53**	**52**	**27**	**4**

Vegetation transects from Camp Two (2050 m) in the Northern Cordillera de Vilcabamba, Department of Junín, Peru

APPENDIX 3

Brad Boyle, Mónica Arakaki, and Hamilton Beltrán

For all transects, subsamples are based on size, as follows: Herbs, < 0.5 m tall; Shrubs, > 0.5 m tall and < 1.0 cm dbh. Abundance for Herbs subsample is the number of subunits in which species was recorded. Abundance for Shrubs subsample is number of individuals. "Habit" is the normal growth form of an adult plant: H = terrestrial herb, E = epiphyte, Hem = hemiepiphyte, L = climber (includes herbaceous vines), S = shrub, T = tree. "nc" = specimen not collected.

A. Transect 14. Ridge crest forest. Abundance of species in Herbs subsample (see Appendices 2B-F for remaining subsamples from this transect). Herbs subsample 0.5 x 2 m (x 10; = 10 m^2), shrubs subsample 1 m x 2 m (= 20 m^2).

FAMILY	GENUS, SPECIES	HABIT	ABUNDANCE
URTICACEAE	*Pilea* sp. 1	H	9
POACEAE	*Chusquea* sp. 4	S	8
PTERIDOPHYTA	*Elaphoglossum lechlerianum* (Mett.) Moore	H	8
PTERIDOPHYTA	*Peltapteris moorei* (E.G.Britt.) Gomez	E	8
MELASTOMATACE	*Miconia* sp. 9	S	7
PTERIDOPHYTA	*Elaphoglossum* sp. 2	H	7
MELASTOMATACE	*Miconia* sp. 10	T	5
PTERIDOPHYTA	*Elaphoglossum raywaense* (Jenm.) Alston	E	5
PTERIDOPHYTA	*Hymenophyllum plumieri* Hk. & Grev.	H	5
ERICACEAE	*Diogenesia octandra* cf. Sleumer	Hem	4
BROMELIACEAE	Indet. sp. 2	H	3
CAMPANULACEAE	*Siphocampylus angustiflorus* Schltdl.	L	3
GENTIANACEAE	*Tapeinostemon zamoranum* Steyerm.	H	3
PTERIDOPHYTA	*Elaphoglossum* sp. 4	H	3
CELASTRACEAE	*Celastrus* sp. 1	L	2
COMMELINACEAE	Indet. sp. 1 (nc)	H	2
MELASTOMATACE	*Clidemia* sp. 1	S	2
PTERIDOPHYTA	*Hymenophyllum verecundum* Morton	E	2
ARACEAE	*Anthurium nigrescens* Engl.	Hem	1
ASTERACEAE	*Munnozia* sp. 2	L	1
CLUSIACEAE	*Clusia* sp. 1	Hem	1
DIOSCORIACEAE	*Dioscorea calcensis* ? Knuth	L	1
MELASTOMATACE	Indet. sp. 1	T	1
MELASTOMATACE	Indet. sp. 3	Hem	1
MELASTOMATACE	*Miconia* sp. 8	T	1
MYRSINACEAE	*Cybianthus pastensis* (Mez) Agost.	S	1
MYRSINACEAE	*Cybianthus peruvianus* (A.DC.) Miq.	T	1
ORCHIDACEAE	*Brachionidium* sp. 1	H	1
OXALIDACEAE	*Oxalis* sp. 1	H	1
PIPERACEAE	*Peperomia glabella* (Sw.) A. Dietr.	H	1
PTERIDOPHYTA	*Elaphoglossum paleaceum* cf. (Hook. & Grev.) Sledge	E	1
PTERIDOPHYTA	*Grammitis lanigera* (Desv.) C.V. Morton	E	1
PTERIDOPHYTA	*Nephrolepis pectinata* (Willd.) Schott	H	1
PTERIDOPHYTA	*Sphaeropteris elongata* (Hooker) Tryon	T	1
RUBIACEAE	*Amaouia* sp. 1	T	1
	Total:		**103**

B. Transect 14. Ridge crest forest. Abundance of species in Shrubs subsample (see Appendices 2A and 2C-F for remaining subsamples from this transect). Herbs subsample 0.5 x 2 m (x 10; = 10 m^2), shrubs subsample 1 m x 20 m (= 20 m^2).

FAMILY	GENUS, SPECIES	HABIT	ABUNDANCE
POACEAEAE	*Chusquea* sp. 4	S	10
MELASTOMATACEAE	*Miconia* sp. 10	T	5
MELASTOMATACEAE	*Miconia* sp. 9	S	5
MELASTOMATACEAE	*Miconia* sp. 7	T	4
MELASTOMATACEAE	*Miconia* sp. 8	T	4
MELASTOMATACEAE	Indet. sp. 1	T	3
MELASTOMATACEAE	Indet. sp. 2	S	3
ASTERACEAE	*Pentacalia* sp. 2	L	2
CLUSIACEAE	*Clusia weberbaueri* Engler	T	2
MELASTOMATACEAE	*Clidemia* sp. 1	S	2
PIPERACEAE	*Piper lanceolatum* R. & P.	S	2
PIPERACEAE	*Piper lanceolatum* R. & P.	S	2
RUBIACEAE	*Psychotria* sp. 2 (nc)	T	2
ARACEAE	*Anthurium* sp. 2 (sect. Calomystrium)	Hem	1
ARALIACEAE	*Dendropanax* sp. 1	T	1
CAMPANULACEAE	*Siphocampylus angustiflorus* Schltdl.	L	1
CLUSIACEAE	*Clusia* sp. 1	Hem	1
CUNONIACEAE	*Weinmannia* sp. 1	T	1
DIOSCORIACEAE	*Dioscorea calcensis* ? Knuth	L	1
ERICACEAE	*Cavendishia bracteata* cf. (R. & P. ex J. St.-Hill) Hoerold	Hem	1
ERICACEAE	*Disterigma* ? sp. 1	Hem	1
ERICACEAE	*Psammisia ulbrichiana* cf. Hoer.	Hem	1
EUPHORBIACEAE	*Alchornea* sp. 1	T	1
LAURACEAE	*Rhodostemonodahne*?/*Ocotea* sp. 1	T	1
MELASTOMATACEAE	Indet. sp. 3	Hem	1
MYRSINACEAE	*Cybianthus peruvianus* (A.DC.) Miq.	T	1
MYRTACEAE	*Myrcia* sp. 1	T	1
PIPERACEAE	*Piper* sp. 2	T	1
POACEAE	*Guadua* sp. 1	S	1
PTERIDOPHYTA	*Blechnum binervatum* C.V. Morton & Lellinger	H	1
RUBIACEAE	*Psychotria steinbachii* Standl.	S	1
RUBIACEAE	*Psychotria/Palicourea* sp. 2	T	1
RUBIACEAE	*Psychotria/Palicourea* sp. 3	S	1
THEACEAE	*Freziera* ? sp. 1	T	1
VITACEAE	*Cissus* sp. 1	L	1
	Total:		**68**

C. Transect 14. Ridge crest forest. Abundance of species in 1-2.5 cm dbh size class (see Appendices 2A-B and 2D-F for remaining subsamples from this transect). Counts are numbers of individuals. Subsamples 1-2.5 cm dbh 2 m x 80 m (= 160 m^2).

FAMILY	GENUS, SPECIES	HABIT	INDIVIDUALS 1-2.5 cm DBH
MELASTOMATACEAE	*Miconia* sp. 10	T	5
ARECACEAE	*Geonoma* sp. 1	T	4
EUPHORBIACEAE	*Alchornea* sp. 1	T	4
ARACEAE	*Anthurium nigrescens* Engl.	Hem	3
CLUSIACEAE	*Clusia* sp. 1	Hem	3
POACEAE	*Chusquea* sp. 4	S	3
LAURACEAE	Indet. sp. 2	T	2
LAURACEAE	Indet. sp. 4	T	2
MELASTOMATACEAE	Indet. sp. 2	S	2
SABIACEAE	*Meliosma* sp. 1	T	2
APOCYNACEAE	*Tabernaemontana vanheurckii* cf. Muell.	T	1
ARACEAE	*Anthurium* sp. 2 (sect. Calomystrium)	Hem	1
ASTERACEAE	Indet. sp. 2	T	1
ASTERACEAE	*Pentacalia* sp. 3	L	1
CUNONIACEAE	*Weinmannia sorbifolia* Kunth	T	1
ERICACEAE	*Psammisia ulbrichiana* cf. Hoer.	Hem	1
EUPHORBIACEAE	*Alchornea* sp. 2	T	1
LAURACEAE	Indet. sp. 1	T	1
LAURACEAE	Indet. sp. 7 (nc)	T	1
LAURACEAE	*Rhodostemonodahne* ?/*Ocotea* sp. 1	T	1
MARCGRAVIACEAE	*Norantea* sp. 1	Hem	1
MELASTOMATACEAE	*Miconia* sp. 11	S	1
MELASTOMATACEAE	*Miconia* sp. 7	T	1
POLYGALACEAE	*Monnina* sp. 1	L	1
RUBIACEAE	Indet. sp. 1	L	1
RUBIACEAE	*Amaouia* sp. 1	T	1
RUBIACEAE	*Psychotria steinbachii* Standl.	S	1
THEACEAE	*Gordonia fruticosa* (Schrader) H. Keng	T	1
	Total individuals:		**48**

D. Transect 14. Ridge crest forest. Abundance of species in 2.5-10 cm dbh size classes (see Appendices 2A-C and 2E-F for remaining subsamples from this transect). Counts are numbers of individuals. Subsamples 2.5-10 cm dbh 2 m x 110 m (= 220 m^2).

FAMILY	GENUS, SPECIES	HABIT	INDIVIDUALS 2.5-10 cm DBH
ARACEAE	*Anthurium* sp. 2 (sect. Calomystrium)	Hem	5
MELASTOMATACEAE	*Miconia* sp. 10	T	5
MARCGRAVIACEAE	*Norantea* sp. 1	Hem	4
PTERIDOPHYTA	*Cyathea pallescens* (Sodiro) Domin	T	4
CELASTRACEAE	*Celastrus* sp. 1	L	3
MELASTOMATACEAE	Indet. sp. 2	S	3
CUNONIACEAE	*Weinmannia sorbifolia* Kunth	T	2
ERICACEAE	*Cavendishia bracteata* cf. (R. & P. ex J. St.-Hill) Hoerold	Hem	2
POLYGALACEAE	*Moutabea aculeata* cf. (R. & P.) Poepp. & Endl.	L	2
RUBIACEAE	*Faramea multiflora* A. Rich. ex DC.	T	2
SAPOTACEAE	*Pouteria* sp. 1	T	2
ARECACEAE	*Geonoma* sp. 1	T	1
ASTERACEAE	*Pentacalia* sp. 3	L	1
ERICACEAE	*Diogenesia octandra* cf. Sleumer	Hem	1
EUPHORBIACEAE	*Alchornea* sp. 2	T	1
LAURACEAE	Indet. sp. 2	T	1
LAURACEAE	Indet. sp. 3	T	1
LAURACEAE	*Rhodostemonodahne ?/Ocotea* sp. 1	T	1
MELASTOMATACEAE	*Miconia* sp. 7	T	1
MYRSINACEAE	*Cybianthus peruvianus* (A.DC.) Miq.	T	1
MYRTACEAE	*Myrcia* sp. 1	T	1
POLYGALACEAE	*Monnina* sp. 1	L	1
RUBIACEAE	Indet. sp. 1	L	1
RUBIACEAE	*Amaouia* sp. 1	T	1
THEACEAE	*Freziera* ? sp. 1	T	1
THEACEAE	*Gordonia fruticosa* (Schrader) H. Keng	T	1
VITACEAE	*Cissus* sp. 1	L	1
	Total individuals:		**50**

E. Transect 14. Ridge crest forest. Abundance of species in 10-30 cm and >30 cm dbh size classes (see Appendices 2A-D and 2F for remaining subsamples from this transect). Counts are numbers of individuals. Subsamples 10-30 cm dbh 4 m x 140 m (= 560 m^2), > 30 cm dbh 8 m x 120 m (= 960 m^2).

		INDIVIDUALS IN SIZE CLASS (DBH)		
FAMILY	**GENUS, SPECIES**	**HABIT**	**10-30 cm**	**>30 cm**
MELASTOMATACEAE	*Miconia* sp. 10	T	8	.
EUPHORBIACEAE	*Alchornea* sp. 2	T	5	.
THEACEAE	*Gordonia fruticosa* (Schrader) H. Keng	T	5	.
CLUSIACEAE	*Clusia weberbaueri* Engler	T	3	.
CUNONIACEAE	*Weinmannia sorbifolia* Kunth	T	3	.
RUBIACEAE	*Amaouia* sp. 1	T	3	.
LAURACEAE	Indet. sp. 2	T	2	1
CLUSIACEAE	*Clusia* sp. 1	Hem	2	.
CUNONIACEAE	*Weinmannia* sp. 1	T	2	.
MYRSINACEAE	*Geissanthus* sp. 1	T	2	.
MYRTACEAE	*Siphoneugena occidentalis* cf. Legrand	T	2	.
PTERIDOPHYTA	*Sphaeropteris elongata* (Hooker) Tryon	T	2	.
RUBIACEAE	*Faramea multiflora* A. Rich. ex DC.	T	2	.
SAPOTACEAE	*Pouteria* sp. 1	T	2	.
PODOCARPACEAE	*Podocarpus oleifolius* D. Don	T	1	1
CUNONIACEAE	*Weinmannia pubescens* H.B.K.	T	1	.
FABACEAE-MIM	*Inga adenophylla* Pittier	T	1	.
LAURACEAE	*Rhodostemonodahne* ?/*Ocotea* sp. 1	T	1	.
MELASTOMATACEAE	Indet. sp. 2	S	1	.
MELIACEAE	*Guarea kunthiana* A. Juss.	T	1	.
PTERIDOPHYTA	*Cyathea pallescens* (Sodiro) Domin	T	1	.
SAPOTACEAE	Indet. sp. 1	T	1	.
CLETHRACEAE	*Clethra revoluta* (R. & P.) Sp.	T	.	2
STYRACACEAE	*Styrax foveolaria* Perkins	T	1	.
SYMPLOCACEAE	*Symplocos* sp. 1	T	.	1
	Total individuals:		**51**	**5**

F. Transect 14. Ridge crest forest. Abundance of species in Epiphytes subsample (see Appendices 2A-E for remaining subsamples from this transect). Counts are total number of host trees (out of six) in which species was present.

FAMILY	GENUS, SPECIES	HABIT	TOTAL HOST TREES WITH SPECIES (OUT OF SIX)
PTERIDOPHYTA	*Elaphoglossum raywaense* (Jenm.) Alston	E	5
PTERIDOPHYTA	*Hymenophyllum plumieri* Hk. & Grev.	H	4
ORCHIDACEAE	*Maxillaria* sp. 4	E	4
ERICACEAE	*Cavendishia bracteata* cf. (R. & P. ex J. St.-Hill) Hoerold	Hem	3
PTERIDOPHYTA	*Polypodium caceresii* Sodiro	E	3
ERICACEAE	*Diogenesia octandra* cf. Sleumer	Hem	3
PTERIDOPHYTA	*Peltapteris moorei* (E.G.Britt.) Gomez	E	3
ORCHIDACEAE	*Elleanthus* sp. 2	E	3
PTERIDOPHYTA	*Grammitis lanigera* (Desv.) C.V. Morton	E	3
PTERIDOPHYTA	*Grammitis* sp. 2 (nc)	E	3
BROMELIACEAE	*Tillandsia* sp. 2	E	3
ORCHIDACEAE	*Stelis* sp. 4	E	3
PTERIDOPHYTA	*Elaphoglossum* sp. 2	H	3
ORCHIDACEAE	*Maxillaria acuminata* aff. Lindl.	E	2
ORCHIDACEAE	Indet. sp. 1 (nc)	E	2
ORCHIDACEAE	*Maxillaria* sp. 1	E	2
ERICACEAE	*Psammisia ulbrichiana* cf. Hoer.	Hem	2
ORCHIDACEAE	*Elleanthus* sp. 1 (nc)	E	2
ORCHIDACEAE	*Elleanthus* sp. 1	E	2
ORCHIDACEAE	*Maxillaria* sp. 2	E	2
ASTERACEAE	*Pentacalia* sp. 4	L	2
ORCHIDACEAE	*Pleurothallis imraei* Lindl.	E	2
URTICACEAE	*Pilea* sp. 1	H	2
OXALIDACEAE	*Oxalis* sp. 1	H	2
PTERIDOPHYTA	*Polypodium* sp. 1	E	2
ASTERACEAE	*Munnozia* sp. 2	L	2
ARACEAE	*Anthurium nigrescens* Engl.	Hem	2
ORCHIDACEAE	*Odontoglossum teretifolius* S. Dalstrom	E	2
ARACEAE	*Philodendron* sp. 1	Hem	1
MENISPERMACEAE	*Abuta aristeguietae* cf. Krukoff & Barneby	L	1
MARCGRAVIACEAE	*Sorobea* sp. 1	Hem	1
MARCGRAVIACEAE	*Marcgravia affinis* cf.	Hem	1
DIOSCORIACEAE	*Dioscorea calcensis* ? Knuth	L	1
ERICACEAE	*Sphyrospermum cordifolium* Benth.	Hem	1
ASCLEPIADACEAE	Indet. sp. 1	L	1
ERICACEAE	*Disterigma* ? sp. 1	Hem	1
ASTERACEAE	*Pentacalia* sp. 3	L	1
ORCHIDACEAE	*Lockhartia parthenocomos* cf. Rchf. f.	E	1
GESNERIACEAE	*Paradrymonia metamorphophylla* (J.D.Sm.) Wiehl.	E	1
ORCHIDACEAE	*Maxillaria quitensis* (Rchf. f.) C. Schueinf	E	1
PTERIDOPHYTA	*Nephrolepis pectinata* (Willd.) Schott	H	1
PTERIDOPHYTA	*Hymenophyllum verecundum* Morton	E	1
PTERIDOPHYTA	*Enterosora parietina* (Klotzsch) Bishop	E	1
PTERIDOPHYTA	*Elaphoglossum lechlerianum* (Mett.) Moore	H	1
ORCHIDACEAE	*Stelis* sp. 3	E	1
ORCHIDACEAE	*Pleurothallis* sp. 5 (nc)	E	1
ORCHIDACEAE	*Lepanthes* sp. 2	E	1
ORCHIDACEAE	*Pleurothallis* sp. 1	E	1
ORCHIDACEAE	*Brassia* sp. 1	E	1
ORCHIDACEAE	*Maxillaria augustae-victoriae* Lehm & Kranzl	E	1

FAMILY	GENUS, SPECIES	HABIT	TOTAL HOST TREES WITH SPECIES (OUT OF SIX)
ORCHIDACEAE	*Maxillaria* sp. 3	E	1
VITACEAE	*Cissus* sp. 1	L	1
ORCHIDACEAE	*Lepanthes* sp. 3	E	1
ACANTHACEAE	*Mendoncia glabra* (P. & E.) Nees	L	1
ORCHIDACEAE	*Lepanthes* sp. 1	E	1
ORCHIDACEAE	*Epidendrum* sp. 4 (nc)	E	1
ORCHIDACEAE	*Dichaea morrisii* Fawe & Rendl.	E	1
ORCHIDACEAE	*Pleurothallis* sp. 2	E	1
	Total:		**103**

G. Transect 15. Tall humid forest. Abundance of species in Herbs subsample (see Appendices 2H -24L for remaining subsamples from this transect). Herbs subsample 0.5 x 2 m (x 17; = 17 m^2), shrubs subsample 1 m x 2 m (x 15; = 30 m^2).

FAMILY	GENUS, SPECIES	HABIT	ABUNDANCE
URTICACEAE	*Pilea diversifolia* Wedd.	S	14
PTERIDOPHYTA	*Campyloneurum fuscosquamatum* Lellinger	H	8
ACANTHACEAE	Indet. sp. 1	S	7
MELASTOMATACEAE	*Miconia* sp. 13	T	5
PIPERACEAE	*Peperomia alata* R. & P.	H	4
PTERIDOPHYTA	*Ctenitis sloanei* (Sprengel) Morton	H	4
URTICACEAE	*Pilea* sp. 2	H	4
ARACEAE	*Anthurium anemonum* Kunth	H	3
HYDRANGEACEAE	*Hydrangea* sp. 1	L	3
OXALIDACEAE	*Oxalis* sp. 1	H	3
PTERIDOPHYTA	*Blechnum binervatum* C.V. Morton & Lellinger	H	3
PTERIDOPHYTA	*Polybotrya* sp. 3	H	3
URTICACEAE	*Pilea* sp. 5	H	3
MELASTOMATACEAE	Indet. sp. 4	S	2
MELIACEAE	*Ruagea ?/Trichilia* ? sp. 1	T	2
PIPERACEAE	*Piper crassinervium* cf. H.B.K.	T	2
PTERIDOPHYTA	*Asplenium auriculatum* Sw.	H	2
PTERIDOPHYTA	*Elaphoglossum lechlerianum* (Mett.) Moore	H	2
ACANTHACEAE	*Mendoncia glabra* (P. & E.) Nees	L	1
ARACEAE	*Philodendron* sp. 2 (sp. nov.)	Hem	1
ARECACEAE	*Geonoma interrupta/obtusifolia*	T	1
ASTERACEAE	*Jungia* sp. 1	L	1
ASTERACEAE	*Munnozia* sp. 2	L	1
BEGONIACEAE	*Begonia exaltata* cf. C. DC.	H	1
CAMPANULACEAE	*Siphocampylus angustiflorus* Schltdl.	L	1
CLUSIACEAE	*Clusia* sp. 2	Hem	1
CYCLANTHACEAE	*Evodianthus ?/Asplundia* ? sp. 1	H	1
EUPHORBIACEAE	*Alchornea* sp. 1	T	1
GESNERIACEAE	*Besleria* sp. 1	H	1
MELASTOMATACEAE	*Miconia* sp. 12	T	1
MONIMIACEAE	*Mollinedia* sp. 1	S	1
PIPERACEAE	*Piper* sp. 2	T	1
POACEAE	*Chusquea* sp. 5	S	1
PTERIDOPHYTA	*Alsophila engelii* Tryon	T	1
PTERIDOPHYTA	*Asplenium* sp. 1	H	1
PTERIDOPHYTA	*Elaphoglossum* sp. 5	H	1

FAMILY	GENUS, SPECIES	HABIT	ABUNDANCE
PTERIDOPHYTA	*Hymenophyllum plumieri* Hk. & Grev.	H	1
PTERIDOPHYTA	*Polypodium loriceum* L.	E	1
PTERIDOPHYTA	*Polypodium triseriale* Sw.	H	1
PTERIDOPHYTA	*Pteris* sp. 1	H	1
PTERIDOPHYTA	*Terpsichore taxifolia* (L.) A.R.Sm.	E	1
PTERIDOPHYTA	*Vittaria* sp. 1	H	1
URTICACEAE	*Pilea* sp. 4	H	1
URTICACEAE	*Pilea haenkei* Killip ex char	S	1
	Total:		**100**

H. Transect 15. Tall humid forest. Abundance of species in Shrubs subsample (see Appendices 2G and 2I-L for remaining subsamples from this transect. Abundance is number of subunits (out of 15) in which species was recorded. Herbs subsample 0.5 x 2 m (x 17; = 17 m^2), shrubs subsample 1 m x 15 m (= 30 m^2).

FAMILY	GENUS, SPECIES	HABIT	ABUNDANCE
ACANTHACEAE	Indet. sp. 1	S	8
URTICACEAE	*Pilea diversifolia* Wedd.	S	8
CYCLANTHACEAE	*Evodianthus ?/Asplundia* ? sp. 1	H	6
PTERIDOPHYTA	*Ctenitis sloanei* (Sprengel) Morton	H	5
MELASTOMATACEAE	*Miconia* sp. 13	T	4
POACEAE	*Chusquea* sp. 5	S	4
ASTERACEAE	*Mikania* sp. 2	L	3
SOLANACEAE	*Cestrum megalophyllum* Dunal	T	3
URTICACEAE	*Pilea haenkei* Killip ex char	S	3
AMARANTHACEAE	*Iresine diffusa* cf. H. & B. ex Willd.	L	2
ARACEAE	*Philodendron* ? subgen. *Pteromischum*	Hem	2
MELASTOMATACEAE	Indet. sp. 5	S	2
MELASTOMATACEAE	*Miconia* sp. 12	T	2
MYRTACEAE	Indet. sp. 1	T	2
PTERIDOPHYTA	Indet. sp. 1	S	2
PTERIDOPHYTA	Indet. sp. 2	E	2
PTERIDOPHYTA	*Blechnum binervatum* C.V. Morton & Lellinger	H	2
PTERIDOPHYTA	*Campyloneurum fuscosquamatum* Lellinger	H	2
PTERIDOPHYTA	*Polybotrya* sp. 2	E	2
PTERIDOPHYTA	*Polybotrya* sp. 3	H	2
RUBIACEAE	*Manettia* sp. 2	L	2
RUBIACEAE	*Psychotria* sp. 1	S	2
SOLANACEAE	*Solanum anceps* R. & P.	S	2
ACANTHACEAE	*Mendoncia glabra* (P. & E.) Nees	L	1
ARACEAE	*Anthurium nigrescens* Engl.	Hem	1
ARACEAE	*Philodendron* sp. 2 (sp. nov.)	Hem	1
ARECACEAE	*Geonoma interrupta/obtusifolia*	T	1
ASTERACEAE	*Jungia* sp. 1	L	1
ASTERACEAE	*Munnozia* sp. 2	L	1
CAMPANULACEAE	*Centropogon* sp. 1	L	1
CAMPANULACEAE	*Siphocampylus angustiflorus* Schltdl.	L	1
CUCURBITACEAE	*Psiguria trihylla* cf.	L	1
ERICACEAE	*Psammisia ulbrichiana* cf. Hoer.	Hem	1
HYDRANGEACEAE	*Hydrangea* sp. 1	L	1

FAMILY	GENUS, SPECIES	HABIT	ABUNDANCE
LAURACEAE	Indet. sp. 5	T	1
LAURACEAE	Indet. sp. 6	T	1
MELASTOMATACEAE	*Miconia* sp. 14	T	1
MELASTOMATACEAE	*Miconia* sp. 15	H	1
MELASTOMATACEAE	*Miconia* sp. 17	S	1
MONIMIACEAE	*Mollinedia* sp. 1	S	1
MORACEAE	*Ficus apollinaris* cf. Dug.	T	1
MYRSINACEAE	*Cybianthus* sp. 1	T	1
PIPERACEAE	*Peperomia alata* R. & P.	H	1
PIPERACEAE	*Piper crassinervium* cf. H.B.K	T	1
PIPERACEAE	*Piper* sp. 6 (nc)	S	1
PTERIDOPHYTA	*Asplenium* sp. 1	H	1
PTERIDOPHYTA	*Diplazium expansum* Willd	H	1
PTERIDOPHYTA	*Pteris podophyla* Sw.	H	1
RUBIACEAE	*Palicourea* sp. 1	T	1
RUBIACEAE	*Psychotria steinbachii* Standl.	S	1
SOLANACEAE	*Solanum pectinatum* aff. Dunal	S	1
THEACEAE	*Freziera* ? sp. 2	S	1
	Total:		**101**

I. Transect 15. Tall humid forest. Abundance of species in 1-2.5 cm dbh size class (see Appendices 2G-H and 2J-L for remaining subsamples from this transect). Counts are numbers of individuals. Subsamples 1-2.5 cm dbh 2 m x 80 m (= 160 m^2).

FAMILY	GENUS, SPECIES	HABIT	INDIVIDUALS 1-2.5 cm DBH
RUBIACEAE	*Elaeagia utilis* cf. Karst.	T	4
ACTINIDIACEAE	*Saurauia biserrata* (R. & P.) Spreng	T	3
CYCLANTHACEAE	*Evodianthus* ?/*Asplundia* ? sp. 1	H	3
PTERIDOPHYTA	*Polybotrya* sp. 1	H	3
PTERIDOPHYTA	*Pteris livida* Mett.	T	3
RUBIACEAE	*Palicourea* sp. 1	T	3
SOLANACEAE	*Cestrum megalophyllum* Dunal	T	3
AMARANTHACEAE	*Iresine diffusa* cf. H. & B. ex Willd.	L	2
ICACINACEAE	*Calatola venezuelana* Pittier	T	2
ARACEAE	*Anthurium nigrescens* Engl.	Hem	1
ASTERACEAE	*Mikania* sp. 2	L	1
ASTERACEAE	*Mikania* sp. 3	L	1
CUCURBITACEAE	*Gurania* sp. 1	L	1
ERICACEAE	*Psammisia ulbrichiana* cf. Hoer.	Hem	1
HYDRANGEACEAE	*Hydrangea* sp. 1	L	1
LAURACEAE	Indet. sp. 5	T	1
MELASTOMATACEAE	*Miconia* sp. 13	T	1
MELASTOMATACEAE	*Miconia* sp. 14	T	1
MELASTOMATACEAE	*Miconia* sp. 15	H	1
MELASTOMATACEAE	*Miconia* sp. 16	T	1
MORACEAE	*Ficus* sp. 2	Hem	1
MYRSINACEAE	*Cybianthus* sp. 1	T	1
MYRTACEAE	*Plinia* ? sp. 1	T	1
PIPERACEAE	*Piper crassinervium* cf. H.B.K.	T	1
POACEAE	*Chusquea* sp. 5	S	1
ROSACEAE	*Prunus pleiantha* Pilger	T	1
RUBIACEAE	*Palicourea* sp. 2	T	1
RUBIACEAE	*Psychotria steinbachii* Standl.	S	1
URTICACEAE	*Pilea diversifolia* Wedd.	S	1
URTICACEAE	*Urera eggersii* Hieron	T	1
	Total individuals:		**47**

J. Transect 15. Tall humid forest. Abundance of species in 2.5-10 cm dbh size class (see Appendices 2G-I and 2K-L for remaining subsamples from this transect). Counts are numbers of individuals. Subsamples 2.5-10 cm dbh 2 m x 140 (= 280 m^2).

FAMILY	GENUS, SPECIES	HABIT	INDIVIDUALS 2.5-10 cm DBH
SOLANACEAE	*Cestrum megalophyllum* Dunal	T	9
PTERIDOPHYTA	*Cyathea* sp. 2 (nc)	T	7
CYCLANTHACEAE	*Evodianthus ?/Asplundia ?* sp. 1	H	5
POACEAE	*Chusquea* sp. 5	S	5
RUBIACEAE	*Elaeagia utilis* cf. Karst.	T	3
RUBIACEAE	*Palicourea* sp. 2	T	3
ANNONACEAE	*Guatteria* sp. 1	T	2
MELASTOMATACEAE	*Miconia* sp. 13	T	2
MORACEAE	*Ficus* sp. 1	T	2
PTERIDOPHYTA	*Dicksonia sellowiana*	T	2
ARACEAE	*Philodendron* ? subgen. *Pteromischum*	Hem	1
ARALIACEAE	*Oreopanax* sp. 3	T	1
ASTERACEAE	*Pentacalia* sp. 3	L	1
BORAGINACEAE	*Cordia* sp. 1	T	1
CUCURBITACEAE	*Gurania* sp. 1	L	1
ERICACEAE	*Psammisia ulbrichiana* cf. Hoer.	Hem	1
EUPHORBIACEAE	*Alchornea* sp. 3	T	1
FLACOURTIACEAE	*Banara guianensis* Aubl.	T	1
LAURACEAE	Indet. sp. 5	T	1
MELIACEAE	*Ruagea ?/Trichilia ?* sp. 1	T	1
MORACEAE	*Cecropia multiflora* cf. Snethlage	T	1
MYRSINACEAE	*Cybianthus* sp. 1	T	1
MYRTACEAE	*Eugenia muricata* DC.	T	1
MYRTACEAE	*Plinia* ? sp. 1	T	1
PTERIDOPHYTA	*Alsophila engelii* Tryon	T	1
PTERIDOPHYTA	*Diplazium* sp. 1	T	1
RUBIACEAE	*Faramea* sp. 1	T	1
RUBIACEAE	*Psychotria/Palicourea* sp. 4	T	1
SAPOTACEAE	Indet. sp. 2	T	1
	Total individuals:		**59**

K. Transect 15. Tall humid forest. Abundance of species in 10-30 cm and >30 cm dbh size classes (see Appendices 2G-J and 2L for remaining subsamples from this transect). Counts are numbers of individuals. Subsamples 10-30 cm dbh 4 m x 140m (= 560 m^2), > 30 cm dbh 8 m x 160 m (= 1280 m^2).

		INDIVIDUALS IN SIZE CLASS (DBH)		
FAMILY	**GENUS, SPECIES**	**HABIT**	**10-30 cm**	**>30 cm**
PTERIDOPHYTA	*Alsophila engelii* Tryon	T	31	.
MELIACEAE	*Guarea kunthiana* A. Juss.	T	7	3
PTERIDOPHYTA	*Cyathea pallescens* (Sodiro) Domin	T	4	.
MELIACEAE	*Ruagea* ?/*Trichilia* ? sp. 1	T	2	1
RUBIACEAE	*Faramea multiflora* A. Rich. ex DC.	T	2	.
RUBIACEAE	*Elaeagia utilis* cf. Karst.	T	1	7
PTERIDOPHYTA	*Dicksonia sellowiana*	T	1	1
CHLORANTHACEAE	*Hedyosmum cuatrecasanum* Occhioni	T	1	.
EUPHORBIACEAE	*Alchornea* sp. 3	T	1	.
EUPHORBIACEAE	*Mabea* ? sp. 1	T	1	.
FLACOURTIACEAE	*Banara guianensis* Aubl.	T	1	.
MORACEAE	*Ficus apollinaris* cf. Dug.	T	1	.
PTERIDOPHYTA	*Cyathea* sp. 1	T	1	.
ROSACEAE	*Prunus pleiantha* Pilger	T	1	.
RUBIACEAE	*Psychotria* sp. 1	S	1	.
SAPOTACEAE	Indet. sp. 2	T	1	.
CUNONIACEAE	*Weinmannia sorbifolia* Kunth	T	.	3
LAURACEAE	*Nectandra lineatifolia* (R. & P.) Mez.	T	.	2
CLETHRACEAE	*Clethra revoluta* (R. & P.) Sp.	T	.	1
LAURACEAE	*Ocotea tesmannii* O. Schmidt	T	.	1
MORACEAE	*Ficus* sp. 2	Hem	.	1
SYMPLOCACEAE	*Symplocos* sp. 2	T	.	1
	Total individuals:		**57**	**21**

L. Transect 15. Tall humid forest. Abundance of species in Epiphytes subsample (see Appendices 2G-K for remaining subsamples from this transect). Counts are total number of host trees (out of seven) in which the species was present.

FAMILY	GENUS, SPECIES	HABIT	TOTAL HOST TREES WITH SPECIES (OUT OF SEVEN)
PTERIDOPHYTA	*Elaphoglossum paleaceum* cf. (Hook. & Grev.) Sledge	E	6
PTERIDOPHYTA	*Polypodium loriceum* L.	E	6
PIPERACEAE	*Peperomia alata* R. & P.	H	5
POACEAE	*Chusquea* sp. 5	S	5
ARACEAE	*Anthurium nigrescens* Engl.	Hem	4
ERICACEAE	*Psammisia ulbrichiana* cf. Hoer.	Hem	4
ARALIACEAE	*Schefflera* sp. 1	Hem	3
BROMELIACEAE	Indet. sp. 3 (nc)	E	3
ERICACEAE	*Sphyrospermum cordifolium* Benth.	Hem	3
HYDRANGEACEAE	*Hydrangea* sp. 1	L	3
MELASTOMATACEAE	*Blakea* sp. 1	Hem	3
ORCHIDACEAE	*Elleanthus* sp. 3 (nc)	E	3
ORCHIDACEAE	*Stelis* sp. 5	E	3
PTERIDOPHYTA	*Campyloneurum fuscosquamatum* Lellinger	H	3
PTERIDOPHYTA	*Niphidium crassifolium* (L.) Lell.	E	3
ARACEAE	*Anthurium weberbaueri* Engl.	E	2
ASTERACEAE	*Munnozia* sp. 2	L	2
CAMPANULACEAE	*Siphocampylus angustiflorus* Schltdl.	L	2
CLUSIACEAE	*Clusia* sp. 2	Hem	2
CYCLANTHACEAE	*Evodianthus ?/Asplundia ?* sp. 1	H	2
MELASTOMATACEAE	Indet. sp. 4	S	2
MORACEAE	*Ficus* sp. 2	Hem	2
URTICACEAE	*Pilea* sp. 4	H	2
ARACEAE	*Anthurium soukupii* Croat	Hem	1
ARACEAE	*Philodendron* sp. 1	Hem	1
ARACEAE	*Philodendron* sp. 2 (sp. nov.)	Hem	1
ARACEAE	*Philodendron ? subgen. Pteromischum*	Hem	1
ASCLEPIADACEAE	*Tassadia obovata* cf. Decne	L	1
ASTERACEAE	*Mikania* sp. 3	L	1
BROMELIACEAE	*Pitcairnia* sp. 1	E	1
BROMELIACEAE	*Tillandsia* sp. 2	E	1
CAMPANULACEAE	*Centropogon* sp. 1	L	1
CLUSIACEAE	*Clusia* sp. 3	Hem	1
CUCURBITACEAE	*Gurania* sp. 1	L	1
CUCURBITACEAE	*Psiguria trihylla* cf.	L	1
ERICACEAE	*Cavendishia bracteata* cf. (R. & P. ex J. St.-Hill) Hoerold	Hem	1
ERICACEAE	*Diogenesia octandra* cf. Sleumer	Hem	1
ORCHIDACEAE	*Cryptocentrum* sp. 1	E	1
ORCHIDACEAE	*Maxillaria* sp. 2	E	1
ORCHIDACEAE	*Maxillaria augustae-victoriae* Lehm & Kranzl	E	1
ORCHIDACEAE	*Pleurothallis* sp. 6 (nc)	E	1
ORCHIDACEAE	*Scaphyglottis* sp. 1 (nc)	E	1
ORCHIDACEAE	*Xylobium* sp. 1	E	1
PTERIDOPHYTA	*Asplenium cuspidatum* Lam.	E	1
PTERIDOPHYTA	*Blechnum binervatum* C.V. Morton & Lellinger	H	1
PTERIDOPHYTA	*Elaphoglossum raywaense* (Jenm.) Alston	E	1
PTERIDOPHYTA	*Grammitis lanigera* (Desv.) C.V. Morton	E	1
PTERIDOPHYTA	*Polybotrya* sp. 1	H	1
PTERIDOPHYTA	*Polybotrya* sp. 3	H	1

FAMILY	GENUS, SPECIES	HABIT	TOTAL HOST TREES WITH SPECIES (OUT OF SEVEN)
PTERIDOPHYTA	*Polypodium* sp. 1	E	1
PTERIDOPHYTA	*Terpsichore taxifolia* (L.) A.R.Sm.	E	1
SMILACACEAE	*Smilax* sp. 1	L	1
URTICACEAE	*Pilea* sp. 2	H	1
URTICACEAE	*Pilea diversifolia* Wedd.	S	1
URTICACEAE	*Pilea haenkei* Killip ex char	S	1
	Total:		**105**

M. Transect 16. Pajonal (*Chusquea-Sphagnum* bog). Species abundance in all subsamples. Abundance for Herbs and Shrubs subsamples is the number of subunits in which species was recorded; abundance for remaining subsample is number of individuals. Herbs subsample 0.5 x 2 m (x 13; = 13 m^2); shrubs subsample 1 m x 2 m (x 17; = 34 m^2); 10-30 cm dbh 4 m x 170 m (= 680 m^2).

			ABUNDANCE IN SIZE CLASS		
FAMILY	GENUS, SPECIES	HABIT	HERBS	SHRUBS	10-30 cm DBH
PTERIDOPHYTA	*Blechnum auratum* (Fee) Tryon & Stolze	T	12	1	.
CUNONIACEAE	*Weinmannia crassifolia* vel. sp. aff. R. & P.	S	11	10	.
PTERIDOPHYTA	*Elaphoglossum* sp. 6	H	11	.	.
ROSACEAE	*Hesperomeles heterophylla* (R. & P.) Hook	S	9	10	.
MYRSINACEAE	*Cybianthus peruvianus* (A.DC.) Miq.	T	6	5	.
PTERIDOPHYTA	*Trichomanes diaphanum* H.B.K.	H	5	.	.
ERICACEAE	*Befaria aestuans* L.	S	4	8	.
ERICACEAE	*Sphyrospermum cordifolium* Benth.	Hem	3	.	.
POACEAE	*Chusquea* sp. 1	S	1	15	.
ERICACEAE	*Gaultheria* sp. 1	S	.	2	.
GENTIANACEAE	*Macrocarpaea* sp. 1	S	.	1	.
CLUSIACEAE	*Clusia weberbaueri* Engler	T	.	.	1
	Totals:		**62**	**52**	**1**

N. Transect 17. *Pajonal* (*Chusquea-Sphagnum* bog). Species abundance in all subsamples. Size criteria for Herbs and Shrubs subsamples as follows: Herbs, < 0.5 m tall; Shrubs, > 0.5 m tall and < 1.0 cm dbh. Abundance for Herbs and Shrubs subsamples is the number of subunits in which species was recorded. Abundance for remaining subsample is number of individuals. Herbs subsample 0.5 x 2 m (x 10; = 10 m^2); shrubs subsample 1 m x 2 m (x 15; = 30 m^2); 10-30 cm dbh 4 m x 150 m (= 600 m^2).

			ABUNDANCE IN SIZE CLASS		
FAMILY	GENUS, SPECIES	HABIT	HERBS	SHRUBS	10-30 cm DBH
PTERIDOPHYTA	*Elaphoglossum* sp. 6	H	10	.	.
PTERIDOPHYTA	*Blechnum auratum* (Fee) Tryon & Stolze	T	9	10	.
CUNONIACEAE	*Weinmannia crassifolia* vel. sp. aff. R. & P.	S	7	13	.
MYRSINACEAE	*Cybianthus peruvianus* (A.DC.) Miq.	T	6	8	.
ROSACEAE	*Hesperomeles heterophylla* (R. & P.) Hook	S	4	4	.
ERICACEAE	*Gaultheria* sp. 1	S	3	9	.
XYRIDACEAE	*Xyris* sp. 2	H	3	.	.
GENTIANACEAE	*Macrocarpaea* sp. 1	S	2	1	.
POACEAE	*Cortaderia* sp. 1	H	2	.	.
PTERIDOPHYTA	*Trichomanes diaphanum* H.B.K.	H	2	.	.
ERICACEAE	*Befaria aestuans* L.	S	1	2	.
PTERIDOPHYTA	*Trichomanes* sp. 1	H	1	.	.
XYRIDACEAE	*Xyris subulata* R. & P. var *breviscapa* Idobro &	H	1	.	.
POACEAE	*Chusquea* sp. 1	S	.	4	.
CLUSIACEAE	*Clusia weberbaueri* Engler	T	.	.	1
	Totals:		**51**	**51**	**1**

O. Transect 18. Dwarf *Clusia* forest. Abundance of species in Herbs subsample (see Appendices 2P-R for remaining subsamples from this transect). Herbs subsample 0.5 x 2 m (x 10; = 10 m^2).

FAMILY	GENUS, SPECIES	HABIT	ABUNDANCE
MELASTOMATACEAE	*Clidemia* sp. 1	S	5
PTERIDOPHYTA	*Elaphoglossum* sp. 7	E	4
ERICACEAE	*Vaccinium* sp. 1	S	3
ORCHIDACEAE	*Maxillaria augustae-victoriae* Lehm & Kranzl	E	3
ORCHIDACEAE	*Pleurothallis* sp. 3	E	3
PTERIDOPHYTA	Indet. sp. 3	E	3
PTERIDOPHYTA	*Blechnum binervatum* C.V. Morton & Lellinger	H	3
URTICACEAE	*Pilea* sp. 1	H	3
VITACEAE	*Cissus* sp. 1	L	3
BROMELIACEAE	*Tillandsia* sp. 3	E	2
DIOSCORIACEAE	*Dioscorea acanthogene* Rusby	L	2
ORCHIDACEAE	*Trichosalpinx/Lepanthes* sp. 1	E	2
OXALIDACEAE	*Oxalis* sp. 1	H	2
POACEAE	*Chusquea* sp. 4	S	2
PTERIDOPHYTA	*Elaphoglossum cuspidatum* (Willd.) Moore	E	2
PTERIDOPHYTA	*Grammitis* sp. 1	E	2
PTERIDOPHYTA	*Hymenophyllum* sp. 2	E	2
PTERIDOPHYTA	*Hymenophyllum plumieri* Hk. & Grev.	H	2
RUBIACEAE	*Manettia* sp. 2	L	2
CLUSIACEAE	*Clusia weberbaueri* Engler	T	1
MELASTOMATACEAE	Indet. sp. 6	T	1
ORCHIDACEAE	*Erythrodes* ? sp. 1	H	1
PASSIFLORACEAE	*Passiflora pascoensis* cf. L. Escobar	L	1
PTERIDOPHYTA	*Elaphoglossum raywaense* (Jenm.) Alston	E	1
RUBIACEAE	*Manettia* sp. 3	H	1
RUBIACEAE	*Nertera granadensis* (L.f.) Druce	H	1
URTICACEAE	*Pilea* sp. 2	H	1
URTICACEAE	*Pilea haenkei* Killip ex char	S	1
	Total:		**59**

P. Transect 18. Dwarf *Clusia* forest. Abundance of species in Shrubs subsample (see Appendices 2O and 2Q-R for remaining subsamples from this transect). Shrubs subsample 1 m x 2 m (x 10; = 20 m^2).

FAMILY	GENUS, SPECIES	HABIT	ABUNDANCE
POACEAE	*Chusquea* sp. 4	S	7
POACEAE	*Chusquea* sp. 6	S	5
ERICACEAE	*Vaccinium* sp. 1	S	4
PTERIDOPHYTA	*Blechnum* sp. 2	H	4
SOLANACEAE	*Lycianthes acutifolia* (R. & P.) Bitter	S	4
ERICACEAE	*Diogenesia octandra* cf. Sleumer	Hem	3
MELASTOMATACEAE	*Clidemia* sp. 1	S	3
VITACEAE	*Cissus* sp. 1	L	3
ASTERACEAE	*Mikania* sp. 4	L	2
ASTERACEAE	*Munnozia* sp. 3	L	2
ERICACEAE	*Cavendishia bracteata* cf. (R. & P. ex J. St.-Hill) Hoerold	Hem	2
PASSIFLORACEAE	*Passiflora pascoensis* cf. L. Escobar	L	2
URTICACEAE	*Pilea haenkei* Killip ex char	S	2
ASTERACEAE	*Jungia* sp. 1	L	1
ASTERACEAE	*Mikania* sp. 5	L	1
CHLORANTHACEA	*Hedyosmum* sp. 2	S	1
CLUSIACEAE	*Clusia weberbaueri* Engler	T	1
CYCLANTHACEAE	*Sphaeradenia* sp. 1	H	1
ERICACEAE	*Psammisia ulbrichiana* cf. Hoer.	Hem	1
EUPHORBIACEAE	*Alchornea* sp. 2	T	1
MELASTOMATACEAE	Indet. sp. 4	S	1
MELASTOMATACEAE	*Miconia* sp. 18	T	1
MYRSINACEAE	*Cybianthus peruvianus* (A.DC.) Miq.	T	1
ORCHIDACEAE	*Erythrodes* ? sp. 1	H	1
ORCHIDACEAE	*Sobralia/Elleanthus* sp. 2 (nc)	E	1
POACEAE	*Chusquea* sp. 1	S	1
PTERIDOPHYTA	*Eriosorus cheilanthoides* cf. (Sw.) A.F.Tryon	H	1
ROSACEAE	*Prunus pleiantha* Pilger	T	1
RUBIACEAE	*Manettia* sp. 2	L	1
RUBIACEAE	*Psychotria/Palicourea* sp. 7 (nc)	T	1
VALERIANACEAE	*Valeriana clematitis* Kunth	L	1
	Total:		**61**

Q. Transect 18. Dwarf *Clusia* forest. Abundance of species in size classes >1 cm dbh (see Appendices 2O-P and 2R for remaining subsamples from this transect). Counts are numbers of individuals. Subsamples 1-2.5 cm dbh 2 m x 100 m (= 200 m^2); 2.5-10 cm dbh 2 m x 100 m (= 200 m^2); 10-30 cm dbh 4 m x 100 m (= 400 m^2).

			INDIVIDUALS IN SIZE CLASS (DBH)		
FAMILY	**GENUS, SPECIES**	**HABIT**	**1-2.5 cm**	**2.5-10 cm**	**10-30 cm**
ERICACEAE	*Cavendishia bracteata* cf. (R. & P. ex J. St.-Hill)Hoerold	Hem	3	2	.
POACEAE	*Chusquea* sp. 6	S	3	2	.
POACEAE	*Chusquea* sp. 4	S	3	.	.
EUPHORBIACEAE	*Alchornea* sp. 2	T	2	.	.
MELASTOMATACEAE	*Miconia* sp. 18	T	2	.	.
ROSACEAE	*Prunus pleiantha* Pilger	T	2	.	.
CLUSIACEAE	*Clusia weberbaueri* Engler	T	1	6	6
MYRSINACEAE	*Cybianthus peruvianus* (A.DC.) Miq.	T	1	1	.
CAPRIFOLIACEAE	*Viburnum* sp. 1	T	1	.	.
MELASTOMATACEAE	Indet. sp. 6	T	1		.
POACEAE	*Chusquea* sp. 1	S	1	.	.
POACEAE	*Guadua* sp. 1	S	1	.	.
PTERIDOPHYTA	*Lycopodiella alopecuroides* (L.) Cranfill	Hem	1	.	.
RUBIACEAE	*Psychotria/Palicourea* sp. 2	T	1	.	.
RUBIACEAE	*Psychotria/Palicourea* sp. 4	T	1	.	.
PODOCARPACEAE	*Podocarpus oleifolius* D. Don	T	.	2	.
MYRTACEAE	Indet. sp. 1	T	.	2	.
PTERIDOPHYTA	*Cyathea* sp. 3 (nc)	T	.	2	.
SAPOTACEAE	Indet. sp. 1	T	.	2	.
LAURACEAE	*Nectandra* sp. 1	T	.	1	2
LAURACEAE	Indet. sp. 2	T	.	1	.
LORANTHACEAE	*Gaiadendron punctatum* (R. & P.) G. Don	S	.	1	.
MYRTACEAE	*Myrcia* sp. 5	T	.	1	.
PTERIDOPHYTA	*Blechnum binervatum* C.V. Morton & Lellinger	H	.	1	.
PTERIDOPHYTA	*Sphaeropteris elongata* (Hooker) Tryon	T	.	.	3
	Total individuals:		**24**	**24**	**11**

R. Transect 18. Dwarf *Clusia* forest. Abundance of species in Epiphytes subsample (see Appendices 2O-Q for remaining subsamples from this transect). Counts are total number of host trees (out of three) in which species was present.

FAMILY	GENUS, SPECIES	HABIT	TOTAL HOST TREES WITH SPECIES (out of three)
ERICACEAE	*Cavendishia bracteata* cf. (R. & P. ex J. St.-Hill) Hoerold	Hem	3
PTERIDOPHYTA	*Elaphoglossum cuspidatum* (Willd.) Moore	E	3
ERICACEAE	*Vaccinium* sp. 1	S	2
ORCHIDACEAE	*Elleanthus* sp. 1	E	2
ORCHIDACEAE	*Elleanthus* sp. 1 (nc)	E	2
PTERIDOPHYTA	*Elaphoglossum* sp. 7	E	2
PTERIDOPHYTA	*Grammitis* sp. 1	E	2
PTERIDOPHYTA	*Hymenophyllum verecundum* Morton	E	2
ASTERACEAE	*Munnozia* sp. 3	L	1
BROMELIACEAE	*Tillandsia* sp. 2	E	1
BROMELIACEAE	*Tillandsia* sp. 3	E	1
ERICACEAE	*Diogenesia octandra* cf. Sleumer	Hem	1
LORANTHACEAE	*Dendrophthora* sp. 1	E	1
LORANTHACEAE	*Dendrophthora* sp. 2	E	1
LORANTHACEAE	*Gaiadendron punctatum* (R. & P.) G. Don	S	1
MELASTOMATACEAE	Indet. sp. 4	S	1
MELASTOMATACEAE	Indet. sp. 6	T	1
ORCHIDACEAE	*Cryptocentrum* sp. 1	E	1
ORCHIDACEAE	*Maxillaria augustae-victoriae* Lehm & Kranzl	E	1
ORCHIDACEAE	*Maxillaria* sp. 5 (nc)	E	1
ORCHIDACEAE	*Pleurothallis* sp. 3	E	1
ORCHIDACEAE	*Pleurothallis* sp. 4	E	1
ORCHIDACEAE	*Stelis* sp. 5	E	1
ORCHIDACEAE	*Stelis* sp. 6	E	1
PTERIDOPHYTA	Indet. sp. 4	S	1
PTERIDOPHYTA	*Elaphoglossum* sp. 6	H	1
PTERIDOPHYTA	*Elaphoglossum paleaceum* cf. (Hook. & Grev.) Sledge	E	1
PTERIDOPHYTA	*Eriosorus cheilanthoides* cf. (Sw.) A.F.Tryon	H	1
PTERIDOPHYTA	*Hymenophyllum* sp. 2	E	1
PTERIDOPHYTA	*Pecluma* ? sp. 1	E	1
URTICACEAE	*Pilea* sp. 2	H	1
URTICACEAE	*Pilea haenkei* Killip ex char	S	1
VALERIANACEAE	*Valeriana clematitis* Kunth	L	1
	Total:		**43**

APPENDIX 4

Density and species richness, by subsample, of plant transects from the Northern Cordillera de Vilcabamba, Peru

Brad Boyle

With the exception of "Epiphytes", subsamples were based on size class: "Herbs", < 0.5 m tall; "Shrubs", > 0.5 m tall and < 1 cm dbh; other size classes as listed. "Epiphytes" subsample included all non-self-supporting individuals not included under the preceding categoties (see text for explanation). "Individuals" for subsamples marked with asterisk (*) are counts of species presence only in transect subunits. "# Individuals per 100 m²" is given only for subsamples where actual number of individuals was recorded. Note that transects 9 and 13 are incomplete: "Epiphytes" not sampled in transect 9; "Epiphytes", "Herbs" or "Shrubs" not sampled in transect 13.

Site Transect	Subsample	Total # Individuals	# Individuals per 100 m^2	Total # Species	# Species per 50 individuals	# Species per 25 individuals
Camp 1 (3300-3400 m)						
1	Herbs*	113	.	22	15	13
	Shrubs	3	3.0	1	.	.
2	Herbs*	96	.	20	17	11
	Shrubs	2	2.0	1	.	.
3	Herbs*	111	.	26	20	16
	Shrubs	7	7.0	2	.	.
4	Herbs*	114	.	27	21	17
	Shrubs	17	17.0	6	.	.
5	Herbs*	86	.	17	17	15
	Shrubs	2	2.0	1	.	.
6	Herbs*	87	.	24	20	16
	Shrubs	7	7.0	1	.	.
7	Herbs*	117	.	22	16	13
	Shrubs	11	11.0	2	.	.
8	Herbs*	106	.	28	23	17
	Shrubs	47	47.0	8	.	4
	1-10 cm dbh	4	4.0	2	.	.
9	Herbs*	66	.	28	22	16
	Shrubs*	37	.	20	.	15
	1-10 cm dbh	35	17.5	13	.	9
	10-30 cm dbh	16	4.0	5	.	.
	> 30 cm dbh	9	1.1	2	.	.
10	Herbs*	96	.	27	20	15
	Shrubs	1	1.0	1	.	.
11	Herbs*	163	.	39	26	18
	Shrubs*	19	.	8	.	.
	1-10 cm dbh	33	11.0	3	.	3
	10-30 cm dbh	20	3.3	2	.	.
	> 30 cm dbh	3	0.3	2	.	.
	Epiphytes*	47	.	19	.	13
12	Herbs*	105	.	33	24	16
	Shrubs*	51	.	20	20	14
	1-2.5 cm dbh	51	17.0	12	11	11
	2.5-10 cm dbh	50	16.6	15	15	13
	10-30 cm dbh	47	3.6	13	.	11
	> 30 cm dbh	50	2.2	10	10	8

Site Transect	Subsample	Total # Individuals	# Individuals per 100 m²	Total # Species	# Species per 50 individuals	# Species per 25 individuals
	Epiphytes*	79	.	34	21	15
13	1-2.5 cm dbh	53	75.7	23	23	14
	2.5-10 cm dbh	52	43.3	19	19	15
	10-30 cm dbh	27	11.2	9	.	9
	> 30 cm dbh	4	0.8	4	.	.
Camp 2 (2050 m)						
14	Herbs*	103	.	35	20	16
	Shrubs*	67	.	35	30	17
	1-2.5 cm dbh	50	31.2	29	29	17
	2.5-10 cm dbh	51	23.1	28	27	16
	10-30 cm dbh	52	9.3	23	22	17
	> 30 cm dbh	6	0.5	5	.	.
	Epiphytes*	101	.	57	39	25
15	Herbs*	102	.	45	28	17
	Shrubs*	103	.	54	36	24
	1-2.5 cm dbh	49	30.6	31	.	21
	2.5-10 cm dbh	60	21.4	30	24	14
	10-30 cm dbh	57	10.1	16	15	11
	> 30 cm dbh	22	1.7	11	.	.
	Epiphytes*	106	.	56	38	24
16	Herbs*	62	.	9	9	7
	Shrubs*	52	.	8	8	6
	10-30 cm dbh	1	0.1	1	.	.
17	Herbs*	51	.	13	13	11
	Shrubs*	51	.	8	8	7
	10-30 cm dbh	1	0.2	1	.	.
18	Herbs*	59	.	28	26	14
	Shrubs*	61	.	31	26	15
	1-2.5 cm dbh	27	13.5	17	.	17
	2.5-10 cm dbh	28	14.0	14	.	13
	10-30 cm dbh	12	3.0	4	.	.
	Epiphytes*	41	.	32	.	20

APPENDIX 5 Species richness, by habit category, of plant transects from the Northern Cordillera de Vilcabamba, Peru

Brad Boyle

All categories refer to growth habits and not to size classes (i.e., species from all subsamples (size classes) are included in each habit category). * = Incomplete. "Epiphytes" subsample omitted in transect 9. "Epiphytes", "Herbs" and "Shrubs" subsample omitted in transect 13 (i.e., sample includes only plants > 1 cm dbh).

Site Transect	Total # Species	Herb	Shrub	Tree	Epiphyte	Hemi-epiphyte	Liana
Camp 1 (3300 - 3400 m)							
1	22	19	3	0	0	0	0
2	20	16	4	0	0	0	0
3	27	22	5	0	0	0	0
4	29	18	10	1	0	0	0
5	17	12	5	0	0	0	0
6	24	19	5	0	0	0	0
7	22	14	7	1	0	0	0
8	32	15	14	2	1	0	0
9*	47	15	16	13	3	0	0
10	27	20	7	0	0	0	0
11	49	27	10	5	5	1	1
12	73	15	6	19	20	1	12
13*	30	1	6	21	0	1	1
Camp 2 (2050 m)							
14	129	13	10	44	35	13	14
15	148	28	16	53	22	14	15
16	12	2	6	3	0	1	0
17	15	6	6	3	0	0	0
18	80	13	12	19	23	4	

Species overlap among plant transects from the Northern Cordillera de Vilcabamba, Peru

APPENDIX 6

Brad Boyle

All subsamples combined. Upper diagonal: number of species shared between pairs of transects. Lower diagonal: proportion of species shared. Proportion shared was calculated using the Sorensen Index, $I_s = 2S_c / (S_1 + S_2)$, where S_1 = number of species in sample 1, S_2 = number of species in sample 2, and S_c = number of species common to both samples. I_s varies between 0 (no overlap) and 1 (total overlap). Note that transects 9 and 13 are incomplete (9: no Epiphytes subsample; 13: no Epiphytes, Herbs or Shrubs subsamples), thus, overlap with other transects should be intepreted with caution. Overlap values between Trees subsamples only for forest transects are given in Appendix 6. See Appendices 2 and 3 for transect habitats and subsample dimensions.

Transet	1	2	3	4	5	6	7	8	9	10	11	12	13	14	15	16	17	18
1		12	16	12	10	10	9	10	7	13	8	0	0	0	0	1	2	1
2	.57		14	14	15	12	10	10	7	14	10	1	0	0	0	0	0	0
3	.65	.60		16	11	12	10	15	9	18	15	1	0	0	0	1	2	1
4	.47	.57	.57		14	11	20	17	10	19	13	2	0	0	0	3	4	1
5	.51	.81	.50	.60		14	10	11	8	11	8	1	0	0	0	1	1	1
6	.43	.54	.47	.41	.68		7	15	10	12	15	1	0	0	0	1	1	2
7	.40	.47	.40	.78	.51	.30		11	5	15	6	0	0	0	0	2	3	0
8	.37	.38	.50	.55	.44	.53	.40		15	15	18	2	2	0	0	2	3	1
9	.20	.20	.24	.26	.25	.28	.14	.37		9	18	14	17	1	0	2	2	2
10	.53	.60	.66	.67	.50	.47	.61	.50	.24		13	1	0	0	0	1	2	1
11	.22	.28	.39	.33	.24	.41	.16	.44	.37	.34		8	2	2	2	3	2	3
12	0	.02	.02	.03	.02	.02	0	.03	.23	.02	.13		8	3	6	1	1	3
13	0	0	0	0	0	0	0	.06	.43	0	.05	.15		0	0	0	0	1
14	0	0	0	0	0	0	0	0	.01	0	.02	.03	0		29	3	2	25
15	0	0	0	0	0	0	0	0	0	0	.02	.05	0	.21		1	0	20
16	.06	0	.05	.14	.06	.05	.11	.09	.06	.05	.10	.02	0	.04	.01		11	4
17	.10	0	.10	.18	.06	.05	.16	.12	.06	.10	.06	.02	0	.02	0	.81		4
18	.02	0	.01	.01	.02	.03	0	.01	.03	.01	.04	.04	.01	.24	.18	.08	.08	

APPENDIX 7 Species overlap among forest transects from the Northern Cordillera de Vilcabamba, Peru

Brad Boyle

Size classes ≥ 1cm dbh only. Upper diagonal: number of species shared between pairs of transects. Lower diagonal: proportion of species shared. Proportion shared was calculated using the Sorensen Index, $I_s = 2S_c / (S_1 + S_2)$, where S_1 = number of species in sample 1, S_2 = number of species in sample 2, and S_c = number of species common to both samples. I_s varies between 0 (no overlap) and 1 (total overlap). Transects 9, 11, 12 and 13 are from Camp 1 at approximately 3300 m, transects 14 and 15 are from Camp 2 at 2050 m. See Appendices 2 and 3 for transect habitats and dimensions.

Transect	9	11	12	13	14	15	18	
9			1	3	12	0	0	0
11		0.12		0	1	0	0	0
12		0.16	0		7	0	1	0
13		0.53	0.06	0.26		0	0	1
14		0	0	0	0		9	9
15		0	0	0.02	0	0.15		2
18		0	0	0	0.03	0.22	0.04	

Plant transect data from RAP Camp Three (600 – 1400 m): The east central Cordillera de Vilcabamba, Department of Cusco, Peru

APPENDIX 8

Bruce K. Holst, Mónica Arakaki, and Hamilton Beltrán

Each transect sampled 50 trees greater than 10 cm dbh.

A. Transect 1. Medium-height ridge forest with numerous small diameter trees. 950 m elevation. Transect area: 51 x 10 m. Number of species in transect: 29. Number of individuals > 30 cm dbh: 2.

Family	Genus, species	# Individuals > 10 cm dbh	# Trees > 30 cm dbh
Myristicaceae	*Virola*	4	
Rubiaceae	*Faramea*	4	
Euphorbiaceae	*Hevea brasiliensis*	3	1
Moraceae		3	1
Moraceae	*Pseudolmedia* cf.	3	
Myrtaceae	*Myrcia fallax*	3	
Apocynaceae	*Aspidosperma*	2	
Caryocaraceae	*Caryocar*	2	
Lecythidaceae	*Eschweilera* cf. *coriacea*	2	
Meliaceae	*Trichilia*	2	
Myristicaceae	*Virola sebifera*	2	
Quiinaceae	*Quiina*	2	
?		1	
?		1	
?		1	
?		1	
Chrysobalanaceae		1	
Chrysobalanaceae		1	
Clusiaceae	*Vismia*	1	
Elaeocarpaceae	*Sloanea* sp. 1	1	
Elaeocarpaceae	*Sloanea* sp. 2	1	
Euporbiaceae		1	
Fabaceae		1	
Fabaceae-Mim.	*Pithecellobium/Zygia*	1	
Fabaceae-Pap.	*Ormosia* ?	1	
Meliaceae	*Guarea*	1	
Moraceae	*Brosimum*	1	
Myrisicaceae	*Iryanthera*	1	
Myrtaceae	*Myrcia splendens*	1	

B. Transect 2. Tall, dense forest on ridge belwo camp. 850 m elevation. Transect area: 3 x 10 m. Number of species in transect: 32. Number of individuals > 30 cm dbh: 3.

Family	Genus, species	# Individuals > 10 cm dbh	# Trees > 30 cm dbh
Moraceae	*Brosimum*	6	
Lecythidaceae	*Eschweilera* cf. *coriacea*	4	
Euphorbiaceae	*Hevea brasiliensis*	3	
Melastomataceae		3	
Burseraceae	*Protium*	2	2
Clusiaceae	*Tovomita*	2	
Fabaceae-Caes.	*Tachigali*	2	
Myristicaceae	*Virola sebifera*	2	
Myrtaceae	*Myrcia fallax*	2	
Rubiaceae	*Faramea*	2	
?	sp. 1	1	
?	sp. 2	1	
?	sp. 3	1	
?	sp. 4	1	
Anacardiaceae?		1	
Cecropiaceae	*Pourouma*	1	
Chrysobalanaceae		1	
Clusiaceae	*Calophyllum brasiliense*	1	
Clusiaceae	*Garcinia*	1	
Fabaceae-Mim.	*Inga*	1	
Fabaceae-Mim.	*Zygia*	1	
Lauraceae		1	
Melastomataceae	*Miconia* sp. 1	1	
Melastomataceae	*Miconia* sp. 2	1	
Meliaceae		1	
Meliaceae	*Trichilia* sp. 1	1	
Meliaceae	*Trichilia* sp. 2	1	
Moraceae		1	
Moraceae	*Ficus*	1	
Sapotaceae		1	
Sapotaceae	*Chrysophyllum*	1	
Sapotaceae	*Pouteria*	1	

C. Transect 3 – Wide, steep ridge with generally deep soil and little exposed rock. 1000 m elevation. Transect area: 59 x 10 m. Number of species in transect: 36. Number of individuals > 30 cm dbh:

Family	Genus, species	# Individuals > 10 cm dbh	# Trees > 30 cm dbh
Arecaceae	*Iriartea deltoidea*	4	
Bombacaceae	*Quararibea*	3	1
Meliaceae	*Guarea*	3	
Moraceae	*Poulsenia armata*	3	2
Myristicaceae	*Virola sebifera*	3	
Hippocrateaceae	*Salacia?*	2	
Lecythidaceae	*Eschweilera*	2	
Sapotaceae	*Pouteria*	2	
Anacardiaceae		1	
Araliaceae	*Dendropanax*	1	1
Bombacaceae	*Quararibea*	1	
Cecropiaceae	*Pourouma mollis*	1	
Chrysobalanaceae	*Hirtella*	1	
Clusiaceae	*Garcinia*	1	
Euphorbiaceae		1	
Euphorbiaceae	*Tetrorchidium*	1	1
Fabaceae-Mim.	*Zygia*	1	
Flacourtiaceae		1	
Lauraceae	*Pleurothyrium?*	1	
Meliaceae	*Guarea pterorhachis*	1	
Moraceae		1	
Moraceae	*Ficus* cf. *maroma*	1	
Moraceae	*Pseudolmedia laevigata*	1	
Myristicaceae	*Iryanthera*	1	
Myrsinaceae		1	
Ochnaceae	*Ouratea?*	1	
Olacaceae		1	1
Rosaceae	*Prunus*	1	
Rubiaceae	*Coussarea*	1	
Rubiaceae	*Pentagonia* sp. 1	1	
Rubiaceae	*Pentagonia* sp. 2	1	
Sapindaceae	*Allophyllus*	1	
Sapotaceae	*Pouteria* sp. 1	1	1
Sapotaceae	*Pouteria* sp. 2	1	
Sapotaceae	*Pouteria* sp. 3	1	
Sterculiaceae	*Pterygota amazonica*	1	1

D. Transect 4 – Rocky limestone ridge forest. 1000 m elevation. Transect area: 51 x 10 m. Number of species in transect: 23. Number of individuals > 30 cm dbh: 2.

Family	Genus, species	# Individuals > 10 cm dbh	# Trees > 30 cm dbh
Arecaceae	*Iriartea deltoidea*	12	
Meliaceae	*Guarea pterorhachis*	8	
Proteaceae	*Roupala montana*	4	1
Cecropiaceae	*Coussapoa* cf. *villosa*	3	
Elaeocarpaceae	*Sloanea*	2	1
Fabaceae-Pap.	*Swartzia* cf. *arborescens*	2	
Meliaceae	*Guarea*	2	
Myrtaceae	*Myrcia splendens*	2	
?		1	
Clusiaceae		1	
Euphorbiaceae	*Hevea brasiliensis*	1	
Flacourtiaceae		1	
Lauraceae		1	
Moraceae		1	
Moraceae	*Sorocea guilleminiana*	1	
Myrsinaceae		1	
Myrtaceae	*Myrcia fallax*	1	
Nyctaginaceae	*Neea*	1	
Olacaceae		1	
Rubiaceae		1	
Sapotaceae	*Pouteria* sp. 1	1	
Sapotaceae	*Pouteria* sp. 2	1	
Sterculiaceae	*Pterygota amazonica*	1	

E. Transect 5 – Steep, tall, open-understory slope forest adjacent to ridge camp. 1000 m elevation. Transect area: 88 x 10 m. Number of species in transect: 36. Number of individuals > 30 cm dbh: 7

Family	Genus, species	# Individuals > 10 cm dbh	# Trees > 30 cm dbh
Arecaceae	*Iriartea deltoide*	5	
Euphorbiaceae	*Hevea brasiliensis*	3	1
Bombacaceae	*Quararibea*	2	
Cecropiaceae	*Pourouma cecropiifolia*	2	
Fabaceae-Mim.	*Inga*	2	1
Flacourtiaceae	*Homalium*	2	
Moraceae	sp. 1	2	
Moraceae	sp. 2	2	
Moraceae	*Ficus*	2	1
Quiinaceae	*Quiina*	2	1
?	sp. 1	1	
?	sp. 2	1	
Araliaceae	*Dendropanax*	1	1
Boraginaceae	*Cordia*	1	
Caryocaraceae	*Anthodiscus*	1	1
Clusiaceae	*Garcinia*	1	
Elaeocarpaceae	*Sloanea*	1	
Euphorbiaceae	*Alchornea*	1	
Euphorbiaceae	*Hyeronima*	1	
Fabaceae		1	
Flacourtiaceae		1	
Flacourtiaceae	*Casearia*	1	
Hippocrateaceae		1	
Icacinaceae		1	
Lauraceae		1	
Moraceae	sp. 3	1	
Moraceae	*Castilla*	1	
Moraceae	*Naucleopsis*	1	
Moraceae	*Sorocea?*	1	
Myristicaceae		1	
Rosaceae	*Prunus*	1	1
Rubiaceae	sp. 1	1	
Rubiaceae	sp. 2	1	
Rubiaceae	sp. 3	1	
Simaroubaceae	*Picramnia*	1	
Sterculiaceae	*Pterygota amazonica*	1	

F. Transect 6 – Tall forest on broad undulating ridge with deep soils. 850 m elevation. Transect area: 64 x 10 m. Number of species in transect: 27. Number of individuals > 30 cm dbh: 13.

Family	Genus, species	# Individuals > 10 cm dbh	# Trees > 30 cm dbh
Euphorbiaceae	*Gavarretia terminalis*	4	4
Sapotaceae		4	
Myristicaceae	*Virola*	3	
Anacardiaceae	sp. 1	2	
Euphorbiaceae	*Hevea brasiliensis*	2	1
Icacinaceae		2	1
Myristicaceae	*Virola sebifera*	2	
Nyctaginaceae	*Neea*	2	
Sapotaceae	*Pouteria*	2	1
Vochysiaceae		2	1
?		1	
Anacardiaceae	sp. 2	1	1
Anacardiaceae	sp. 3	1	
Bombacaceae		1	1
Caryocaraceae	*Anthodiscus*	1	
Caryocaraceae	*Caryocar*	1	1
Cecropiaceae	*Pourouma minor*	1	1
Chrysobalanaceae	sp. 1	1	
Chrysobalanaceae	sp. 2	1	
Chrysobalanaceae	*Hirtella*	1	
Clusiaceae	*Garcinia*	1	
Clusiaceae	*Tovomita*	1	
Elaeocarpaceae	*Sloanea*	1	
Elaeocarpaceae	*Sloanea*	1	
Euphorbiaceae	sp. 1	1	
Euphorbiaceae?	sp. 2	1	1
Fabaceae-Mim.	*Inga*	1	
Moraceae		1	
Moraceae	*Brosimum lactescens*	1	
Moraceae	*Pseudolmedia* cf. *macrophylla*	1	
Myristicaceae	*Iryanthera*	1	
Myristicaceae	*Iryanthera?*	1	
Rubiaceae	*Cinchona?*	1	
Sapotaceae	*Micropholis*	1	
Simaroubaceae	*Simarouba amara*	1	

G. Transect 7 – Tall forest on reddish soils with a thick layer of organic material, on a gentle slope immediately above the steep descent to Tangoshiari. 900 m elevation. Transect area: 52 x 10 m. Number of species in transect: 29. Number of individuals > 30 cm dbh: 5.

Family	Genus, species	# Individuals > 10 cm dbh	# Trees > 30 cm dbh
Myristicaceae	*Virola sebifera*	6	
Chrysobalanaceae		4	
Euphorbiaceae	*Hevea brasiliensis*	4	
Elaeocarpaceae	*Sloanea*	3	2
Moraceae		3	
Moraceae	*Ficus*	2	
Sapotaceae	sp. 1	2	1
Sapotaceae	sp. 2	2	1
Sapotaceae	sp. 3	2	
Sapotaceae	sp. 4	2	
?		1	
Annonaceae	*Xylopia?*	1	
Burseraceae	*Protium*	1	
Cecropiaceae	*Pourouma*	1	
Chrysobalanaceae		1	
Euphorbiaceae	*Aparisthmium*	1	
Euphorbiaceae?		1	1
Fabaceae	sp. 1	1	
Fabaceae	sp. 2	1	
Fabaceae	sp. 3	1	
Icacinaceae	*Dendrobangia?*	1	
Lauraceae		1	
Lauraceae	*Nectandra?*	1	
Myristicaceae	*Iryanthera*	1	
Myristicaceae	*Virola*	1	
Myrtaceae	*Myrciaria floribunda*	1	
Nyctaginaceae		1	
Nyctaginaceae	*Neea*	1	
Sapotaceae	sp. 5	1	
Sapotaceae	sp. 6	1	

H. Transect 8 – Tall forest on reddish soils in gently sloping, protected valley. 900 m elevation. Transect area: 84 x 10 m. Number of species in transect: 34. Number of individuals > 30 cm dbh: 6.

Family	Genus, species	# Individuals > 10 cm dbh	# Trees > 30 cm dbh
Clusiaceae	*Tovomita*	3	
Euphorbiaceae	*Hevea brasiliensis*	3	
Myristicaceae	*Virola sebifera*	3	
Sapotaceae	sp. 1	3	
Chrysobalanaceae		2	1
Clusiaceae	*Calophyllum brasiliense*	2	1
Euphorbiaceae	*Alchorneopsis floribunda*	2	1
Fabaceae	sp. 1	2	
Nyctaginaceae	*Neea*	2	
Rubiaceae	sp. 1	2	
Rubiaceae	sp. 2	2	
Sapotaceae	sp. 2	2	
Anacardiaceae		1	
Burseraceae	*Protium*	1	
Burseraceae?		1	
Caryocaraceae	*Caryocar*	1	
Clusiaceae	*Garcinia*	1	
Euphorbiaceae		1	
Euphorbiaceae	*Conceveiba guianensis*	1	
Euphorbiaceae	*Pera*	1	
Fabaceae-Caes.	*Cedrelinga cateniformis*	1	
Fabaceae-Mim.		1	
Fabaceae-Pap.		1	
Lauraceae		1	
Lauraceae	*Nectandra?*	1	
Lecythidaceae	*Couratari guianensis*	1	1
Moraceae		1	1
Ochnaceae	*Cespedezia spathulata*	1	1
Rubiaceae	*Pagamea*	1	
Sapotaceae	sp. 3	1	
Sapotaceae	sp. 4	1	
Sapotaceae	sp. 5	1	
Sapotaceae	*Chrysophyllum*	1	
Verbenaceae	*Vitex*	1	

I. Transect 9 – Montane forest on broad, gently rolling summit of limestone hill. 1350 m elevation. Transect area: 51 x 10 m. Number of species in transect: 32. Number of individuals > 30 cm dbh: 11.

Family	Genus, species	# Individuals > 10 cm dbh	# Trees > 30 cm dbh
Rubiaceae	*Bathysa obovata*	5	
Moraceae	*Perebea tessmannii*	4	
Burseraceae	*Protium*	3	1
Cecropiaceae	*Pourouma* cf. *tomentosa*	3	2
Euphorbiaceae	*Hyeronima* sp. 1	3	
Pteridophyte		3	
Araliaceae	*Schefflera morototoni*	2	1
Rubiaceae	*Faramea*	2	
Sapotaceae	*Micropholis*	2	1
?		1	1
Apocynaceae	*Lacmellea*	1	
Arecaceae	*Iriartea deltoidea*	1	1
Burseraceae	*Protium*	1	
Elaeocarpaceae	*Sloanea*	1	
Euphorbiaceae	*Hyeronima* sp. 2	1	
Fabaceae		1	
Icacinaceae?		1	1
Lauraceae		1	
Melastomataceae	*Miconia?*	1	1
Meliaceae	*Guarea*	1	
Meliaceae	*Trichilia*	1	
Monimiaceae	*Mollinedia*	1	
Moraceae	*Pseudolmedia laevis*	1	
Myristicaceae	*Iryanthera*	1	
Myrtaceae	*Myrcia* cf. *sylvatica*	1	1
Myrtaceae	*Myrcia fallax*	1	
Quiinaceae		1	
Rosaceae	*Prunus*	1	
Rubiaceae		1	
Sapindaceae		1	
Sapotaceae		1	1
Simaroubaceae	*Simarouba amara*	1	

J. Transect 10 – Tall, slightly disturbed forest on undulating hills. 650 m elevation.
Transect area: 79 x 10 m. Number of species in transect: 37. Number of individuals > 30 cm dbh: 13.

Family	Genus, species	# Individuals > 10 cm dbh	# Trees > 30 cm dbh
Arecaceae	*Iriartea deltoidea*	4	
Arecaceae	*Socratea salazarii*	2	
Cecropiaceae	*Pourouma minor*	2	2
Clusiaceae	*Chrysochlamys*	2	
Euphorbiaceae	*Glycydendron amazonicum*	2	
Flacourtiaceae	*Laetia?*	2	1
Meliaceae	*Guarea*	2	
Myristicaceae	*Virola sebifera*	2	
Myrtaceae	*Myrcia fallax*	2	
Rubiaceae	sp. 1	2	2
Rubiaceae	sp. 2	2	
?		1	1
Anacardiaceae		1	1
Annonaceae		1	
Annonaceae	*Guatteria*	1	
Apocynaceae	*Himatanthus*	1	1
Chrysobalanaceae	*Hirtella*	1	1
Fabaceae-Caes.	*Cedrelinga cateniformis*	1	1
Fabaceae-Mim.	*Inga*	1	
Lauraceae	sp. 1	1	1
Lauraceae	sp. 2	1	
Monimiaceae	*Mollinedia*	1	
Moraceae	sp. 1	1	1
Moraceae	sp. 2	1	1
Moraceae	sp. 3	1	
Moraceae	sp. 4	1	
Moraceae	*Naucleopsis*	1	
Moraceae	*Perebea guianensis*	1	
Moraceae	*Sorocea*	1	
Nyctaginaceae	*Neea*	1	
Olacaceae	*Minquartia guianensis*	1	
Rubiaceae	sp. 3	1	
Rubiaceae	sp. 4	1	
Sapotaceae		1	
Sapotaceae	*Chrysophyllum*	1	
Simaroubaceae	*Simarouba amara*	1	
Violaceae	*Leonia glycycarpa*	1	

Plant species found in Whittaker Plots at Llactahuaman (1710 m), Southern Cordillera de Vilcabamba, Peru

APPENDIX 9

Percy Nuñez, William Nauray, Rafael de la Colina and Severo Baldeon

FAMILY	GENUS AND SPECIES
PTERIDOPHYTA	*Adiantum poeppigii*
PTERIDOPHYTA	*Campyloneurum* sp.1
PTERIDOPHYTA	*Cyathea pubescens*
PTERIDOPHYTA	*Elaphoglossum* sp.1
PTERIDOPHYTA	*Grammitis* sp.1
PTERIDOPHYTA	*Hymenophyllum* sp.1
PTERIDOPHYTA	Undet. sp.1
PTERIDOPHYTA	Undet. sp.2
PTERIDOPHYTA	Undet. sp.3
PTERIDOPHYTA	Undet. sp.4
PTERIDOPHYTA	Undet. sp.5
PTERIDOPHYTA	*Nephrolepis* sp.1
PTERIDOPHYTA	*Polypodium* sp.1
PTERIDOPHYTA	*Pteridium* sp.1
PTERIDOPHYTA	*Pteridium aquilinum*
PTERIDOPHYTA	*Pteris* sp.1
PTERIDOPHYTA	*Thelypteris* sp.1
ACANTHACEAE	Undet. sp.1
ANACARDIACEAE	Undet. sp.1
ANNONACEAE	*Anaxagorea* sp.1
ANNONACEAE	*Guatteria* sp1
APIACEAE	Undet. sp.1
APOCYNACEAE	*Aspidosperma* sp.1
ARACEAE	*Anthurium* sp.1
ARACEAE	*Caladium* sp.1
ARACEAE	Undet. sp.1
ARACEAE	Undet. sp.2
ARACEAE	*Philodendron* sp.1
ARECACEAE	*Dictyocaryum lamarckianum*
ARECACEAE	*Euterpe* sp.1
ARECACEAE	*Euterpe precatoria*
ARECACEAE	*Wettinia* sp.1
ASTERACEAE	Undet. sp.1
BIGNONIACEAE	Undet. sp.1
BIGNONIACEAE	Undet. sp.2

Plant species found in Whittaker Plots at Llactahuaman (1710 m)

FAMILY	GENUS AND SPECIES
BROMELIACEAE	Undet. sp.1
BROMELIACEAE	*Tillandsia* sp.1
BURSERACEAE	Undet. sp.1
BURSERACEAE	*Trattinnickia* sp.1
CECROPIACEAE	*Pourouma* sp.1
CECROPIACEAE	*Pourouma* minor
CHLORANTHACEAE	*Hedyosmum scabrum*
CLETHRACEAE	*Clethra* sp.1
CLUSIACEAE	*Clusia trochiformis*
CLUSIACEAE	*Quapoya* sp.1
COLUMELLIACEAE	*Columellia* sp.1
COMMELINACEAE	Undet. sp.1
COMMELINACEAE	Undet. sp.2
CUCURBITACEAE	Undet. sp.1
CUNONIACEAE	*Weinmannia* sp.1
CUNONIACEAE	*Weinmannia* sp.2
DIOSCOREACEAE	*Dioscorea* sp.1
ERICACEAE	Undet. sp.1
ERICACEAE	*Psammisia* sp.1
EUPHORBIACEAE	*Alchornea* sp.1
EUPHORBIACEAE	Undet. sp.1
EUPHORBIACEAE	*Mabea* sp.1
EUPHORBIACEAE	*Sapium* sp.1
FLACOURTIACEAE	Undet. sp.1
GESNERIACEAE	Undet. sp.1
GESNERIACEAE	Undet. sp.2
LAURACEAE	*Nectandra* sp.1
LAURACEAE	*Ocotea* sp.1
LAURACEAE	*Ocotea* sp.2
LAURACEAE	*Pleurothyrium* sp.1
LEGUMINOSAE	Undet. sp.1
LEGUMINOSAE	*Platymiscium* sp.1
MELASTOMATACEAE	*Blakea* sp.1
MELASTOMATACEAE	*Clidemia* sp.1
MELASTOMATACEAE	Undet. sp.1
MELASTOMATACEAE	*Miconia* sp.1
MELASTOMATACEAE	*Miconia* sp.2
MELASTOMATACEAE	*Miconia* sp.3
MELASTOMATACEAE	*Miconia paleacea*
MELASTOMATACEAE	*Mouriri* sp.1
MELIACEAE	Undet. sp.1
MONIMIACEA	*Siparuna* sp.1
MORACEAE	*Brosimum* sp.1
MORACEAE	*Ficus* sp.1
MORACEAE	*Helianthostylis* sp.1
MORACEAE	Undet. sp.1
MORACEAE	*Pseudolmedia* sp.1
MYRISTICACEAE	*Iryanthera* sp.1
MYRSINACEAE	Undet. sp.1
MYRSINACEAE	*Myrsine* sp.1
MYRTACEAE	*Calyptranthes* sp.1
MYRTACEAE	*Eugenia* sp.1
MYRTACEAE	*Myrcia* sp.1
MYRTACEAE	*Myrciaria* sp.2

FAMILY	GENUS AND SPECIES
ORCHIDACEAE	*Dichaea muricata*
ORCHIDACEAE	*Elleanthus* sp.1
ORCHIDACEAE	*Elleanthus bambusaceus*
ORCHIDACEAE	*Epidendrum* sp.1
ORCHIDACEAE	*Erythrodes* sp.1
ORCHIDACEAE	Undet. sp.1
ORCHIDACEAE	Undet. sp.2
ORCHIDACEAE	*Maxillaria graminifolia*
ORCHIDACEAE	*Odontoglossum* sp.1
ORCHIDACEAE	*Pleurothallis* sp.1
ORCHIDACEAE	*Stelis* sp.1
ORCHIDACEAE	*Trichosalpinx* sp.1
OXALIDACEAE	*Oxalis* sp.1
PIPERACEAE	*Peperomia* sp.1
PIPERACEAE	*Piper* sp.1
POACEAE	*Chusquea* sp.1
POACEAE	*Olyra* sp.1
POACEAE	*Olyra latifolia*
PODOCARPACEAE	*Podocarpus* sp.1
RUBIACEAE	*Cinchona* sp.1
RUBIACEAE	*Psychotria* sp.1
RUBIACEAE	*Psychotria* sp.2
SAPINDACEAE	Undet. sp.1
SAPINDACEAE	*Serjania* sp.1
SAPOTACEAE	*Chrysophyllum* sp.1
SAPOTACEAE	Undet. sp.1
SAPOTACEAE	*Micropholis* sp.1
TILIACEAE	*Apeiba* sp.1
ULMACEAE	*Celtis schippii*
URTICACEAE	*Parietaria* sp.1
VITACEAE	*Cissus* sp.1
VITACEAE	Undet. sp.1

APPENDIX 10 Additional Plant Species recorded from Llactahuaman (1710 m), Southern Cordillera de Vilcabamba, Peru

Percy Nuñez, William Nauray, Rafael de la Colina and Severo Baldeon

FAMILY	GENUS AND SPECIES
MARCHANTIACEAE	*Marchantia* sp.1
FERN ALLIES	*Lycopodium* sp.1
FERN ALLIES	*Lycopodium cernuum*
FERN ALLIES	*Musci* sp.1
FERN ALLIES	*Polytrichum* sp.1
FERN ALLIES	*Sphagnum* sp.1
PTERIDOPHYTA	*Cyathea* sp.1
PTERIDOPHYTA	*Elaphoglossum* sp.2
PTERIDOPHYTA	*Lomariopsis* sp.1
PTERIDOPHYTA	*Polypodium* sp.2
PTERIDOPHYTA	*Polypodium* sp.3
PTERIDOPHYTA	*Polypodium* sp.4
ACANTHACEAE	*Drymonia* sp.1
ACANTHACEAE	*Justicia* sp.1
ACANTHACEAE	*Ruellia* sp.1
ACANTHACEAE	*Sanchezia* sp.1
AMARANTHACEAE	*Iresine* sp.1
APIACEAE	*Hydrocotyle* sp.1
ARACEAE	*Anthurium* sp.2
ARACEAE	*Anthurium* sp.3
ARACEAE	*Anthurium* sp.4
ARACEAE	*Philodendron* sp.2
ARACEAE	*Philodendron* sp.3
ARALIACEAE	*Schefflera* sp.1
ARALIACEAE	*Schefflera* sp.2
ARALIACEAE	*Schefflera* sp.3
ARECACEAE	*Geonoma* sp.1
ASTERACEAE	*Baccharis* sp.1
ASTERACEAE	*Baccharis* sp.2
ASTERACEAE	*Erato* sp.1
ASTERACEAE	*Mikania* sp.1
ASTERACEAE	*Munnozia* sp.1
ASTERACEAE	*Munnozia* sp.2
ASTERACEAE	*Vernonia* sp.1
ASTERACEAE	*Vernonia* sp.2
BEGONIACEAE	*Begonia* sp.1
BEGONIACEAE	*Begonia* sp.2
BROMELIACEAE	*Puya* sp.1
BROMELIACEAE	*Tillandsia* sp.2
BROMELIACEAE	*Tillandsia* sp.3
BROMELIACEAE	*Tillandsia* sp.4
CAMPANULACEAE	*Centropogon* sp.1
CAMPANULACEAE	*Centropogon* sp.2
CAMPANULACEAE	*Centropogon* sp.3
CAMPANULACEAE	*Centropogon* sp.4
CECROPIACEAE	*Coussapoa* sp.1
CHLORANTHACEAE	*Hedyosmum* sp.1
CHLORANTHACEAE	*Hedyosmum* sp.2
CHLORANTHACEAE	*Hedyosmun* sp.3
CHLORANTHACEAE	*Hedyosmun* sp.4

FAMILY	GENUS AND SPECIES
CLUSIACEAE	*Clusia* sp.1
CLUSIACEAE	*Clusia* sp.2
CLUSIACEAE	*Clusia* sp.3
CLUSIACEAE	*Clusia* sp.4
CLUSIACEAE	*Clusia* sp.5
CLUSIACEAE	*Clusia* sp.6
CLUSIACEAE	*Clusia* sp.7
CUNONIACEAE	*Weinmannia* sp.2
CUNONIACEAE	*Weinmannia* sp.3
CUNONIACEAE	*Weinmannia* sp.4
DIOSCOREACEAE	*Dioscorea* sp.2
ELAEOCARPACEAE	*Vallea* sp.1
ERICACEAE	*Demosthenesia* sp.1
ERICACEAE	*Psammisia* sp.2
ERICACEAE	*Sphyrospermum* sp.1
EUPHORBIACEAE	*Acalypha* sp.1
EUPHORBIACEAE	*Hyeronima* sp.1
EUPHORBIACEAE	*Hyeronima* sp.2
EUPHORBIACEAE	*Mabea* sp.2
EUPHORBIACEAE	*Mabea* sp.3
GESNERIACEAE	*Besleria* sp.1
GESNERIACEAE	*Gloxinia* sp.1
HELICONIACEAE	*Heliconia* sp.1
LAURACEAE	*Nectandra* sp.2
LAURACEAE	*Pleurothyrium* sp.2
LAURACEAE	*Pleurothyrium* sp.3
LEGUMINOSAE	*Inga* sp.1
LORANTHACEAE	*Gaiadendron* sp.1
LORANTHACEAE	*Phoradendron* sp.1
MARCGRAVIACEAE	*Marcgravia* sp.1
MARCGRAVIACEAE	*Sarcopera* sp.1
MARCGRAVIACEAE	*Sarcopera* sp.2
MELASTOMATACEAE	*Brachyotum* sp.1
MELASTOMATACEAE	*Brachyoum* sp.2
MELASTOMATACEAE	*Miconia* sp.4
MELASTOMATACEAE	*Miconia* sp.5
MELASTOMATACEAE	*Miconia* sp.6
MELASTOMATACEAE	*Miconia* sp.7
MELASTOMATACEAE	*Miconia* sp.8
MELASTOMATACEAE	*Miconia* sp.9
MELASTOMATACEAE	*Miconia* sp.10
MELASTOMATACEAE	*Miconia* sp.11
MELASTOMATACEAE	*Miconia* sp.12
MELASTOMATACEAE	*Miconia* sp.13
MORACEAE	*Castilla* sp.1
MORACEAE	*Helianthostylis* sp.1
MORACEAE	*Helycostylis* sp.1
MORACEAE	*Perebea* sp.1
MYRTACEAE	*Myrciaria* sp.3
ONAGRACEAE	*Fuchsia* sp.1
ORCHIDACEAE	*Dichaea* sp.1
ORCHIDACEAE	*Elleanthus* sp.2
ORCHIDACEAE	*Elleanthus* sp.3
ORCHIDACEAE	*Epidendrum* sp.2
ORCHIDACEAE	*Maxillaria* sp.1
ORCHIDACEAE	*Scaphyglottis* sp.1
OXALIDACEAE	*Oxalis* sp.2
OXALIDACEAE	*Oxalis* sp.3
PASSIFLORACEAE	*Passiflora* sp.1

Additional plant species recorded from Llactahuaman (1710 m)

FAMILY	GENUS AND SPECIES
PIPERACEAE	*Peperomia* sp.2
PIPERACEAE	*Peperomia* sp.3
PIPERACEAE	*Peperomia* sp.4
POACEAE	*Chusquea* sp.2
POACEAE	*Chusquea* sp.3
POACEAE	*Guadua* sp.1
PODOCARPACEAE	*Podocarpus* sp.2
PODOCARPACEAE	*Podocarpus* sp.3
POLYGALACEAE	*Monnina* sp.1
POLYGALACEAE	*Monnina* sp.2
ROSACEAE	*Oreocallis* sp.1
RUBIACEAE	*Faramea* sp.1
RUBIACEAE	*Palicourea* sp.2
RUBIACEAE	*Remijia* sp.1
SAPOTACEAE	*Manilkara* sp.1
SMILACACEAE	*Smilax* sp.1
SMILACACEAE	*Smilax* sp.2
SMILACACEAE	*Smilax* sp.3
SMILACACEAE	*Smilax* sp.4
SOLANACEAE	*Solanum* sp.1
SOLANACEAE	*Solanum* sp.2
SYMPLOCACEAE	*Symplocos* sp.1
SYMPLOCACEAE	*Symplocos* sp.2
THEACEAE	*Goronia* sp.1
VOCHYSIACEAE	*Vochysia* sp.1

Plant species found in Whittaker Plots at Wayrapata (2445 m). Southern Cordillera de Vilcabamba, Peru

APPENDIX 11

Percy Nuñez, William Nauray, Rafael de la Colina and Severo Baldeon

FAMILY	GENUS AND SPECIES
MARCHANTIACEAE	*Marchantia polimorfica*
FERN ALLIES	*Cora pavonia*
FERN ALLIES	*Huperzia saururus*
PTERIDOPHYTA	*Adiantum poeppigii*
PTERIDOPHYTA	*Adiantum* sp.1
PTERIDOPHYTA	*Asplenium auritum*
PTERIDOPHYTA	*Blechnum* sp.1
PTERIDOPHYTA	*Campyloneurum* sp.1
PTERIDOPHYTA	*Cyathea* sp.1
PTERIDOPHYTA	*Danaea* sp.1
PTERIDOPHYTA	*Elaphoglossum* sp.1
PTERIDOPHYTA	*Grammitis* sp.1
PTERIDOPHYTA	Undet. sp.1
PTERIDOPHYTA	Undet. sp.2
PTERIDOPHYTA	Undet. sp.3
PTERIDOPHYTA	*Polypodium* sp.1
PTERIDOPHYTA	*Polypodium* sp.2
PTERIDOPHYTA	*Pteridium aquilinum*
PTERIDOPHYTA	*Pteris* sp.1
PTERIDOPHYTA	*Thelypteris* sp.1
AMARYLLIDACEAE	*Bomarea* sp.1
AMARYLLIDACEAE	*Bomarea* sp.2
ANNONACEAE	*Guatteria* sp.1
AQUIFOLIACEAE	*Ilex* sp.1
ARACEAE	*Anthurium huanucense*
ARACEAE	*Anthurium* sp.1
ARACEAE	*Monstera* sp.1
ARALIACEAE	*Schefflera* sp.1
ARALIACEAE	*Schefflera* sp.2
ARECACEAE	*Ceroxylon* sp.1
ARECACEAE	*Chamaedorea* sp.1
ASTERACEAE	*Pentacalia senecionidis*
ASTERACEAE	*Pentacalia* sp.1
ASTERACEAE	*Senecio* sp.1
ASTERACEAE	*Vernonia* sp.1
BROMELIACEAE	*Guzmania seemannii*
BROMELIACEAE	*Guzmania* sp.1
BROMELIACEAE	*Pitcairnia* sp.1
BROMELIACEAE	*Tillandsia* sp.1
BROMELIACEAE	*Tillandsia* sp.2
CAMPANULACEAE	*Centropogon* sp.1
CHLORANTHACEAE	*Hedyosmum* sp.1
CLUSIACEAE	*Clusia* sp.2
CLUSIACEAE	*Clusia trochiformis*
CUNONIACEAE	*Weinmannia* sp.1
CUNONIACEAE	*Weinmannia* sp.2
CUNONIACEAE	*Weinmannia* sp.3
CUNONIACEAE	*Weinmannia* sp.4
CUNONIACEAE	*Weinmannia* sp.5
CUNONIACEAE	*Weinmannia* sp.6
CUNONIACEAE	*Weinmannia* sp.7

Plant species found in the Whittaker Plots at Wayrapata (2445 m)

FAMILY	GENUS AND SPECIES
CYCLANTHACEAE	*Asplundia* sp.1
CYCLANTHACEAE	*Cyclanthus* sp.1
DIOSCOREACEAE	*Dioscorea* sp.1
ERICACEAE	*Bejaria aestuans*
ERICACEAE	*Cavendishia* sp.1
ERICACEAE	*Demosthenesia* sp.1
ERICACEAE	*Gaultheria bracteata*
ERICACEAE	*Gaultheria* sp.1
ERICACEAE	*Psammisia* sp.1
ERICACEAE	*Sphyrospermum* sp.1
ERICACEAE	*Vaccinium floribundum*
EUPHORBIACEAE	*Hyeronima oblonga*
FLACOURTIACEAE	*Prockia* sp.1
GENTIANACEAE	*Irlbachia* sp.1
GESNERIACEAE	*Alloplectus* sp.1
GESNERIACEAE	*Columnea* sp.1
LAURACEAE	*Nectandra* sp.1
LAURACEAE	*Persea* sp.1
LORANTHACEAE	*Dendrophthora* sp.1
LORANTHACEAE	*Phoradendron* sp.1
MELASTOMATACEAE	*Brachyotum* sp.1
MELASTOMATACEAE	*Miconia paleacea*
MELASTOMATACEAE	*Miconia* sp.1
MELASTOMATACEAE	*Miconia* sp.10
MELASTOMATACEAE	*Miconia* sp.2
MELASTOMATACEAE	*Miconia* sp.3
MELASTOMATACEAE	*Miconia* sp.4
MELASTOMATACEAE	*Miconia* sp.6
MELASTOMATACEAE	*Miconia* sp.7
MELASTOMATACEAE	*Miconia* sp.8
MELASTOMATACEAE	*Miconia* sp.9
MYRSINACEAE	*Cybianthus* sp.1
MYRSINACEAE	*Myrsine* sp.5
MYRTACEAE	*Myrcia* sp.1
MYRTACEAE	*Myrciaria* sp.1
MYRTACEAE	*Myrteola* sp.1
ORCHIDACEAE	*Brachionidium* sp.1
ORCHIDACEAE	*Elleanthus aurantiacus*
ORCHIDACEAE	*Elleanthus bambusaceus*
ORCHIDACEAE	*Elleanthus* sp.1
ORCHIDACEAE	*Epidendrum gramineum*
ORCHIDACEAE	*Epidendrum* sp.1
ORCHIDACEAE	*Epistephium amplexicaule*
ORCHIDACEAE	*Lepanthes* sp.1
ORCHIDACEAE	*Lepanthes* sp.2
ORCHIDACEAE	*Lepanthes* sp.3
ORCHIDACEAE	*Maxillaria alpestris*
ORCHIDACEAE	*Maxillaria aurea*
ORCHIDACEAE	*Maxillaria floribunda*
ORCHIDACEAE	*Maxillaria quitensis*
ORCHIDACEAE	*Maxillaria* sp.1
ORCHIDACEAE	*Maxillaria* sp.2
ORCHIDACEAE	*Maxillaria* sp.3
ORCHIDACEAE	*Maxillaria* sp.4
ORCHIDACEAE	*Pleurothallis* sp.1
ORCHIDACEAE	*Sauroglossum* sp.1
ORCHIDACEAE	*Stelis boliviensis*
ORCHIDACEAE	*Stelis* sp.1

FAMILY	GENUS AND SPECIES
OXALIDACEAE	*Oxalis sp.1*
PIPERACEAE	*Peperomia* sp.1
POACEAE	*Arundinella* sp.1
POACEAE	*Aulonemia* sp.1
POACEAE	*Chusquea scandens*
POACEAE	*Olyra* sp.1
ROSACEAE	*Hesperomeles* sp.1
ROSACEAE	*Hesperomeles* sp.2
ROSACEAE	*Prunus* sp.1
RUBIACEAE	*Cinchona* sp.1
RUBIACEAE	*Psychotria* sp.1
RUBIACEAE	*Psychotria* sp.2
SABIACEAE	*Meliosma peytonii*
SMILACACEAE	*Smilax* sp.1
SYMPLOCACEAE	*Symplocos* sp.1
URTICACEAE	*Pilea* sp.1
ZINGIBERACEAE	*Renealmia* sp.1

APPENDIX 12

Additional Plant Species recorded from Wayrapata (2445 m), Southern Cordillera de Vilcabamba, Peru

Percy Nuñez, William Nauray, Rafael de la Colina and Severo Baldeon

FAMILY	GENUS AND SPECIES
MARCHANTIACEAE	*Marchantia* sp.1
FERN ALLIES	*Octoblepharum* sp.1
FERN ALLIES	*Sphagnum* sp.1
FERN ALLIES	*Sphagnum* sp.2
PTERIDOPHYTA	*Asplenium* sp.1
PTERIDOPHYTA	*Blechnum* sp.2
PTERIDOPHYTA	*Cyathea* sp.2
PTERIDOPHYTA	*Elaphoglossum* sp.2
PTERIDOPHYTA	*Eriosorus* sp.1
PTERIDOPHYTA	*Grammitis* sp.2
PTERIDOPHYTA	*Grammitis* sp.3
PTERIDOPHYTA	*Grammitis* sp.4
PTERIDOPHYTA	*Grammitis* sp.5
PTERIDOPHYTA	*Grammitis* sp.6
PTERIDOPHYTA	*Grammitis* sp.7
PTERIDOPHYTA	*Huperzia* sp.1
PTERIDOPHYTA	*Huperzia* sp.2
PTERIDOPHYTA	*Hymenophyllum* sp.1
PTERIDOPHYTA	*Hymenophyllum* sp.2
PTERIDOPHYTA	*Hymenophyllum* sp.3
PTERIDOPHYTA	*Hymenophyllum* sp.4
PTERIDOPHYTA	*Polypodium* sp.3
PTERIDOPHYTA	*Polypodium* sp.4
PTERIDOPHYTA	*Polypodium* sp.5
PTERIDOPHYTA	*Polypodium* sp.6
PTERIDOPHYTA	*Polypodium* sp.7
PTERIDOPHYTA	*Polypodium* sp.8
PTERIDOPHYTA	*Polypodium* sp.9
PTERIDOPHYTA	*Pteris* sp.2
PTERIDOPHYTA	*Tectaria* sp.1
ACTINIDACEAE	*Saurauia natalicia*
AMARYLLIDACEAE	*Bomarea* sp.3
AMARYLLIDACEAE	*Bomarea* sp.4
AMARYLLIDACEAE	*Bomarea* sp.5
AMARYLLIDACEAE	*Bomarea* sp.6
ANNONACEAE	*Guatteria* sp.2
ANNONACEAE	*Guatteria* sp.3
AQUIFOLIACEAE	*Ilex* sp.2
AQUIFOLIACEAE	*Ilex* sp.3
AQUIFOLIACEAE	*Ilex* sp.4
AQUIFOLIACEAE	*Ilex* sp.5
AQUIFOLIACEAE	*Ilex* sp.6
AQUIFOLIACEAE	*Ilex* sp.7
ARACEAE	*Anthurium* sp.2
ARACEAE	*Anthurium* sp.3
ARACEAE	*Anthurium* sp.4
ARACEAE	*Anthurium* sp.5
ARACEAE	*Anthurium* sp.6
ARACEAE	*Philodendron* sp.1

FAMILY	GENUS AND SPECIES
ARALIACEAE	*Oreopanax* sp.1
ARALIACEAE	*Oreopanax* sp.2
ARALIACEAE	*Schefflera* sp.3
ARALIACEAE	*Schefflera* sp.4
ARALIACEAE	*Schefflera* sp.5
ARALIACEAE	*Schefflera* sp.6
ARALIACEAE	*Schefflera* sp.7
ARALIACEAE	*Schefflera* sp.8
ARALIACEAE	*Schefflera* sp.9
ARALIACEAE	*Schefflera* sp.10
ARECACEAE	*Ceroxylon* sp.2
ARECACEAE	*Ceroxylon* sp.3
ARECACEAE	*Ceroxylon* sp.4
ARECACEAE	*Chamaedorea* sp.2
ARECACEAE	*Chamaedorea* sp.3
ARECACEAE	*Geonoma* sp.1
ARECACEAE	*Geonoma* sp.2
ASTERACEAE	*Gynoxys* sp.1
ASTERACEAE	*Mikania* sp.1
ASTERACEAE	*Mikania* sp.2
ASTERACEAE	*Munnozia* sp.1
ASTERACEAE	*Munnozia* sp.2
ASTERACEAE	*Munnozia* sp.3
ASTERACEAE	*Munnozia* sp.4
ASTERACEAE	*Pentacalia nunezii*
ASTERACEAE	*Pentacalia* sp.2
ASTERACEAE	*Pentacalia* sp.3
ASTERACEAE	*Senecio* sp.2
ASTERACEAE	*Senecio* sp.3
ASTERACEAE	*Vernonia patens*
BEGONIACEAE	*Begonia* sp.1
BROMELIACEAE	*Guzmania* sp.2
BROMELIACEAE	*Tillandsia* sp.3
BRUNELLIACEAE	*Brunellia* sp.1
BRUNELLIACEAE	*Brunellia* sp.2
BRUNELLIACEAE	*Brunellia* sp.3
CAMPANULACEAE	*Centropogon* sp.2
CAMPANULACEAE	*Centropogon* sp.3
CAPRIFOLIACEAE	*Viburnum* sp.1
CAPRIFOLIACEAE	*Viburnum* sp.2
CECROPIACEAE	*Cecropia* sp.1
CELASTRACEAE	*Maytenus* sp.1
CHLORANTHACEAE	*Hedyosmum* sp.2
CHLORANTHACEAE	*Hedyosmum* sp.3
CHLORANTHACEAE	*Hedyosmum* sp.4
CLETHRACEAE	*Clethra* sp.1
CLUSIACEAE	*Clusia* sp.1
CLUSIACEAE	*Clusia* sp.3
CLUSIACEAE	*Clusia* sp.4
CLUSIACEAE	*Clusia* sp.5
CUNONIACEAE	*Weinmannia* sp.8
CUNONIACEAE	*Weinmannia* sp.9

Additional plant species recorded from Wayrapata (2445 m)

FAMILY	GENUS AND SPECIES
CUNONIACEAE	*Weinmannia* sp.10
CUNONIACEAE	*Weinmannia* sp.11
CUNONIACEAE	*Weinmannia* sp.12
CUNONIACEAE	*Weinmannia* sp.13
CUNONIACEAE	*Weinmannia* sp.14
CUNONIACEAE	*Weinmannia* sp.15
CUNONIACEAE	*Weinmannia* sp.16
CUNONIACEAE	*Weinmannia* sp.17
DIOSCOREACEAE	*Dioscorea* sp.2
DIOSCOREACEAE	*Dioscorea* sp.3
DIOSCOREACEAE	*Dioscorea* sp.4
ERICACEAE	*Befaria* sp.1
ERICACEAE	*Psammisia* sp.2
ERICACEAE	*Psammisia* sp.3
ERICACEAE	*Psammisia* sp.4
ERICACEAE	*Sphyrospermum* sp.2
ERICACEAE	*Sphyrospermum* sp.3
ERICACEAE	*Sphyrospermum* sp.4
ERICACEAE	*Sphyrospermum* sp.5
ERICACEAE	*Sphyrospermum* sp.6
ERICACEAE	*Sphyrospermum* sp.7
ERICACEAE	*Vaccinium* sp.1
ERICACEAE	*Vaccinium* sp.2
EUPHORBIACEAE	*Alchornea grandis*
EUPHORBIACEAE	*Alchornea* sp.1
EUPHORBIACEAE	*Alchornea* sp.2
EUPHORBIACEAE	*Alchornea* sp.3
EUPHORBIACEAE	*Hyeronima* sp.1
FLACOURTIACEAE	*Prockia* sp.2
GENTIANACEAE	*Irlbachia* sp.2
GENTIANACEAE	*Irlbachia* sp.3
GESNERIACEAE	*Alloplectus* sp.2
GESNERIACEAE	*Besleria* sp.1
GESNERIACEAE	*Columnea* sp.2
LAURACEAE	*Persea ferruginea*
LEGUMINACEA	*Inga* sp.1
LORANTHACEAE	*Tristerix* sp.1
MARCGRAVIACEAE	*Marcgravia* sp.1
MELASTOMATACEAE	*Brachyotum* sp.2
MELASTOMATACEAE	*Meriana cuzcoana*
MELASTOMATACEAE	*Miconia* sp.11
MELASTOMATACEAE	*Miconia* sp.12
MELASTOMATACEAE	*Miconia* sp.13
MELASTOMATACEAE	*Miconia* sp.14
MELASTOMATACEAE	*Miconia* sp.15
MELASTOMATACEAE	*Miconia* sp.16
MELASTOMATACEAE	*Miconia* sp.17
MELASTOMATACEAE	*Miconia* sp.18
MELASTOMATACEAE	*Miconia* sp.19
MELASTOMATACEAE	*Miconia* sp.20
MELASTOMATACEAE	*Miconia* sp.21
MELASTOMATACEAE	*Miconia* sp.22
MELASTOMATACEAE	*Miconia* sp.23
MELASTOMATACEAE	*Miconia* sp.24
MELASTOMATACEAE	*Miconia* sp.25
MONIMIACEAE	*Mollinedia* sp.1
MONIMIACEAE	*Mollinedia* sp.2
MONIMIACEAE	*Siparuna* sp.1
MYRICACEAE	*Myrica pubescens*

FAMILY	GENUS AND SPECIES
MYRICACEAE	*Myrica* sp.1
MYRSINACEAE	*Cybianthus* sp.2
MYRSINACEAE	*Cybianthus* sp.3
MYRSINACEAE	*Myrsine* sp.1
MYRSINACEAE	*Myrsine* sp.2
MYRSINACEAE	*Myrsine* sp.3
MYRSINACEAE	*Myrsine* sp.4
MYRSINACEAE	*Stylogyne* sp.1
MYRSINACEAE	*Stylogyne* sp.2
MYRTACEAE	*Myrteola* sp.2
MYRTACEAE	*Myrteola* sp.3
OLACACEAE	*Schoepfia flexuosa*
ONAGRACEAE	*Fuchsia* sp.1
ORCHIDACEAE	*Brachionidium* sp.2
ORCHIDACEAE	*Cyclopogon* sp.1
ORCHIDACEAE	*Elleanthus* sp.2
ORCHIDACEAE	*Epidendrum aniculatum*
ORCHIDACEAE	*Epidendrum* sp.2
ORCHIDACEAE	*Epistephium* sp.1
ORCHIDACEAE	*Lepanthes* sp.4
ORCHIDACEAE	*Maxillaria* sp.5
ORCHIDACEAE	*Maxillaria* sp.6
ORCHIDACEAE	*Maxillaria* sp.7
ORCHIDACEAE	*Neobardia* sp.1
ORCHIDACEAE	*Pleurothallis cordifolia*
ORCHIDACEAE	*Pleurothallis* sp.2
ORCHIDACEAE	*Pleurothallis* sp.3
ORCHIDACEAE	*Pleurothallis* sp.4
ORCHIDACEAE	*Sobralia* sp.1
OXALIDACEAE	*Oxalis* sp.2
OXALIDACEAE	*Oxalis* sp.3
OXALIDACEAE	*Oxalis* sp.4
PASSIFLORACEAE	*Passiflora* sp.1
PASSIFLORACEAE	*Passiflora* sp.2
PIPERACEAE	*Peperomia* sp.2
PIPERACEAE	*Peperomia* sp.3
PIPERACEAE	*Peperomia* sp.4
PIPERACEAE	*Peperomia* sp.5
PIPERACEAE	*Peperomia* sp.6
PIPERACEAE	*Piper* sp.1
PIPERACEAE	*Piper* sp.2
PIPERACEAE	*Piper* sp.3
PIPERACEAE	*Piper* sp.4
PIPERACEAE	*Piper* sp.5
POACEAE	*Arundinella* sp.2
POACEAE	*Aulonemia* sp.2
POACEAE	*Chusquea* sp.1
POACEAE	*Chusquea* sp.2
POACEAE	*Chusquea* sp.3
POACEAE	*Rhipidocladum* sp.1
POLYGALACEAE	*Monnina* sp.1
POLYGALACEAE	*Monnina* sp.2
POLYGONACEAE	*Muehlenbeckia* sp.1
ROSACEAE	*Prunus* sp.2

FAMILY	GENUS AND SPECIES
ROSACEAE	*Rubus* sp.1
ROSACEAE	*Rubus* sp.2
ROSACEAE	*Rubus* sp.3
RUBIACEAE	*Cephaelis flaviflora*
RUBIACEAE	*Cinchona* sp.2
RUBIACEAE	*Cinchona* sp.3
RUBIACEAE	*Faramea* sp.1
RUBIACEAE	*Palicourea* sp.1
RUBIACEAE	*Palicourea* sp.2
RUBIACEAE	*Pentagonia* sp.1
RUBIACEAE	*Psychotria* sp.3
RUBIACEAE	*Psychotria* sp.4
SABIACEAE	*Meliosma dudleyi*
SABIACEAE	*Meliosma* sp.1
SMILACACEAE	Smilax sp.2
SMILACACEAE	*Smilax* sp.3
SMILACACEAE	*Smilax* sp.4
SOLANACEAE	*Solanum* sp.1
SOLANACEAE	*Solanum* sp.2
STYRACACEAE	*Pamphilia vilcabambae*
STYRACACEAE	*Styrax* sp.1
SYMPLOCACEAE	*Symplocos* sp.2
SYMPLOCACEAE	*Symplocos* sp.3
SYMPLOCACEAE	*Symplocos* sp.4
SYMPLOCACEAE	*Symplocos* sp.5
SYMPLOCACEAE	*Symplocos* sp.6
SYMPLOCACEAE	*Symplocos* sp.7
SYMPLOCACEAE	*Symplocos* sp.8
SYMPLOCACEAE	*Symplocos* sp.9
SYMPLOCACEAE	*Symplocos* sp.10
SYMPLOCACEAE	*Symplocos* sp.11
SYMPLOCACEAE	*Symplocos* sp.12
THEACEAE	*Freziera* sp.1
THEACEAE	*Freziera* sp.2
URTICACEAE	*Pilea* sp.2
VERBENACEAE	*Aegiphila* sp.1
VITACEAE	*Cissus* sp.1
VOCHYSIACEAE	*Vochysia* sp.1

Preliminary list of the birds at three sites in the Northern Cordillera de Vilcabamba, Peru

APPENDIX 13

Thomas S. Schulenberg, Lawrence López, Grace Servat, and Armando Valdes

Codes for Avian Data:

HABITATS

Fm Montane evergreen forest
Elf Elfin forest
Pol *Polylepis* forest
Fe Forest edge
P *Pajonales* (paramo grasslands)
Fsm Forest stream margins
B Bamboo
O Overhead

ABUNDANCE

F Fairly common
U Uncommon
R Rare

EVIDENCE

sp Specimen
t Tape
si Species identification by sight or sound
ph Photograph

Bird species not reported previously from the Cordillera de Vilcabamba (Weske 1972) are shown in bold face.

	Habitat	Camp One 3350 m	Camp Two 1800-2050 m	Ridge Camp 750-1150 m	Evidence
TINAMIDAE (4)					
Tinamus tao	Fm			R	si
Crypturellus obsoletus	Fm			U	t
Crypturellus soui	Fm			R	si
Crypturellus variegatus	Fm			U	si
CATHARTIDAE (1)					
Cathartes aura	O			U	si
ACCIPITRIDAE (4)					
Accipiter striatus	Fm		R		t
Buteo albigula	O	R			
Buteo leucorrhous	Fm		R		t
Buteo polyosoma	O	R			si
Spizaetus tyrannus	Fm			R	si
FALCONIDAE (2)					
Daptrius americanus	Fm			R	t
Micrastur ruficollis	Fm			U	t
PHASIANIDAE (1)					
Odontophorus balliviani	Fm	R	U		t
SCOLOPACIDAE (1)					
Gallinago jamesoni	P	U			t
COLUMBIDAE (3)					
Columba fasciata	Fm	U			si
Columba plumbea	Fm		U	U	t
Geotrygon frenata	Fm		U	U	si

	Habitat	Camp One 3350 m	Camp Two 1800-2050 m	Ridge Camp 750-1150 m	Evidence
PSITTACIDAE (3)					
Ara couloni	Fm			R	t
Pyrrhura picta	Fm		R	U	t
Bolborhynchus orbygnesius	Elf, Fe, P	F			t
Pionus menstruus	Fm			U	si
Pionus tumultuosus	Fm		U		t
Amazona mercenaria	Fm		U	U	t
CUCULIDAE (1)				R	
Piaya cayana	Fm		U	U	t
STRIGIDAE (6)					
Otus albogularis	Fm	U	U		t
Otus guatemalae	Fm			U	t
Otus ingens	Fm		U		t
Pulsatrix melanota	Fm			U	t
Glaucidium jardinii	Fm, Elf	U	U		t
Ciccaba albitarsus	Fm		U		t
CAPRIMULGIDAE (2)					
Lurocalis rufiventris	Fm		R		t
Uropsalis segmentata	P, Pol	U	U		t, sp
APODIDAE (3)					
Streptoprocne rutilus	O			U	si
Streptoprocne zonaris	O		F	U	si
Chaetura brachyura	O			U	t
TROCHILIDAE (17)					
Phaethornis guy	Fm			U	si
Phaethornis koepckeae	Fm			F	si
Phaethornis sp. (*stuarti* ?)	Fm			U	si
Phaethornis superciliosus	Fm			U	si
Eutoxeres condamini	Fm			R	
Colibri thalassinus	Fe		F		t
Thalurania furcata	Fm			U	t
Adelomyia melanogenys	Fm		F		t
Heliodoxa branickii	Fm			F	t
Heliodoxa rubinoides	Fm		U		t
Coeligena coeligena	Fm		F		t
Coeligena violifer	Fm, Pol	U			t, sp
Boissonneaua matthewsii	Fm		U		Si
Metallura tyrianthina	Elf, P, Pol	F			t
Aglaiocercus kingi	Fm		U		si
Heliothryx aurita	Fm			R	si
Acestrura mulsant	Fe		R		si
TROGONIDAE (4)					
Pharomachrus auriceps	Fm		F		t
Trogon collaris	Fm			U	t
Trogon curucui	Fm			U	t
Trogon personatus	Fm	R	U		t

	Habitat	Camp One 3350 m	Camp Two 1800-2050 m	Ridge Camp 750-1150 m	Evidence
MOMOTIDAE (1)					
Electron platyrhynchum	Fm			U	t
GALBULIDAE (2)					
Galbula cyanescens	Fm			U	si
Jacamerops aurea	Fm			R	t
BUCCONIDAE (1)					
Nystalus striolatus	Fm			U	t
CAPITONIDAE (1)					
Capito niger	Fm			U	si
RAMPHASTIDAE (5)					
Aulacorhynchus coeruleicinctus	Fm		U		t
Aulacorhynchus (*prasinus*)	Fm			U	si
Pteroglossus azara	Fm			R	si
Selenidera reinwardtii	Fm			U	si
Andigena hypoglauca	Fm		U		t
PICIDAE (11)					
Picumnus aurifrons	Fm			R	si
Veniliornis affinis	Fm			U	t
Veniliornis dignus	Fm		R		si
Veniliornis nigriceps	Fm	R			si
Piculus chrysochloros	Fm			U	t
Piculus (*leucolaemus*)	Fm			R	t
Piculus rivolii	Fm	U			si
Piculus rubiginosus	Fm		U		si
Celeus grammicus	Fm			U	t
Campephilus haematogaster	Fm		U		t
Campephilus melanoleucos	Fm			U	t
DENDROCOLAPTIDAE (9)					
Dendrocincla fuliginosa	Fm			R	si
Deconychura longicauda	Fm			R	si
Sittasomus griseicapillus	Fm			U	t
Glyphorynchus spirurus	Fm			F	t
Xiphocolaptes promeropirhynchus	Fm		U	R	t
Xiphorhynchus ocellatus	Fm			F	t
Xiphorhynchus triangularis	Fm		U		t
Lepidocolaptes affinis	Fm		U		si
Campylorhamphus trochilirostris	B, Fm			U	t
FURNARIIDAE (20)					
Schizoeaca vilcabambae	Elf, P, Pol	F			t, sp
Synallaxis cabanisi	B			R	t
Synallaxis azarae	Fe		F		t
Cranioleuca gutturata	Fm			U	si
Cranioleuca marcapatae weskei	Fm	U			sp
Margarornis squamiger	Fm, Elf, Pol	F	U		t
Premnornis guttuligera	Fm		U		t
Premnplex brunnescens	Fm		F		t

	Habitat	Camp One 3350 m	Camp Two 1800-2050 m	Ridge Camp 750-1150 m	Evidence
Pseudocolaptes boissonneautii	Fm	R	F		t
Hyloctistes subulatus	Fm			U	t
Simoxenops ucayalae	B			R	t
Philydor erythrocercus ochrogaster	Fm			U	si
Philydor ruficaudatus	Fm			U	Si
Automolus dorsalis	B			R	t
Automolus ochrolaemus	Fm			F	t
Automolus rubiginosus	Fm			R	t
Thripadectes holostictus	Fm		U		si
Xenops minutus	Fm			U	si
Xenops rutilans	Fm			R	t
Sclerurus albigularis	Fm			U	t
THAMNOPHILIDAE (26)					
Cymbilaimus lineatus	Fm			U	si
Thamnophilus caerulescens	Fm		F		t
Thamnophilus palliatus	Fm			U	t
Thamnophilus schistaceus	Fm			R	t
Dysithamnus mentalis	Fm			F	t
Thamnomanes schistogynus	Fm			R	t
Myrmotherula brachyura	Fm			R	si
Myrmotherula erythrura	Fm			U	si
Myrmotherula spodionota	Fm			U	t
Myrmotherula menetriesii	Fm			U	si
Herpsilochmus motacilloides	Fm		R		t
Herpsilochmus rufimarginatus	Fm			F	t
Microrhopias quixensis	B			U	t
Drymophila caudata	B		F		t
Cercomacra cinerascens	Fm			R	si
Pyriglena leuconota	Fm		U		t
Myrmoborus leucophrys	Fm			U	t
Myrmoborus myotherinus	Fm			F	t
Hypocnemis cantator	B			U	t
Percnostola lophotes	B			R	t
Myrmeciza fortis	Fm			F	t
Myrmeciza goeldii	B			R	t
Myrmeciza hemimelaena	Fm			F	t
Rhegmatorhina melanosticta	Fm			R	si
Hylophylax naevia	Fm			U	t
Hylophylax poecilonota	Fm			U	t
FORMICARIIDAE (6)					
Formicarius analis	Fm			U	t
Formicarius rufipectus	Fm		U		
Chamaeza mollissima	Fm		F		t
Grallaria erythroleuca	Fm		F		t, sp
Grallaria rufula	Elf, Pol	F			t, sp
Myrmothera campanisona	Fm			U	t

	Habitat	Camp One 3350 m	Camp Two 1800-2050 m	Ridge Camp 750-1150 m	Evidence
Grallaricula flavirostris	Fm		U		t, sp
RHINOCRYPTIDAE (2)					
Scytalopus atratus	Fm		U		t, sp
Scytalopus parvirostris	Fm, Elf, Pol	F	F		t, sp
TYRANNIDAE (42)					
Zimmerius bolivianus	Fm		F		
Zimmerius gracilipes	Fm			U	si
Myiopagis gaimardii	Fm			F	t
Elaenia pallatangae	Fe		R		si
Elaenia sp.	Fm			R	si
Mecocerculus leucophrys	Elf, Pol, Fm	F			t
Mionectes olivaceus	Fm			U	si
Mionectes striaticollis	Fm		F		si
Leptopogon superciliaris	Fm			F	t
Leptopogon taczanowskii	Fm		U		t
Phylloscartes ophthalmicus	Fm			R	si
Phylloscartes orbitalis	Fm			U	si
Phylloscartes parkeri	Fm			U	si
Phylloscartes ventralis	Fm		F		t
Pseudotriccus ruficeps	Fm	U	F		t
Hemitriccus rufigularis	Fm			U	si
Tolmomyias assimilis	Fm			U	t
Tolmomyias poliocephalus	Fm			R	t
Platyrinchus mystaceus	Fm			R	si
Myiotriccus ornatus	Fm			F	t
Terenotriccus erythrurus	Fm			U	si
Myiobius (*villosus*)	Fm			R	si
Myiophobus fasciatus	B			R	t
Myiophobus flavicans	Fm		U		t
Myiophobus sp.	Fm		R		si
Pyrrhomyias cinnamomea	Fm		F	U	t
Mitrephanes olivaceus	Fm		U		t
Contopus fumigatus	Fm		F		si
Lathrotriccus euleri	B			R	t
Ochthoeca cinnamomeiventris	Fm, Fsm		U		t
Ochthoeca frontalis	Elf	R			si
Ochthoeca fumicolor	Elf, P, Pol	F			t, sp

	Habitat	Camp One 3350 m	Camp Two 1800-2050 m	Ridge Camp 750-1150 m	Evidence
Ochthoeca pulchella	Fm		R		
Ochthoeca rufipectoralis	Elf, P, Pol	F			t
Myiotheretes fumigatus	Fm	U			t
Myiotheretes striaticollis	Fe		R		si
Attila spadiceus	Fm			U	t
Laniocera hypopyrra	Fm			U	si
Myiarchus tuberculifer	Fm		F	U	si
Pachyramphus polychopterus	Fm			U	si
Pachyramphus validus	Fm			R	si
Pachyramphus versicolor	Fm		U		si
PIPRIDAE (6)					
Schiffornis turdinus	Fm			U	t
Piprites chloris	Fm			F	t
Chiroxiphia boliviana	Fm			R	t
Pipra coeruleocapilla	Fm			U	si
Pipra chloromeros	Fm			F	t
Pipra pipra	Fm			F	t
COTINGIDAE (11)					
Ampelion rubrocristatus	Elf, Fm, Pol	U			si
Ampelion rufaxilla	Fm		F		t
Pipreola chlorolepidota	Fm			U	t
Pipreola intermedia	Fm		R		
Pipreola pulchra	Fm		F		t
Ampelioides tschudii	Fm			U	si
Iodopleura isabellae	Fm			R	t
Lipaugus subalaris	Fm			U	t
Cotinga cayana	Fm			U	si
Rupicola peruviana	Fm		R	U	t
Oxyruncus cristatus	Fm			U	t
HIRUNDINIDAE (2)					
Notiochelidon flavipes	Fm		F		t
Notiochelidon murina	P	F			si
CORVIDAE (2)					
Cyanolyca viridicyana	Fm		U		t
Cyanocorax yncas	Fm		R		t
TROGLODYTIDAE (7)					
Campylorhynchus turdinus	Fe			R	t
Cinnycerthia fulva	Fm		F		t
Thryothorus genibarbis	B			R	t
Troglodytes solstitialis	Fm		F		t
Henicorhina leucophrys	Fm		F		t
Microcerculus marginatus	Fm			F	t
Cyphorhinus aradus	Fm			R	si
TURDIDAE (6)					
Myadestes ralloides	Fm		F		t

	Habitat	Camp One 3350 m	Camp Two 1800-2050 m	Ridge Camp 750-1150 m	Evidence
Entomodestes leucotis	Fm		F		t
Catharus fuscater	Fm		U		t, sp
Turdus albicollis	Fm			U	si
Turdus fuscater	Elf, P, Pol	F			t
Turdus serranus	Fm		F		t
VIREONIDAE (7)					
Cyclarhis gujanensis	Fe			U	t
Vireolanius leucotis	Fm			F	t
Vireo leucophrys	Fm		R		si
Vireo olivaceus	Fm			F	si
Hylophilus hypoxanthus	Fm			U	t
Hylophilus ochraceiceps	Fm			R	si
Hylophilus thoracicus	Fm			U	t
EMBERIZINAE (5)					
Zonotrichia capensis	P, Pol	R			si
Sporophila (luctuosa)	Fe		R		si
Catamenia homochroa	P	R			si
Atlaptetes brunneinucha	Fm		U		si
Atlapetes tricolor	Fe		U		t
CATAMBYRHYNCHINAE (1)					
Catamblyrhynchus diadema	Elf, Fm	R	R		si
CARDINALINAE (2)					
Pitylus grossus	Fm			F	t
Saltator maximus	Fm			U	si
THRAUPINAE (44)					
Cissopis leveriana	Fe			U	t
Chlorornis riefferii	Elf, Fm		F		t
Chlorospingus ophthalmicus	Fm		F		t
Cnemoscopus rubrirostris	Fm		U		si
Hemispingus atropileus	Fm	U			si
Hemispingus frontalis	Fm		R		si
Hemispingus melanotis	Fm		U		si
Hemispingus xanthophthalmus	Elf, Pol	U			si
Chlorothraupis carmioli	Fm			U	t
Lanio versicolor	Fm			U	t
Creurgops dentata	Fm		R		si
Tachyphonus luctuosus	Fm			U	si
Tachyphonus rufiventer	Fm			U	si
Piranga flava	Fm			R	si
Thraupis cyanocephala	Fe, Fm		F		t
Thraupis palmarum	Fe, Fm			F	si
Buthraupis montana	Fm		R		
Anisognathus igniventris	Elf, Pol, Fm	F			t
Anisognathus lacrymosus	Fm	R	U		si
Iridosornis analis	Fm		U		t

	Habitat	Camp One 3350 m	Camp Two 1800-2050 m	Ridge Camp 750-1150 m	Evidence
Iridosornis reinhardti	Fm		U		si
Delothraupis castaneoventris	Elf	U	R		si
Euphonia xanthogaster	Fm			F	t
Chlorochrysa calliparaea	Fm		R		si
Tangara arthus	Fm			F	si
Tangara chilensis	Fm			F	t
Tangara cyanicollis	Fm			R	si
Tangara gyrola	Fm			F	si
Tangara nigroviridis	Fm		U		si
Tangara parzudakii	Fm		F		si
Tangara punctata	Fm			R	si
Tangara schrankii	Fm			F	si
Tangara vassorii	Fm		U		si
Tangara xanthocephala	Fm		U		si
Tangara xanthogastra	Fm			U	
Dacnis cayana	Fm			U	si
Dacnis lineata	Fm			U	Si
Chlorophanes spiza	Fm			F	t
Cyanerpes caeruleus	Fm			U	si
Diglossa albilatera	Fe, Fm		F		t, sp
Diglossa cyanea	Fm		F		t
Diglossa glauca	Fm		R		si
Diglossa mystacalis	Elf, Pol	F			t
PARULIDAE (11)					
Parula pitiayumi	Fm			F	t
Myioborus melanocephalus	Fm, Pol	U	F		t
Myioborus miniatus	Fm		U	F	t
Basileuterus chrysogaster	Fm			U	t
Basileuterus coronatus	Fm		F		t
Basileuterus luteoviridis	Fm, Pol	U	U		t
Basileuterus signatus	Fm		R		t
Basileuterus tristriatus	Fm		F		t
Conirostrum albifrons	Fm		F		si
Conirostrum ferrugineiventre	Elf, Pol	U			t
Conirostrum sitticolor	Elf, Pol	U			si
ICTERIDAE (4)					
Icterus cayanensis	Fm			R	t
Psarocolius atrovirens	Fm		U		t
Cacicus holosericeus	Fm	R	R		si
Cacicus leucoramphus	Fm		U		t
FRINGILLIDAE (1)					
Carduelis (olivacea)	Fm, Fe		R		t

Bird Species Observed at Llactahuaman (1710 m), Southern Cordillera de Vilcabamba, Peru

APPENDIX 14

Tatiana Pequeno, Edwin Salazar, and Constantino Aucca

Family	Species	Common Name	Type of Record				
			Captured	Visual	Taped	Photo	Song
Tinamidae	*Crypturellus obsoletus*	Brown Tinamou	X				
Accipitridae	*Buteo polyosoma*	Red backed Hawk		X	X		
	Oroaetus isidoris	Black & chesnut Eagle		X	X		
Columbidae	*Columba subvinacea*	Ruddy Pigeon			X		
Psittacidae	*Pionus tumultuosus*	Plum-crowned Parrot					X
Cuculidae	*Piaya cayana*	Squirrel Cuckoo					X
Strigidae	*Otus choliba*	Tropical screech-Owl					X
	Pulsatrix perspicilata	Spectacle Owl			X		
Caprimulgidae	*Caprimulgus longirostris*	Band winged Nightjar	X	X			
Trochilidae	*Doryfera ludoviciae*	Green-fronted Lancebill	X				
	Eutoxeres condamini	Buff-tailed Sickelebill	X				
	Colibri thalassinus	Green violetear	X	X			
	Colibri coruscans	Sparkling Violetear	X	X			
	Amazilia viridicauda	Green and white Hummingbird	X				
	Adelomyia melanogenys	Specktacle Hummingbird	X	X			
	Pterophanes cyanoptera	Great Sapphirewing		X			
	Coeligena coeligena	Bronzy Inca	X				
	Heliangelus amethysticollis	Amethyst-throated Sunangel	X	X			
Trogonidae	*Trogon collaris*	Collared Trogon		X		X	X
	Trogon personatus	Masked Trogon	X				
Momotidae	*Electron platyrhynchus*	Broad-billed Motmot	X				X
	Momotus momota	Blue-crowned Motmot	X			X	
Bucconidae	*Malacoptila fulvogularis*	Black-streaked Puffbird	X				
Capitonidae	*Eubucco versicolor*	Versicolored Barbet	X				
Ramphastidae	*Aulacorhynchus prasinus*	Emerald Toucanet	X	X			
Picidae	*Picumnus dorbygnianus*	Ocellated Piculet	X			X	
	Chrysoptilus atricollis	Black-necked Woodpecker	X	X			
	Campephilus melanoleucus	Crimson-crested Woodpecker					X
	Campephilus polens	Powerful Woodpecker					X

Family	Species	Common Name	Type of Record				
			Captured	Visual	Taped	Photo	Song
Dendrocolaptidae	*Xiphorhynchus triangularis*	Olive-backed Woodcreeper	X				
	Campylorhamphus trochilirostris	Red-billed Scythebill	X				
Furnaridae	*Synallaxis azarae*	Azara´s Spinetail	X	X			X
	Premnoplex brunnescens	Spotted Barbtail	X	X			
	Syndactyla subalaris	Lineated Foliage-gleaner	X				
	Anabacertia striaticollis	Montane Foliage-gleaner	X				
	Xenops rutilans	Streaked Xenops					
	Xenops minutus	Plain Xenops	X	X			
	Lochmias nematura	Sharp-tailed Streamcreeper	X				
Formicariidae	*Thamnophilus aethiops*	White-shouldered Antshrike		X	X	X	X
	Thamnophilus caerulescens	Variable Antshrike	X				
	Dysithamnus mentalis	Plain Antvireo	X				
	Myrmotherula schisicolor	Slaty Antwren	X				
	Drymophila caudata	Long-tailed Antbird	X			X	X
	Cercomacra nigrescens	Blackish Antbird	X				
	Cercomacra serva	Black Antbird	X				
	Pyriglena leuconota	White-backed Fire-eye	X				
	Chamaeza campanisona	Short-tailed Antthrush	X				X
	Grallaria guatimalensis	Scaled Antpitta	X				
	Grallaricula flavirostris	Ochre-breasted Antpitta	X				X
	Conopophaga castaneiceps	Chestnut-crowned Gnateater	X				
Rhinocryptidae	*Scytalopus unicolor*	Unicolored Tapaculo	X		X		X
Tyrannidae	*Phyllomyas cinereiceps*	Ashy-headed Tylanulet	X				
	Camptostoma obsoletum	Southern Bearless Tyranulet		X			
	Mionectes olivaceus	Olive-striped Flycatcher	X	X		X	
	Leptopogon superciliaris	Slaty-capped Flycatcher	X				
	Phylloscartes ventralis	Mottle-cheeked Tyrannulet	X				
	Pseudotriccus pelzelni	Bronze-olive Pygmy-Tyrant	X				
	Lophotriccus pileatus	Scale-crested Pygmy-Tyrant	X				
	Rhynchocyclus fulvipectus	Fulvous-brested Flatbill	X	X	X		
	Platyrinchus mystaceus	White-throated Spadebill	X			X	
	Myiotrichus ornatus	Ornate Flycatcher	X	X		X	
	Myobius barbatus	Sulphur-rumped Flycatcher	X	X			
	Pyrromyias cinnamomea	Cinnamon Flycatcher	X	X			
	Lathrotriccus euleri	Euler´s Flycatcher	X				
	Knipolegus poecilurus	Rufous-tailed Tyrant	X				
	Miarchus ferox	Short-crested Flycatcher		X			
	Miarchus cephalotes	Pale-edged Flycatcher	X			X	
	Myarchus tuberculifer	Dusky-capped Flycatcher	X				
Pipridae	*Chloropipo unicolor*	Jet Manakin	X				
	Pipra pipra	White-crowned Manakin	X	X		X	
	Pipra coronata	Blue-crowned Manakin	X				
Cotingidae	*Pipreola pulchra*	Masked Fruiteater	X			X	
	Rupicola peruviana	Andean Cock-of-the-Rock		X			

Family	Species	Common Name	Type of Record				
			Captured	Visual	Taped	Photo	Song
Troglodytidae	*Thryothorus einsenmanni*	Inca Wren	X			X	
	Thryothorus coraya	Coraya Wren	X			X	
	Troglodytes solstitialis	Mountain Wren	X				
	Henicorhina leucophrys	Gray-breasted Wood-wren	X	X	X	X	X
Muscicapidae	*Myadestes ralloides*	Andean Solitaire	X				
Emberezidae	*Oryzoborus angolensis*	Leseer Seed-Finch	X			X	
	Atlapetes brunneinucha	Chestnut-capped Brush-Finch	X			X	
	Atlapetes rufinucha	Rufous-naped Brush-Finch	X			X	
	Pheucticus chrysogaster	Golden-bellied Grosbeak	X				
	Chlorospingus ophthalmicus	Common Bush-Tanager	X				
	Hemispingus frontalis	Oleagineus Hemispingus	X				
	Thlypopsis ornata	Rufous-chested Tanager		X			
	Chlorothraupis carmioli	Carmiol's Tanager					X
	Creurgops dentata	Slaty Tanager	X			X	
	Piranga flava	Hepatic Tanager		X			
	Thraupis palmarum	Palm Tanager	X				X
	Thraupis cyanocephala	Blue-capped Tanager	X	X		X	
	Anisognatus flavinuchus	Blue-winged Mountain-Tanager	X	X			
	Iridosornis analis	Yellow-throated Tanager	X	X			
	Iridosornis jelskii	Golden-collared Tanager	X	X			
	Euphonia xanthogaster	Orange-bellied Euphonia	X	X		X	
	Chlorochrysa calliparaea	Orange-eared Tanager	X			X	
	Tangara xanthocephala	Saffron-crowned Tanager	X				
	Tangara parzudakii	Flame-faced Tanager	X	X		X	X
	Chlorophanes spiza	Green Honeycreeper		X			
	Diglossa caerulescens	Bluish Flower-piercer	X				
	Diglossa glauca	Deep-blue Flower-piercer	X				
	Myioborus miniatus	Slate-throuted Redstart	X	X			
	Myioborus melanocephalus	Spevtacle Redstart	X	X			
	Basileuterus tristriatus	Three-striped Warbler	X			X	
	Basileuterus coronatus	Russet-crowned Warbler	X	X		X	
	Conirostrum albifrons	Capped Conebill		X			
Icteridae	*Psaracolius angustifrons*	Russet-backed Oropendola		X	X		X
	Cacicus cela	Yellow-rumped Cacique		X			
	Cacicus leucorhampus	Mountain Cacique		X			
	Amblycercus holocericeus	Yellow-billed Cacique	X				
	Icterus icterus	Troupial		X			
Fringillidae	*Carduelis magellanica*	Hooded Siskin		X			

APPENDIX 15

Bird Species Observed at Wayrapata (2445 m), Southern Cordillera de Vilcabamba, Peru

Tatiana Pequeno, Edwin Salazar, and Constantino Aucca

Family	Species	Common Name	Type of Record				
			Captured	Visual	Taped	Photo	Song
Accipitridae	*Elanoides forficatus*	Swallow tailed Kite		X			
	Geranoaetus melanoleucos	Balk chested Buzzard Eagle		X			
	Buteo polyosoma	Red-backed Hawk		X			
Falconidae	*Milvago chimachima*	Yellow-headed Caracara		X			
	Falco sparverius	American Kestrel		X			
Cracidae	*Penelope montagni*	Andean Guan		X			
Columbidae	*Columba subvinacea*	Ruddy pigeon		X			
Psittacidae	*Bolborhynchus lineola*	Barred Parakeet	X	X	X	X	X
	Amazona mercenaria	Scaly-naped Parrot	X	X			
Strigidae	*Otus choliba*	Tropical screech-Owl	X			X	X
	Glaucidium jardinii	Andean pygmy-Owl	X			X	X
Caprimulgidae	*Uropsalis segmentata*	Swallow-tailed Nightjar	X	X		X	X
Apodidae	*Streptoprogne zonaris*	White-collared Swift		X			
	Cypseloides rutilus	Chesnut-collared swift		X			
Trochilidae	*Doryfera ludoviciae*	Green-fronted Lancebill	X			X	
	Colibri thalassinus	Green Violetear	X	X		X	
	Colibri coruscans	Sparkling Violetear	X	X		X	
	Amazilia viridicauda	Green and white Hummingbird	X	X		X	
	Adelomyia melanogenys	Speckled Hummingbird	X	X		X	
	Oreothrochilus stella	Andean Hillstar	X			X	
	Aglaeactis cupripennis	Shining Sunbeam	X			X	
	Coeligena coeligena	Bronzy Inca	X			X	
	Coeligena torquata	Collared Inca	X			X	
	Heliangelus amethiticollis	Amethys-throated Sunangel	X			X	
	Chalcostigma ruficeps	Rufous-capped Thornbill	X			X	
	Aglaiocercus kingi	Long tailed Sulph	X			X	
Trogonidae	*Trogon persognatus*	Masked Trogon		X	X		X
Ramphastidae	*Andigena hypoglauca*	Gray-breasted mountain Tucan	X			X	
Picidae	*Piculus rivolii*	Crimson-mantled Woodpecker	X			X	
Furnariidae	*Synallaxis azarae*	Azara´s Spinetail	X	X		X	X
	Synallaxis sp.		X				
	Synallaxis unirufa	Rufous Spinetail	X				
	Margarornis squamiger	Pearled Treerunner	X	X		X	X
	Cranioleuca curtata	Ash-browed Spinetail	X				

Family	Species	Common Name	Type of Record				
			Captured	Visual	Taped	Photo	Song
Formicariidae	*Drymophyla caudata*	Long tailed Antbird	X		X		X
	Chamaza campanisona	Short-tailed Antthrush					X
	Grallaricula flavirostris	Ochre-breasted Antpitta					X
Rhynocryptidae	*Scytalopus unicolor*	Unicolored Tapaculo		X	X		X
Tyranidae	*Phyllomyias uropygialis*	Tawny-rumped Tyrannulet	X			X	
	Mionectes olivaceus	Olive-striped Flycatcher	X			X	
	Pseudotriccus ruficeps	Rufous-headed Pygmy-Tyrant	X	X		X	
	Hemitriccus granadensis	Black-throated Tody-Tyrant	X			X	
	Pyrrhomyias cinnamomea	Cinnamon Flycatcher	X	X		X	
	Ochthoeca cinnamomeiventris	Salaty-backed Chat-tyrant	X	X		X	
	Ochthoeca pulchella	Golden-browed Chat-tyrant	X			X	
	Ochthoeca rufipectoralis	Rufous-breasted Chat-tyrant	X				
	Myiotheretes striaticollis	Streak-throated Bush-tyrant	X			X	
	Myiotheretes fuscorufus	Rufous-bellied Bush-tyrant	X			X	
Cotingidae	*Ampelion rubrocristatus*	Red-crested Cotnga	X			X	
	Pipreola intermedia	Band-tailed Fruiteater	X			X	
	Pipreola arcuata	Barred Fruiteater	X			X	
	Pipreola pulchra	Masked Fruiteater	X			X	
	Rupicola peruviana	Andean Cock-of-the-Rock					
Hirundinidae	*Notiochelidon flavipes*	Pale-footed Swallow	X	X		X	X
Troglodytidae	*Cinnycerthia peruana*	Sepia-brown Wren	X			X	
	Troglodytes solstitialis	Mountain Wren	X				
	Henicorhyna leucophrys	Gray-breasted Wood-wren		X	X		X
Muscicapidae	*Entomodestes leucotis*	White-eared Solitaire	X			X	
	Platycichla leucops	Pale-eyed Thrush	X				
	Turdus chiguanco	Chiguanco Thrush	X			X	
	Turdus serranus	Glossy Black Thrush	X	X		X	
Parulidae	*Mioborus melanocephalus*	Slate-throated Redstar	X	X		X	X
	Basileuterus luteoviridis	Citrine Warbler	X	X		X	
Coerebidae	*Diglossa mystacalis*	Moustache Flower-piercer	X			X	
	Diglossa bruneiventris	Black-throated Flower-piercer	X			X	
	Diglossa albilatera	White-sided Flower-piercer	X			X	
	Digolssa cyanea	Masked Flower-piercer	X	X		X	
Emberezidae	*Zonothrichia capensis*	Rufous-collared Sparrow	X			X	
	Phrygilus unicolor	Plumbeous Sierra-finch	X			X	
	Haplospiza rustica	Slaty Finch	X			X	
	Catamenia homocroa	Paramo Seedeater	X			X	
	Catamenia inornata	Pain-colores Seedeater	X			X	
	Catamblyrhyncus diadema	Plus-capped Finch	X			X	
	Pheucticus auroventris	Black-backed Grosbeak	X			X	
	Chlorornis riefferii	Grass-green Tanager	X	X	X	X	X
	Chlorospingus ophthalmicus	Common Bush-tanager	X			X	
	Hemispingus atropileus	Black-capped Hemispingus	X	X		X	
	Hemispingus frontalis	Oleaginous Hemispingus	X	X		X	
	Hemispingus xanthophthalmus	Drab Hemispingus	X			X	
	Thraupis cyanocephala	Blue-capped Tanager	X	X		X	X
	Anisognatus igniventris	Scarlet-bellied Mountain Tanager	X			X	
	Anisognatus lacrimosus	Lacrimose Mountain Tanager	X	X		X	X

Family	Species	Common Name	Type of Record				
			Captured	Visual	Taped	Photo	Song
	Iridosornis analis	Yellow-throated Tanager	X	X		X	
	Iridosornis jelskii	Golden-collared Tanager	X	X		X	X
	Dubusia taeniata	Buff-breasted Mountain Tanager	X			X	
	Euphonia xanthogaster	Orange-bellied Euphonia	X	X		X	
	Chlorophonia cyanea	Blue-naped Chlorophonia	X			X	
	Tangara nigroviridis	Beryl-spangled Tanager	X	X		X	
	Tangara vassorii	Blue and black Tanager	X			X	
Icteridae	*Cacicus cela*	Yellow-rumped Cacique	X				
	Cacicus holocericeus	Yellow-billed Cacique	X			X	
Fringillidae	*Carduelis magellanica*	Hooded Siskin	X			X	

Preliminary list of mammals from three sites in the Northern Cordillera de Vilcabamba, Peru

APPENDIX 16

Louise H. Emmons, Lucia Luna W. and Mónica Romo R.

Key to records: X = specimen collected; O = observed; T = tracks and/or sign; R = sound recorded by ornithologists.

Species	Camp One (3350 m)	Camp Two (2050 m)	Ridge Camp (1000 m)
MARSUPIALS			
Didelphis albiventris		X	
Caluromys lanatus			X
Gracilinamus aceramarcae	X		
Lestoros inca		X	
Marmosops impavidus		X	
Marmosops noctivagus			X
Metachirus nudicaudatus			X
Monodelphis adusta			X
Monodelphis osgoodi		X	
Monodelphis cf. *theresa*		X	
BATS			
Anoura sp.			X
Anoura cultrata			X
Artibeus obscurus			X
Artibeus glaucus			X
Artibeus planirostris			X
Carollia brevicauda			X
Carollia perspicillata			X
Chiroderma villosum			X
Chiroderma salvini			X
Choeroniscus sp.			X
Cormura brevirostris			X
Desmodus rotundus			X
Lionycteris spurrelli			X
Lonchophylla handleyi			X
Mesophylla macconnelli			X
Micronycteris sp.			X
Micronycteris megalotis			X
Platyrrhinus dorsalis		X	
Platyrrhinus helleri			X
Platyrrhinus infuscus			X
Platyrrhinus vittatus		X	X
Platyrrhinus umbratus			X
Phylloderma stenops			X
Sturnira bidens		X	
Sturnira erythromos	X	X	
Sturnira magna			X
Sturnira oporaphilum		X	X

Species	Camp One (3350 m)	Camp Two (2050 m)	Ridge Camp (1000 m)
Sturnira sp.			X
Tonatia saurophila			X
Tonatia silvicola			X
Sturnira bidens		X	
Sturnira erythromos	X	X	
Sturnira magna			X
Sturnira oporaphilum		X	X
Sturnira sp.			X
Tonatia saurophila			X
Tonatia silvicola			X
Trachops cirrhosus			X
Uroderma magnirostrum			X
Vampyressa bidens			X
Vampyressa brocki			X
Myotis keaysi		X	
Myotis simus			X
Myotis sp.			X
Eptesicus andinus		X	
Histiotus montanus		X	
PRIMATES			
Aotus sp.		R	O
Alouatta seniculus			heard
Ateles belzebuth chamek ?		R	heard
Cebus albifrons		?O	O
Cebus apella		R, O	O
Lagothrix lagothricha			O
CARNIVORA			
Bassaricyon gabii			O
Nasua nasua			O
Mustela frenata	X	X	
Potos flavus			O
Tremarctos ornatus	T		
felid (puma or jaguar)	T	T	
TAPIRIDAE			
Tapirus terrestris			T, O
CERVIDAE			
Mazama chunyi	T	?	
RODENTIA			
Akodon torques	X	X	
Akodon cf *aerosus*			X
Microryzomys minutus		X	
Neacomys spinosus			X
Oecomys bicolor			X
Oecomys keaysi		X	
Oryzomys macconnelli			X
Oryzomys megacephalus			X

Species	**Camp One (3350 m)**	**Camp Two (2050 m)**	**Ridge Camp (1000 m)**
Thomasomys aureus	X		
Thomasomys cf *gracilis*	X		
Thomasomys cf *notatus*		X	
Thomasomys sp. near *kalinowski*		X	
Thomasomys sp. near *caudivarius*	X	X	
Cuscomys ashaninka	X		
Cuniculus paca			O
Cavia tschudii	T		
Dasyprocta sp.			T
Dactylomys bolivianus			X
Dactylomys peruanus		X	
Mesomys hispidus			X
Proechimys simonsi			X
TOTAL	**12**	**28**	**58**

APPENDIX 17 Bat species collected by Terborgh and Weske on an elevational transect of the Cordillera de Vilcabamba, Peru

Mónica Romo R.

Specimens were collected during 1966-1968, and are deposited at the American Museum of Natural History, New York. This list is based on data presented by Koopman (1978).

Species	Elevational range
Anoura caudifera	1700 - 2840 m
Anoura cultrata	1660 - 2260 m
Anoura geoffroyi	1700 - 3440 m
Carollia brevicauda	685 - 1800 m
Carollia castanea	685 m
Carollia perspicillata	340 - 950 m
Sturnira bidens	2600 - 3500 m
Sturnira erythromos	1700 - 3540 m
Sturnira lilium	685 m
Sturnira magna	685 - 2260 m
Sturnira oporaphilum	685 - 2260 m
Uroderma bilobatum	685 m
Platyrrhinus dorsalis	2065 - 3540 m
Platyrrhinus infuscus	685 - 950 m
Platyrrhinus nigellus	1535 - 2640 m
Platyrrhinus vittatus	---
Vampyressa bidens	685 m
Chiroderma villosum	685 m
Mesophylla macconnelli	685 - 950 m
Artibeus hartii	1700 - 3540 m
Artibeus planirostris	685 m
Myotis keaysi	3170 - 3540 m
Myotis nigricans	2620 - 3320 m
Myotis riparius	685 m

Mammal species collected in 1915 by E. Heller in the upper Urubamba valley

APPENDIX 18

Louise H. Emmons

List compiled from Thomas (1920) and specimens catalogued in the United States National Museum (USNM). Taxonomic names have been changed to current names in the USNM catalogue. ! signifies a species or subspecies described from the Heller collections.

Species	Elevation and location
Lestoros inca ! (Thomas 1917)	9500-14,000 ft; Torontoy, Machu Picchu, Ocobamba valley
Caluromys lanatus	2000 ft; Río Comberciato
Didelphis albiventris pernigra	10,000 ft; Torontoy, Chospyoc, Machu Picchu
Didelphis marsupialis	3000 ft; Santa Ana
Metachirus nudicaudatus	3000 ft; Río Cosireni
Marmosops impavidus albiventris ! (Tate 1931)	8000 ft; Torontoy
Marmosa murina quichua	Ocobamba (collected by O. Garlepp)
Monodelphis adusta	9100 ft; Ocobamba valley
Monodelphis osgoodi	9100 ft; Ocobamba valley
Peropteryx macrotis ?	Machu Picchu
Glossophaga soricina	3500, 6000 ft; Santa Ana, Idma
Anoura geoffroyi	Ollantaytambo
Carollia brevicauda	6000 ft; Idma
Phyllostomus hastatus	1800 ft; Comberciato
Desmodus rotundus	9500 ft; Puquiura
Myotis oxyotus	9400 ft; Ollantaytambo
Mormopterus phrudus ! Handley 1956	6000 ft; San Miguel Bridge
Tadarida brasiliensis	3500, 6000 ft; Santa Ana, Machu Picchu
Ateles belzebuth chamek	3000 ft; Río Comberciato
Alouatta seniculus	2000-3000 ft; Río Comberciato
Cebus albifrons	4000-5000 ft; Río Cosireni
Lagothrix lagothricha	2000 ft; Río Comberciato
Saimiri saimiri boliviensis	3000 ft; Río Comberciato
Saguinus fuscicollis	3000-4500 ft; Río Comberciato, Río San Miguel
Lycalopex culpaeus	Tocopoqueu, Ocobamba valley
Puma concolor incarum ! Nelson and Goldman 1929	8700 ft; Piscocucho
Leopardus pardalis	3500 ft; Santa Ana
Potus flavus	4000 ft; Río Cosireni
Bassaricyon alleni	3000 ft; Río Cosireni
Mustela frenata	9000 ft; Ollantaytambo

Mammal species collected in 1915 by E. Heller

Species	Elevation and location
Conepatus chinga	Machu Picchu, Chospyoc, Urca, Ocobamba valley
Lontra longicaudis	2000 ft; Río Comberciato
Tremarctos ornatus	Machu Picchu
Odocoileus virginianus	Ollantaytambo, Chopyoc, Paso Panticalla
Mazama americana	Río Cosireni
Mazama chunyi	3480 ft; Santa Ana
Hippocamelus antisensis	14,000 ft; Ollantaytambo
Lama guanicoe	16,000-17,000 ft; La Raya Pass
Vicugna vicugna mensalis ! Thomas 1917	16,000 ft; La Raya Pass
Sciurus spadiceus	2000 ft; Río Comberciato
Sciurus ignitus	2000-4000 ft; Río Comberciato, Río San Miguel
Akodon boliviensis	3500-14,000 ft; 10 localities
Akodon subfuscus	10,000 ft; Chospyoc
Akodon surdus ! Thomas 1917	5000-6000 ft; five localities
Akodon torques ! (Thomas 1917)	9000-14,000; four localities
Auliscomys pictus	11,000-14,000 ft; Huaracondo, Ollantaytambo, La Raya Pass
Bolomys amoenus	11,000 ft; Huarocondo
Calomys sorellus (= *Calomys frida* ! [Thomas 1917])	8000-11,400 ft; four localities
Calomys lepidus	14,000 ft; La Raya Pass
Chroeomys jelskii (= *Chroeomys inornatus*!, Thomas 1917)	Ollantaytambo
Holochilus sciurus	6000-9000 ft; Huaacondo, Chospyoc, Ollantaytambo, Idma
Microryzomys minutus (= *Microryzomys aurillus* !, Thomas 1917)	9500- 14,000 ft; Torontoy, Machu Picchu, Ocobamba valley
Neacomys spinosus	4500 ft; San Miguel
Nectomys squamipes	2000-4500 ft; Río Comberciato, Río Cosireni, Río San Miguel, Santa Ana
Oryzomys keaysi	3400-6000 ft; Machu Picchu, Paltaybamba, Santa Ana
Oligoryzomys destructor	5000-14,000 ft; many localities
Oligoryzomys sp.	Ocobamba Pass
Oryzomys megacephalus	3000 ft; Río Cosireni
Oryzomys nitidus	3000-6000 ft; Río Comberciato, Río Cosireni, Río San Miguel
Oryzomys yunganus	3500 ft; Santa Ana
Oxymycterus paramensis	10,000-13,000 ft; Chospyoc, Ollantaytambo
Phyllotis darwini	9100-13,000 ft; three localities
Phyllotis osilae	9000-14,000 ft; seven localities
Rhipidomys couesi	3000-6000 ft; Machu Picchu, Río San Miguel, Santa Ana
Thomasomys aureus	9000-10,000 ft; Torontoy, Ocobamba

Species	Elevation and location
Thomasomys daphne ! Thomas 1917	9100 ft; Ocobamba valley
Thomasomys gracilis ! Thomas 1917	10,700-14,000 ft; Torontoy, Machu Picchu, Ocobamba valley, Lucma
Thomasomys notatus ! Thomas 1917	8000-9500 ft; Torontoy, Machu Picchu
Thomasomys taczanowskii	Ocobamba valley
Dasyprocta variegata	5000 ft; Pumachaca (collected by E. C. Erdis)
Cavia tschudii	14000 ft.
Lagidium peruanum	13,500-14,000 ft; Ollantaytambo, Puquiura pass, La Raya

APPENDIX 19 Small mammal diversity from several montane forest localities (1300 - 2800 m) on the eastern slope of the Peruvian Andes

Sergio Solari, Elena Vivar, Paul Velazco, and Juan José Rodriquez

Key to Localities and Collections (see Chapter 8 for references)
ABI = Río Abiseo(Leo and Romo 1992; M. Romo pers. com.)
YCH = Yanachaga Chemillen (E. Vivar and S. Solari pers. obs.)
PNM = Manu (Pacheco et al. 1993, Patterson et al. 1998, S. Solari pers. obs.)
MPV = Machu Picchu and Vilcabamba (Koopman 1978, Thomas 1920)
APU = Southern Cordillera de Vilcabamba (this study)

Key to Families
did = Didelphidae
cal = Caluromyidae
mar = Marmosidae
cae = Caenolestidae
emb = Emballonuridae
phy = Phyllostomatidae
ves = Vespertilionidae
mol = Molossidae
mur = Muridae
ech = Echimyidae

		Locality Sampled				
Species	**Family**	**ABI**	**YCH**	**PNM**	**MPV**	**APU**
Didelphimorphia (marsupials)						
Didelphis albiventris	did	X			X	
Didelphis marsupialis	did			X		
Caluromys lanatus	cal					X
Gracilinanus cf. *aceramarcae*	mar	X				
Marmosa murina	mar				X	
Marmosops impavidus	mar	X		X	X	
Marmosops noctivagus	mar			X		X
Micoureus demerarae	mar	X				X
Monodelphis adusta	mar			X	X	
Monodelphis osgoodi	mar					X
Monodelphis theresa	mar		X			X
Paucituberculata						
Lestoros inca	cae			X	X	
Chiroptera (bats)						
Peropteryx macrotis	emb				X	
Micronycteris megalotis	phy	X	X	X	X	
Mimon crenulatum	phy			X	X	
Anoura caudifer	phy	X	X	X	X	X
Anoura cultrata	phy				X	
Anoura geoffroyi	phy	X	X	X	X	X
Anoura latidens	phy		X			X
Anoura sp. nov.	phy			X		

		Locality Sampled				
Species	**Family**	**ABI**	**YCH**	**PNM**	**MPV**	**APU**
Glossophaga soricina	phy				X	
Lonchophylla tomasi	phy			X		
Carollia brevicauda	phy	X	X	X	X	X
Carollia perspicillata	phy			X	X	X
Carollia sp. nov.	phy					X
Dermanura cinerea	phy		X			X
Dermanura glauca	phy		X	X	X	X
Enchistenes hartii	phy			X	X	X
Chiroderma salvini	phy			X		
Platyrrhinus "dorsalis norte"	phy	X				
Platyrrhinus "dorsalis sur"	phy		X	X	X	X
Platyrrhinus infuscus	phy			X		
Platyrrhinus cf. *nigellus*	phy	X		X	X	X
Platyrrhinus vittatus	phy				X	X
Sturnira aratathomasi	phy	X				
Sturnira bidens	phy	X	X		X	X
Sturnira erythromos	phy	X	X	X	X	X
Sturnira lilium	phy			X		
Sturnira magna	phy			X	X	
Sturnira oporaphilum	phy		X	X	X	X
Sturnira tildae	phy					X
Vampyressa macconnelli	phy			X	X	
Vampyressa melissa	phy			X		
Desmodus rotundus	phy			X	X	
Eptesicus andinus	ves		X			
Eptesicus brasiliensis	ves	X	X	X		
Eptesicus furinalis	ves			X		
Histiotus macrotus	ves	X				
Lasiurus borealis	ves	X				
Myotis keaysi	ves	X	X	X	X	X
Myotis nigricans	ves		X	X	X	
Myotis oxyotus	ves	X		X		
Mormopterus phrudus	mol				X	
Tadarida brasiliensis	mol			X	X	
Rodentia (rodents)						
Akodon aerosus	mur			X	X	X
Akodon mollis	mur		X			
Akodon orophilus	mur	X	X			
Akodon subfuscus	mur			X	X	
Akodon surdus	mur				X	
Akodon torques	mur			X	X	X
Microryzomys minutus	mur	X	X	X	X	X
Neacomys spinosus	mur	X		X	X	
Neacomys sp. nov.	mur			X		
Nectomys squamipes	mur			X	X	
Oecomys phaeotis	mur			X		
Oecomys superans	mur					X
Oligoryzomys destructor	mur			X	X	X
Oligoryzomys sp. C	mur			X		
Oryzomys albigularis	mur	X	X			
Oryzomys keaysi	mur			X	X	X
Oryzomys levipes	mur			X		
Oryzomys nitidus	mur				X	
Rhipidomys cf. *couesi*	mur			X		
Rhipidomys leucodactylus	mur				X	

		Locality Sampled				
Species	**Family**	**ABI**	**YCH**	**PNM**	**MPV**	**APU**
Rhipidomys sp.	mur	X				
Rhipidomys sp.	mur	X				
Thomasomys aureus	mur	X	X	X		
Thomasomys incanus	mur	X	X			
Thomasomys ischyrus	mur	X				
Thomasomys notatus	mur		X	X	X	
Thomasomys oreas	mur			X		
Thomasomys taczanowskii	mur				X	
Thomasomys sp.	mur	X				
Thomasomys sp.	mur		X			
Thomasomys sp.	mur		X			
Dactylomys cf. *peruanus*	ech			X		
Total species		**28**	**25**	**49**	**42**	**28**

Preliminary list of amphibians and reptiles at three sites in the Northern Cordillera de Vilcabamba, Peru

APPENDIX 20

Lily Rodríguez and Carlos Rivera

Species	Camp One (3350 m)	Camp Two (2050 m)	Ridge Camp (550-1200 m)
ANURA			
Bufonidae			
Atelopus spumarius Cope 1871			X
Bufo glaberrimus Gunther 1868			X
Bufo marinus (L. 1758)			X
Bufo typhonius complex			X
Bufo sp. A gr. *veraguensis*			X
Bufo sp. nov. gr. *veraguensis*		X	
Centrolenidae			
Centrolene sp. nov. gr. *prosoblepon*		X	
Cochranella sp. nov.		X	
Dendrobatidae			
Colostethus sp. nov.			X
Dendrobates biolat Morales 1992			X
Dendrobates sp.			X
Epipedobates femoralis (Boulenger 1884)			X
Epipedobates haneli			X
Epipedobates macero (Myers and Rodríguez 1993)			X
Hylidae			
Gastrotheca sp. nov. gr. *marsupiata*	X		
Gastrotheca sp. nov. A		X	
Gastrotheca sp. nov. B			X
Hemiphractus jhonsoni (Noble 1917)			X
Hyla fasciata Gunther 1859			X
Hyla lanciformis (Cope 1871)			X
Hyla sp.			X
Scinax rubra (Laurenti 1768)			X
Leptodactylidae			
Adenomera sp.			X
Eleutherodactylus cf *carvalhoi*			X
Eleutherodactylus sp. nov. gr. *conspicillatus*		X	
Eleutherodactylus cf *cruralis*		X	
Eleutherodactylus cf *fenestratus*			X
Eleutherodactylus sp. gr. *lacrimosus*		X	
Eleutherodactylus cf *ockendeni*			X
Eleutherodactylus cf *peruvianus*			X

Species	Camp One (3350 m)	Camp Two (2050 m)	Ridge Camp (550-1200 m)
Eleutherodactylus cf *rhabdolaemus*		X	
Eleutherodactylus skydmainos Flores and Rodríguez 1997			X
Eleutherodactylus toftae Duellman 1978			X
Eleutherodactylus sp. nov. gr. *unistrigatus*	X		
Eleutherodactylus sp. gr. *unistrigatus*		X	
Eleutherodactylus variabilis Lynch 1968			X
Eleutherodactylus sp. A			X
Eleutherodactylus sp. B			X
Ischnocnema quixensis (Jiménez de la Espada 1872)			X
Phyllonastes myrmecoides (Lynch 1976)			X
Phrynopus sp. nov. gr. *peruvianus*		X	
Phrynopus sp. nov.	X		
Telmatobius sp. nov. gr. *ventrimarmoratus*		X	
SAURIA			
Gymnophthalmidae			
Neusticurus cf *ecpleopus*			X
Proctoporus sp. nov.	X		
Hoplocercidae			
Emyalioides palpebralis			X
Polychrotidae			
Anolis ortonii			X
Teiidae			
Kentropyx pelviceps			X
Tropiduridae			
Stenocercus roseiventris			X
SERPENTES			
Boidae			
Corallus caninus			X
Cubridae			
Atractus major			X
Chironius monticola Roze 1952		X	
Dipsas sp. gr. *catesbyi*		X	
Viperidae			
Bothrops atrox			X
Bothrops cf *castelnaudi*			X
Total number of species	4	13	38

Species of amphibians and reptiles recorded at Llactahuaman and Wayrapata, Southern Cordillera de Vilcabamba, Peru

APPENDIX 21

Javier Icochea, Eliana Quispitupac, Alfredo Portilla, and Elias Ponce

Species	Number of Individuals Recorded	
	Llactahuaman (1710 m)	Wayrapata (2445 m)
ANURA (FROGS AND TOADS)		
Bufonidae		
Atelopus sp. nov.		2
Centrolenidae		
Cochranella cf. *pluvialis*	1	
Hylidae		
Hyla?	1 (larva)	
Osteocephalus sp. 1	1 (vocalization)	
Leptodactylidae		
Eleutherodactylus cruralis	5	18
Eleutherodactylus aff. *ockendeni*		1
Eleutherodactylus rhabdolaemus	2	4
Eleutherodactylus toftae	1	
Eleutherodactylus gr. *unistrigatus* A		9
Eleutherodactylus gr. *unistrigatus* B		5
Eleutherodactylus gr. *unistrigatus* C	13	9
Eleutherodactylus aff. *ventrimarmoratus*	6	
Leptodactylus sp.		1
Phrynopus?		1
Phyllonastes aff. *myrmecoides*		1
Telmatobius?		4 (larvae)
SQUAMATA: SAURIA (LIZARDS)		
Gymnophthalmidae		
Euspondylus cf. *rahmi*	3	30
Neusticurus ecpleopus	16	
Proctoporus guentheri	2	2
Tropiduridae		
Stenocercus crassicaudatus	1	
SQUAMATA: SERPENTES (SNAKES)		
Colubridae		
Atractus cf. *peruvianus*	2	
Chironius exoletus	1	
Chironius fuscus	1	
Chironius monticola		2
Liophis andinus (new record for Peru)		1
Liophis taeniurus	1	
Oxyrhopus marcapatae		5
Elapidae		
Micrurus annellatus	1	1
Viperidae		
Bothrops andianus		4
TOTAL (29 species)	**17**	**18**

APPENDIX 22 Fish species recorded in the Río Picha basin, Peru

Fonchii Chang

FAMILY	SPECIES
POTAMOTRYGONIDAE	
	Paratrygon aiereba (Müller & Henle 1841)
	Potamotrygon castexi Castello & Yagalkowski 1969
CHARACIDAE	
	Aphyocharax alburnus (Günther 1869)
	Aphyocharax nattereri (Steindachner 1862)
	Astyanacinus multidens Pearson 1924
	Astyanax bimaculatus (Linnaeus 1758)
	Astyanax fasciatus (Cuvier 1819)
	Brycon erythropterus (Cope 1872)
	Brycon melanopterus (Cope 1872)
	Bryconacidnus ellisi (Pearson 1924)
	Bryconamericus sp.
	Bryconamericus grosvenori Eigenmann 1927
	Bryconamericus pachacuti Eigenmann 1927
	Bryconamericus aff. *phoenicopterus* (Cope 1872)
	Ceratobranchia binghami Eigenmann 1919
	Ceratobranchia obtusirostris Eigenmann 1914
	Charax tectifer (Cope 1870)
	Characidium sp. A
	Characidium sp. B
	Cheirodon fugitiva (Cope 1870)
	Cheirodon piaba Lütken 1874
	Creagrutus sp.
	Ctenobrycon hauxwellianus (Cope 1870)
	Gephyrocharax sp.
	Hemibrycon jelskii (Steindachner 1875)
	Holoshestes heterodon Eigenmann 1915
	Knodus sp. A (aff. *breviceps*)
	Knodus sp. B (aff. *beta*)
	Knodus megalops Myers 1929
	Knodus moenkhausii (Eigenmann & Kennedy 1903)
	Moenkhausia aff. *chrysargyrea* (Günther 1864)
	Mylossoma duriventris (Cuvier 1818)
	Piaractus brachypomus (Cuvier 1818)
	Phenacogaster pectinatus (Cope 1870)
	Prodontocharax melanotus Pearson 1924
	aff. *Rhinopetitia* sp.
	Roeboides affinis (Günther 1868)
	Salminus hilarii Valenciennes 1849
	Scopaeocharax sp. (aff. *atopodus*)
	Serrasalmus rhombeus (Linnaeus 1776)
	Serrasalmus spilopleura Kner 1860
	Tyttocharax sp. (aff. *tambopatensis*)
GASTEROPELECIDAE	
	Thoracocharax stellatus (Kner 1860)
CYNODONTIDAE	
	Rhaphiodon vulpinus Spix 1829
ERYTHRINIDAE	
	Hoplias malabaricus (Bloch 1794)
PARODONTIDAE	
	Apareiodon sp.

FAMILY	SPECIES
PROCHILODONTIDAE	
	Prochilodus nigricans Agassiz 1829
CURIMATIDAE	
	Steindachnerina guentheri Eigenmann & Eigenmann 1889
ANOSTOMIDAE	
	Leporinus friderici (Bloch 1794)
	Schizodon fasciatus (Spix 1829)
GYMNOTIDAE	
	Gymnotus carapo Linnaeus 1758
STERNOPYGIDAE	
	Eigenmannia virescens (Valenciennes 1847)
DORADIDAE	
	Megalodoras irwini Eigenmann 1925
	Pseudodoras niger (Valenciennes 1817)
PIMELODIDAE	
	Brachyplatystoma juruense (Boulenger 1898)
	Hemisorubim platyrhynchos (Valenciennes 1840)
	Microglanis sp.
	Pimelodella gracilis (Valenciennes 1840)
	Pimelodus clarias Bloch 1785
	Pimelodus ornatus Kner 1858
	Pseudoplatystoma fasciatum (Linnaeus 1776)
	Pseudoplatystoma tigrinum (Valenciennes 1840)
	Rhamdia quelen (Quoy & Gaimart 1824)
	Sorubim lima (Schneider 1801)
	Sorubimichthys planiceps (Agassiz 1829)
	Zungaro zungaro (Humboldt 1833)
CETOPSIDAE	
	Cetopsis coecutiens (Lichtenstein 1819)
TRICHOMYCTERIDAE	
	Henonemus sp.
	Trichomycterus sp.
	Stegophylus sp.
	Vandellia plazaii Castelnau 1855
CALLICHTHYIDAE	
	Corydoras sp.
LORICARIIDAE	
	Ancistrus sp. A.
	Ancistrus sp. B
	Ancistrus sp. C
	Ancistrus aff. *teminckii* (Valenciennes 1840)
	Aphanotorulus frankei Isbrücker & Nijssen 1983
	Chaetostoma sp.
	Cochliodon sp.
	Hypostomus sp.
	Rineloricaria lanceolata (Günther 1868)
	Spatuloricaria aff. *evansii* (Boulenger 1892)
	Sturisoma sp.
SYNBRANCHIDAE	
	Synbranchus sp.
CICHLIDAE	
	Bujurquina robusta Kullander 1986
	Crenicichla sedentaria Kullander 1986

APPENDIX 23 Principal species of fish consumed in Matsigenka communities along the Río Picha

Fonchii Chang and Glenn Shepard, Jr.

Species	Spanish name	Matsigenka name
Paratrygon aiereba	aya	mamaro
Potamotrygon castexi	raya	inaro
Brycon erythropterus	sábalo	mamori
Brycon melanopterus	sábalo	mamotsi
Mylossoma duriventris	palometa	chomenta
Piaractus brachypomus	paco	komagiri
Salminus hilarii	sábalo	koviri
Serrasalmus rhombeus	paña	perero
Serrasalmus spilopleura	paña	kaporaro
Hoplias malabaricus	fasaco	tsenkori
Prochilodus nigricans	boquichico	shima
Schizodon fasciatus	lisa	kovana
Megalodoras irwini	cahuara	taya
Pseudodoras niger	turushuqui	togoso
Brachyplatystoma juruense	achuni, mota	kapeshi
Hemisorubim platyrhynchos	toa	kitepatsari
Pimelodus clarias	cunshi	korio
Pimelodus ornatus	cunshi	pariantisama
Pseudoplatystoma fasciatum	doncella	kayonaro
Pseudoplatystoma tigrinum	pumazúngaro	manitigotsi
Sorubim lima	pico de pato	sevitantsi
Sorubimichthys planieps	achacubo	charava
Zungaro zungaro	zúngaro	omani
Cochliodon sp.	carachama	shapona
Hypostomus sp.	carachama	chogeti

Number of Aquatic Invertebrate species per family found in quantitative and qualitative sampling at Llactahuaman and Wayrapata, Southern Cordillera de Vilcabamba, Peru

APPENDIX 24

Raúl Acosta, Max Hidalgo, Edgardo Castro, Norma Salcedo, and Daisy Reyes

LLA = Llactahuaman (1710 m)
WAY = Wayrapata (2445 m)

Class	Order	Family	Genus	Species	LLA	WAY
Turbellaria	Tricladida	Planariidae				6
Phy.Nemátoda						7
Phy.Molusca	Bivalvia	Piladae			X	
Oligochaeta				sp 1		25
Oligochaeta				sp 2		5
Insecta	Colembolla				X	
Insecta	Colembolla	Isotomidae		sp 1		X
Insecta	Colembolla	Isotomidae		sp 2		X
Insecta	Colembolla	Sminthuridae		sp 1		X
Insecta	Colembolla	Sminthuridae		sp 2		X
Insecta	Coleoptera	Carabidae		sp 1	X	
Insecta	Coleoptera	Carabidae		sp 2		1
Insecta	Coleoptera	Dryopidae		sp 1	3	
Insecta	Coleoptera	Dryopidae		sp 2	2	
Insecta	Coleoptera	Dytiscidae		sp 1	X	
Insecta	Coleoptera	Dytiscidae		sp 2		1
Insecta	Coleoptera	Elmidae		sp 1	7	
Insecta	Coleoptera	Elmidae		sp 2	5	
Insecta	Coleoptera	Elmidae		sp 3	1	
Insecta	Coleoptera	Elmidae		sp 4	X	
Insecta	Coleoptera	Elmidae		sp 5	X	
Insecta	Coleoptera	Elmidae		sp 6	X	
Insecta	Coleoptera	Elmidae		sp 7	X	
Insecta	Coleoptera	Elmidae		sp 8	X	
Insecta	Coleoptera	Elmidae	*Heterelmis*		62	9
Insecta	Coleoptera	Elmidae	*Macrelmis*	sp 1	13	
Insecta	Coleoptera	Elmidae	*Macrelmis*	sp 2	5	
Insecta	Coleoptera	Elmidae	*Neocylloepus*		7	
Insecta	Coleoptera	Elmidae	*Neoelmis*		45	219
Insecta	Coleoptera	Elmidae	*Phanoceroides*		4	145
Insecta	Coleoptera	Gyrinidae	*Andogyrus*			2
Insecta	Coleoptera	Gyrinidae	*Gyretes*		X	
Insecta	Coleoptera	Heteroceridae				1
Insecta	Coleoptera	Hydraenidae	*Hydraena*			38
Insecta	Coleoptera	Hydrophilidae		sp 1	X	

Class	Order	Family	Genus	Species	LLA	WAY
Insecta	Coleoptera	Hydrophilidae		sp 2	X	
Insecta	Coleoptera	Hydrophilidae		sp 3	X	
Insecta	Coleoptera	Hydrophilidae	*Derallus*			2
Insecta	Coleoptera	Hydrophilidae	*Helochares*		7	10
Insecta	Coleoptera	Hydrophilidae	*Hydrochus*			2
Insecta	Coleoptera	Hydrophilidae	*Paracymus*	sp 1		1
Insecta	Coleoptera	Hydrophilidae	*Paracymus*	sp 2		1
Insecta	Coleoptera	Lampyridae				X
Insecta	Coleoptera	Limnichidae			10	
Insecta	Coleoptera	Psephenidae			4	
Insecta	Coleoptera	Ptilidae		sp 1		1
Insecta	Coleoptera	Ptilidae		sp 2		1
Insecta	Coleoptera	Ptilidae		sp 3		1
Insecta	Coleoptera	Ptylodactilidae			4	
Insecta	Coleoptera	Scirtidae	*Scirtes*		1	30
Insecta	Coleoptera	Staphylinidae		sp 1	X	
Insecta	Coleoptera	Staphylinidae		sp 2	X	
Insecta	Coleoptera	Staphylinidae		sp 3	X	
Insecta	Coleoptera	Staphylinidae		sp 4	X	
Insecta	Coleoptera	Staphylinidae		sp 5	X	
Insecta	Coleoptera	Staphylinidae		sp 6	X	
Insecta	Coleoptera	Staphylinidae		sp 7	X	
Insecta	Coleoptera	Staphylinidae		sp 8		X
Insecta	Coleoptera	Staphylinidae		sp 9		X
Insecta	Coleoptera	Staphylinidae		sp 10		X
Insecta	Diptera	Ceratopogonidae	*Atrichopogon*	sp 1		3
Insecta	Diptera	Ceratopogonidae	*Atrichopogon*	sp 2		1
Insecta	Diptera	Ceratopogonidae	*Atrichopogon*	sp 3		X
Insecta	Diptera	Ceratopogonidae	*Bezzia*		1	
Insecta	Diptera	Ceratopogonidae	*Ceratopogon?*			22
Insecta	Diptera	Ceratopogonidae	*Forcipomyia*	sp 1		X
Insecta	Diptera	Ceratopogonidae	*Forcipomyia*	sp 2		5
Insecta	Diptera	Ceratopogonidae	*Forcipomyia*	sp 3		4
Insecta	Diptera	Ceratopogonidae	*Forcipomyia*	sp 4		1
Insecta	Diptera	Ceratopogonidae	*Forcipomyia*	sp 5		1
Insecta	Diptera	Ceratopogonidae	*Forcipomyia*	sp 6		X
Insecta	Diptera	Ceratopogonidae	*Palpomyia*	spp		47
Insecta	Diptera	Ceratopogonidae	*Probezzia*			1
Insecta	Diptera	Chironomidae		sp 1		48
Insecta	Diptera	Chironomidae		sp 2		76
Insecta	Diptera	Chironomidae		sp 3		75
Insecta	Diptera	Chironomidae		sp 4		47
Insecta	Diptera	Chironomidae		sp 5		48
Insecta	Diptera	Chironomidae		sp 6		77
Insecta	Diptera	Chironomidae		sp 7		19
Insecta	Diptera	Chironomidae		sp 8		31
Insecta	Diptera	Chironomidae		sp 9		2
Insecta	Diptera	Chironomidae		sp 10		1
Insecta	Diptera	Chironomidae		sp 11	11	

Class	Order	Family	Genus	Species	LLA	WAY
Insecta	Diptera	Chironomidae		sp 12	3	
Insecta	Diptera	Chironomidae		sp 13	23	
Insecta	Diptera	Chironomidae		sp 14	61	
Insecta	Diptera	Dixidae	*Dixella*		X	7
Insecta	Diptera	Empididae	*Chelifera*		X	
Insecta	Diptera	Empididae	*Chelifera*	sp 1	2	8
Insecta	Diptera	Empididae	*Chelifera*	sp 2	2	16
Insecta	Diptera	Ephydridae		sp 1		1
Insecta	Diptera	Ephydridae		sp 2		1
Insecta	Diptera	Ephydridae		sp 3		1
Insecta	Diptera	Ephydridae		sp 4		2
Insecta	Diptera	Ephydridae		sp 5		1
Insecta	Diptera	Ephydridae		sp 6		1
Insecta	Diptera	Muscidae		sp 1	2	
Insecta	Diptera	Muscidae		sp 2	1	
Insecta	Diptera	Psychodidae		sp 1		4
Insecta	Diptera	Psychodidae		sp 2		1
Insecta	Diptera	Psychodidae	*Maruina*			1
Insecta	Diptera	Psychodidae	*Pericoma*			2
Insecta	Diptera	Simuliidae	*Simulium*		1	28
Insecta	Diptera	Stratiomyidae	*Nemotelus*?			1
Insecta	Diptera	Tabanidae				1
Insecta	Diptera	Tipulidae		sp 1		3
Insecta	Diptera	Tipulidae		sp 2		3
Insecta	Diptera	Tipulidae		sp 3		1
Insecta	Diptera	Tipulidae		sp 4		1
Insecta	Diptera	Tipulidae		sp 5		1
Insecta	Diptera	Tipulidae	*Erioptera*			1
Insecta	Diptera	Tipulidae	*Gonomyia*			1
Insecta	Diptera	Tipulidae	*Helius*			1
Insecta	Diptera	Tipulidae	*Hexatoma*	sp 1	4	
Insecta	Diptera	Tipulidae	*Hexatoma*	sp 2	2	
Insecta	Diptera	Tipulidae	*Hexatoma*	sp 3		10
Insecta	Diptera	Tipulidae	*Limnophila*			1
Insecta	Diptera	Tipulidae	*Limonia*		4	1
Insecta	Diptera	Tipulidae	*Molophilus*		8	
Insecta	Diptera	Tipulidae	*Pseudolimnophila*			1
Insecta	Diptera	Tipulidae	*Rhabdomastix*?			1
Insecta	Diptera	Tipulidae	Tipula	sp 1	X	
Insecta	Diptera	Tipulidae	Tipula	sp 2		X
Insecta	Diptera	Tipulidae	Tipula	sp 3		X
Insecta	Diptera	Tipulidae	Tipula	sp 4		X
Insecta	Ephemeroptera				1	
Insecta	Ephemeroptera	Baetidae	*Baetis*	sp 1	38	
Insecta	Ephemeroptera	Baetidae	*Baetis*	sp 2	4	
Insecta	Ephemeroptera	Baetidae	*Baetis*	sp 3	1	
Insecta	Ephemeroptera	Baetidae	*Baetis*	sp 4		10
Insecta	Ephemeroptera	Baetidae	*Baetodes*		2	1
Insecta	Ephemeroptera	Leptohyphidae				1
Insecta	Ephemeroptera	Leptohyphidae	*Haplohyphes*		7	

Class	Order	Family	Genus	Species	LLA	WAY
Insecta	Ephemeroptera	Leptohyphidae	*Leptohyphes*		24	1
Insecta	Ephemeroptera	Leptophlebiidae		sp 1	2	
Insecta	Ephemeroptera	Leptophlebiidae		sp 2	3	
Insecta	Ephemeroptera	Leptophlebiidae		sp 3	1	
Insecta	Ephemeroptera	Leptophlebiidae		sp 4	2	
Insecta	Ephemeroptera	Leptophlebiidae		sp 5	2	
Insecta	Ephemeroptera	Leptophlebiidae		sp 6	7	
Insecta	Ephemeroptera	Leptophlebiidae		sp 7		1
Insecta	Ephemeroptera	Leptophlebiidae	*Thraulodes*		41	
Insecta	Ephemeroptera	Leptophlebiidae	*Traverella*		X	
Insecta	Hemiptera	Gerridae				X
Insecta	Hemiptera	Gerridae	*Eurygerris ?*			X
Insecta	Hemiptera	Gerridae	*Limnoporus*		X	
Insecta	Hemiptera	Gerridae	*Trepobates*		X	
Insecta	Hemiptera	Naucoridae	*Limnocoris*		1	
Insecta	Hemiptera	Veliidae	*Rhagovelia*		X	X
Insecta	Hemiptera	Veliidae	*Microvelia*		X	
Insecta	Megaloptera	Corydalidae	*Corydalus*		X	
Insecta	Lepidoptera			sp 1		1
Insecta	Lepidoptera			sp 2	1	
Insecta	Lepidoptera	Arctiidae				X
Insecta	Lepidoptera	Cossidae				1
Insecta	Lepidoptera	Pyralidae		sp 1	X	
Insecta	Lepidoptera	Pyralidae		sp 2		X
Insecta	Odonata	Aeshnidae	*Aeshna*			9
Insecta	Odonata	Gomphidae	*Progomphus*		X	
Insecta	Odonata	Megapodagrionidae			X	
Insecta	Odonata	Megapodagrionidae	*Megapodagrion*		X	
Insecta	Odonata	Polythoridae		sp 1	17	
Insecta	Odonata	Polythoridae		sp 2		2
Insecta	Plecoptera	Perlidae	*Anacroneuria*		28	59
Insecta	Trichoptera	Calamoceratidae	*Banyallarga*			177
Insecta	Trichoptera	Calamoceratidae	*Phylloicus ?*		42	
Insecta	Trichoptera	Hydroptilidae	*Ochrotrichia*		2	12
Insecta	Trichoptera	Hydrobioscidae	*Atopsyche*		2	9
Insecta	Trichoptera	Hydropsychidae	*Leptonema*	sp 1	2	
Insecta	Trichoptera	Hydropsychidae	*Leptonema*	sp 2	12	
Insecta	Trichoptera	Hydropsychidae	*Macrostemum*		2	
Insecta	Trichoptera	Hydropsychidae	*Smicridea*		30	236
Insecta	Trichoptera	Leptoceridae		sp 1		16
Insecta	Trichoptera	Leptoceridae		sp 2		32
Insecta	Trichoptera	Leptoceridae		sp 3		8
Insecta	Trichoptera	Leptoceridae		sp 4		20
Insecta	Trichoptera	Leptoceridae		sp 5		2
Insecta	Trichoptera	Leptoceridae		sp 6		7

Class	Order	Family	Genus	Species	LLA	WAY
Insecta	Trichoptera	Leptoceridae	*Triplectides*	sp 1	20	11
Insecta	Trichoptera	Leptoceridae	*Triplectides*	sp 2	3	
Insecta	Trichoptera	Odontoceridae	*Marilia*	sp 1	7	
Insecta	Trichoptera	Odontoceridae	*Marilia*	sp 2	8	
Insecta	Trichoptera	Philopotamidae	*Wormaldia*		1	
Insecta	Trichoptera	Philopotamidae	*Chimarra*		2	
Insecta	Trichoptera	Polycentropodidae	*Polycentropus*		1	18
Arachnida	Acarina	N.d	sp1			3
Arachnida	Acarina	N.d	sp2			9
Arachnida	Acarina	N.d	sp3			1
Arachnida	Acarina	N.d	sp4			1
	Sub.ord. Siluriformes	Astroblepidae	Astroblepus		X	
	Sub.ord. Siluriformes	Trichomycteridae	Trichomycterus		X	

APPENDIX 25 Lepidoptera collected at two sites in the Northern Cordillera de Vilcabamba, Peru

Gerardo Lamas and Juan Grados

Species	Camp One (3350 m)	Camp Two (2050 m)
NYMPHALIDAE		
Heliconiinae		
1. *Altinote hilaris hilaris* (Jordan 1910)		X
2. *Dione glycera* (C. Felder & R. Felder 1861)	X	X
Nymphalinae		
3. *Telenassa delphia nana* (Druce 1874)		X
4. *Vanessa braziliensis* (Moore 1883)	X	
Biblidinae		
5. *Orophila diotima* ssp. n.	X	
Satyrinae		
6. *Corades cistene* Thieme 1907	X	
7. *Corades medeba medeba* Hewitson 1850		X
8. *Corades melania* Staudinger 1894	X	
9. *Corades ulema ulema* Hewitson 1850	X	
10. *Corades* sp. n.	X	
11. *Junea doraete gideon* (Thieme 1907)	X	
12. *Lasiophila zapatoza orbifera* Butler 1868		X
13. *Lasiophila* sp. n.	X	
14. *Lymanopoda galactea siviae* Fuchs 1954	X	
15. *Lymanopoda obsoleta* (Westwood 1851)		X
16. *Manerebia cyclopella* Staudinger 1897		X
17. *Manerebia* sp. n.	X	
18. *Pedaliodes phrasa* Grose-Smith & Kirby 1894		X
19. *Pedaliodes phrasicla* (Hewitson 1874)		X
20. *Pedaliodes* sp. n. 1	X	
21. *Pedaliodes* sp. n. 2	X	
22. *Pedaliodes* sp. n. 3	X	
23. *Pedaliodes* sp. n. 4	X	
24. *Pedaliodes* sp. n. 5		X
25. *Physcopedaliodes praxithea* (Hewitson 1870)		X
26. *Steremnia monachella* (Weymer 1911)	X	
27. *Steroma bega andensis* C. Felder & R. Felder 1867		X
28. *Steroma superba* Butler 1868		X
29. *Steroma* sp. n.	X	
30. *Forsterinaria rustica rustica* (Butler 1868)		X

Species	Camp One (3350 m)	Camp Two (2050 m)
Ithomiinae		
31. *Napeogenes harbona domiduca* (Hewitson 1876)		X
PIERIDAE		
32. *Leodonta tagaste* ssp. n.	X	
33. *Lieinix nemesis nemesis* (Latreille [1813])		X
34. *Catasticta similis* Lathy & Rosenberg 1914	X	
35. *Catasticta colla punctata* Lathy & Rosenberg 1914		X
36. *Catasticta chelidonis contrasta* Reissinger 1972		X
37. *Catasticta cinerea* ssp. n.	X	
38. *Catasticta* sp. n.	X	
39. *Tatochila xanthodice paucar* Lamas 1981	X	
LYCAENIDAE		
40. *Penaincisalia candor* (Druce 1907)	X	
41. *Johnsonita* sp. n.	X	
HESPERIIDAE		
Pyrrhopyginae		
42. *Metardaris cosinga catana* Evans 1951	X	
Hesperiinae		
43. *Dalla costala costala* Evans 1955	X	
44. *Dalla* sp. n.	X	
45. *Serdis viridicans* ssp. n.	X	
46. *Thespieus thona* Evans 1955	X	
Pyrginae		
47. *Potamanaxas laoma cosna* Evans 1953		X
MEGALOPYGIDAE		
48. *Podalia* sp.	X	
CERCOPHANIDAE		
49. *Janiodes bethulia* (Druce 1904)	X	
SATURNIIDAE		
50. *Bathyphlebia aglia* R. Felder 1874	X	
SPHINGIDAE		
51. *Agrius cingulatus* (Fabricius 1775)	X	
52. *Euryglottis aper* (Walker 1856)	X	
53. *Euryglottis dognini* Rothschild 1896	X	
54. *Pseudosphinx tetrio* (Linnaeus 1771)	X	
55. *Erinnyis alope alope* (Drury 1773)	X	
56. *Erinnyis ello ello* (Linnaeus 1758)	X	
57. *Erinnyis lassauxii* (Boisduval 1859)	X	
58. *Pachylia ficus* (Linnaeus 1758)	X	
59. *Oryba achemenides* (Cramer [1779])	X	
60. *Perigonia lusca interrupta* Walker 1864	X	
61. *Enyo lugubris lugubris* (Linnaeus 1771)	X	
62. *Enyo ocypete* (Linnaeus 1758)	X	
63. *Xylophanes crotonis* (Walker 1856)	X	

Species	Camp One (3350 m)	Camp Two (2050 m)
64. *Xylophanes pluto* (Fabricius 1776)	X	
ARCTIIDAE		
65. *Neonerita haemasticta* (Dognin 1906)	X	
66. *Elysius atrata* (R. Felder 1874)	X	
67. *Elysius terra* Druce 1906	X	
68. *Hemihyalea watkinsi* (Rothschild 1916)	X	
69. *Amastus aconia* (Herrich-Schäffer 1853)	X	
70. *Amastus* aff. *persimilis* Hampson 1901	X	
71. *Amastus* aff. *subtenuimargo* Rothschild 1916	X	
72. *Amastus* sp. 3	X	
73. *Amastus* sp. 4	X	
74. *Amastus* sp. 5	X	
75. *Amastus* sp. 6	X	
76. *Chlorhoda* sp.	X	

Coleoptera sampled in three study plots at Llactahuaman (1710 m), Southern Cordillera de Vilcabamba, Peru

APPENDIX 26

Jose Santisteban, Roberto Polo, Gorky Valencia, Saida Córdova, Manuel Laime, and Alicia de la Cruz

LL02, LL04, LL05 = Llactahuaman sampling plots 2, 4 and 5
see Chapter 15 for plot descriptions.

		Number of Individuals collected			
Family	**species**	**LL02**	**LL04**	**LL05**	**Total**
Cantharidae	sp01	3	1	28	32
Cantharidae	sp02	2			2
Carabidae	sp01			2	2
Carabidae	sp02		2	1	3
Carabidae	sp03		6	2	8
Carabidae	sp04		1		1
Carabidae	sp05		1		1
Cerambycidae	sp01		1		1
Chrysomelidae	sp01			1	1
Chrysomelidae	sp02			1	1
Chrysomelidae	sp03	1	1	2	4
Chrysomelidae	sp04		1	1	2
Chrysomelidae	sp05			1	1
Chrysomelidae	sp06		18	2	20
Chrysomelidae	sp07		14	1	15
Chrysomelidae	sp08			2	2
Chrysomelidae	sp09			1	1
Chrysomelidae	sp10			1	1
Chrysomelidae	sp11			1	1
Chrysomelidae	sp12		6	2	8
Chrysomelidae	sp13		2	2	4
Chrysomelidae	sp14	5	4		9
Chrysomelidae	sp15			2	2

			Number of Individuals collected		
Family	species	LL02	LL04	LL05	Total
Chrysomelidae	sp16			1	1
Chrysomelidae	sp17		14		14
Chrysomelidae	sp18		2		2
Chrysomelidae	sp19		1		1
Chrysomelidae	sp20		2		2
Chrysomelidae	sp21		1		1
Chrysomelidae	sp22		1		1
Chrysomelidae	sp23	1	2		3
Chrysomelidae	sp24		7		7
Chrysomelidae	sp25		1		1
Chrysomelidae	sp26		10		10
Chrysomelidae	sp27		2		2
Chrysomelidae	sp28		1		1
Chrysomelidae	sp29		8		8
Chrysomelidae	sp30		1		1
Chrysomelidae	sp31	1	1		2
Chrysomelidae	sp32		1		1
Chrysomelidae	sp33		1		1
Chrysomelidae	sp34		1		1
Chrysomelidae	sp35		1		1
Chrysomelidae	sp36	4	4		8
Chrysomelidae	sp37		3		3
Chrysomelidae	sp38		1		1
Chrysomelidae	sp39		1		1
Chrysomelidae	sp40		2		2
Chrysomelidae	sp41		4		4
Chrysomelidae	sp42	5	2		7
Chrysomelidae	sp43		1		1
Chrysomelidae	sp44		1		1
Chrysomelidae	sp45	4			4
Chrysomelidae	sp46	1			1
Chrysomelidae	sp47	1			1
Chrysomelidae	sp48	4			4
Chrysomelidae	sp49	1			1
Chrysomelidae	sp50	1			1
Chrysomelidae	sp51	1			1

		Number of Individuals collected			
Family	**species**	**LL02**	**LL04**	**LL05**	**Total**
Chrysomelidae	sp52	1			1
Coccinellidae	sp01			1	1
Coccinellidae	sp02		1		1
Coccinellidae	sp03	5	1		6
Coleoptera	und		5	3	8
Coleoptera to det	sp01		3		3
Coleoptera to det	sp02		2		2
Coleoptera to det	sp03	2	1		3
Coleoptera to det	sp04	1			1
Coleoptera to det	sp05	1			1
Coleoptera to det	sp06	1			1
Coleoptera to det	sp07	1			1
Coleoptera to det	sp08	2			2
Coleoptera to det	sp09	1			1
Coleoptera to det	sp10	4			4
Coleoptera to det	sp11	1			1
Coleoptera to det	sp12	1			1
Curculionidae	sp01	4			4
Curculionidae	sp02		1		1
Curculionidae	sp03		1		1
Curculionidae	sp04	2			2
Curculionidae	sp05	2			2
Curculionidae	sp06	1			1
Curculionidae	sp07	4			4
Curculionidae	sp08	8			8
Curculionidae	sp09	1			1
Elateridae	sp01	1			1
Erotylidae	sp01		1		1
Histeridae	sp01	1		1	2
Histeridae	sp02	1			1
Lampyridae	sp01		2		2
Lampyridae	sp02		1		1
Lycidae	sp01		1		1
Lycidae	sp02	2			2
Lycidae	sp03	2			2

		Number of Individuals collected			
Family	species	LL02	LL04	LL05	Total
Lycidae	sp04	1			1
Melyridae	sp01		4		4
Mordellidae	sp01			1	1
Nitidulidae	sp01	1		4	5
Nitidulidae	sp02	1		1	2
Nitidulidae	sp03		1		1
Nitidulidae	sp04	2			2
Nitidulidae	sp05	1			1
Nitidulidae	sp06	1			1
Nitidulidae	sp07	1			1
Nitidulidae	sp08	1			1
Pselaphidae	sp01		3		3
Pselaphidae	sp03		1		1
Pselaphidae	sp04	1			1
Pselaphidae	sp05	3			3
Pselaphidae	sp06	1			1
Ptiliidae	sp01	1	1		2
Ptilodactylidae	sp01	2	5	1	8
Scolytidae	sp01			2	2
Scolytidae	sp02	1		1	2
Scolytidae	sp03	1	1		2
Scolytidae	sp04		3		3
Scolytidae	sp05		1		1
Scolytidae	sp06		1		1
Scolytidae	sp07		1		1
Scolytidae	sp08		1		1
Scolytidae	sp09		3		3
Scolytidae	sp10		1		1
Scolytidae	sp11	2	1		3
Scolytidae	sp12		1		1
Scolytidae	sp13	3			3
Scolytidae	sp14	1			1
Scolytidae	sp15	1			1
Scolytidae	sp16	1			1
Scolytidae	sp17	1			1

		Number of Individuals collected			
Family	species	LL02	LL04	LL05	Total
Scolytidae	und		1		1
Staphylinidae	sp01		2	1	3
Staphylinidae	sp02	3	4	2	9
Staphylinidae	sp03	18	17	7	42
Staphylinidae	sp04	3		2	5
Staphylinidae	sp05		1	1	2
Staphylinidae	sp06			2	2
Staphylinidae	sp07		1	1	2
Staphylinidae	sp08		2	1	3
Staphylinidae	sp09		1		1
Staphylinidae	sp10		3		3
Staphylinidae	sp11	1	1		2
Staphylinidae	sp12		4		4
Staphylinidae	sp13		1		1
Staphylinidae	sp14	2	13		15
Staphylinidae	sp15		1		1
Staphylinidae	sp16		1		1
Staphylinidae	sp17	3	1		4
Staphylinidae	sp18		1		1
Staphylinidae	sp19	1	2		3
Staphylinidae	sp20	2	9		11
Staphylinidae	sp21		1	1	2
Staphylinidae	sp22		3		3
Staphylinidae	sp23	1	3		4
Staphylinidae	sp24		1		1
Staphylinidae	sp25		1		1
Staphylinidae	sp26		1		1
Staphylinidae	sp27		2		2
Staphylinidae	sp28		2		2
Staphylinidae	sp29	3	1		4
Staphylinidae	sp30	5			5
Staphylinidae	sp31	4			4
Staphylinidae	sp32	1			1
Staphylinidae	sp33	1			1
Staphylinidae	sp34	1			1
Staphylinidae	und	23	22	9	54
Tenebrionidae	sp01			1	1
TOTAL		**185**	**287**	**97**	**569**

APPENDIX 27

Hymenoptera-Aculeata sampled in three study plots at Llactahuaman (1710 m), Southern Cordillera de Vilcabamba, Peru

Jose Santisteban, Roberto Polo, Gorky Valencia, Saida Córdova, Manuel Laime, and Alicia de la Cruz

LL02 and LL04= Llactahuaman sampling plots 2 and 4;
See Chapter 15 for plot descriptions.

Family	species	LL02	LL04	Total
Apidae	*Apis mellifera*		8	8
Apidae	*Melipona* sp01	1	2	3
Apidae	*Melipona* sp02		1	1
Apidae	*Meliponini* sp01		1	1
Apidae	*Meliponini* sp02		7	7
Apidae	*Nannotrigona* sp01		6	6
Apidae	*Nannotrigona* sp02		10	10
Apidae	*Partamona* sp01		2	2
Apidae	*Partamona* sp02		1	1
Apidae	*Partamona* sp03		2	2
Apidae	*Partamona* sp04		1	1
Apidae	*Partamona* sp05		5	5
Apidae	*Partamona* sp06		1	1
Apidae	sp01		6	6
Apidae	sp12	2		2
Apidae	*Trigona* sp01	23	12	35
Apidae	*Trigona* sp02	2	57	59
Apidae	*Trigona* sp03	9		9
Apidae	*Trigona* sp04	1	5	6
Apidae	*Trigona* sp05		37	37
Apidae	*Trigona* sp06		1	1
Apidae	*Trigona* sp07		40	40
Apidae	*Trigona* sp08		1	1

Family	species	LL02	LL04	Total
Bethylidae	sp01	5	10	15
Bethylidae	sp02	4		4
Bethylidae	sp02		20	20
Bethylidae	sp03	4	1	5
Bethylidae	sp04	1	2	3
Bethylidae	sp05	1		1
Bethylidae	sp05		1	1
Bethylidae	sp07		3	3
Bethylidae	sp08		29	29
Bethylidae	sp09		10	10
Bethylidae	sp10		1	1
Bethylidae	sp11		2	2
Bethylidae	sp12		6	6
Bethylidae	sp13		3	3
Bethylidae	sp14		2	2
Bethylidae	sp16		2	2
Bethylidae	sp17		1	1
Bethylidae	sp18		5	5
Bethylidae	sp19		1	1
Bethylidae	sp20		1	1
Bethylidae	sp21		5	5
Bethylidae	sp22		21	21
Bethylidae	sp23		1	1
Bethylidae	sp24		1	1
Bethylidae	sp25		1	1
Bethylidae	sp26		4	4
Bethylidae	sp27		1	1
Bethylidae	sp28		1	1
Bethylidae	sp29		10	10
Bethylidae	sp30		1	1
Bethylidae	sp31		1	1
Bethylidae	sp32		1	1
Bethylidae	sp33		2	2
Bethylidae	sp34		1	1
Chrysididae	*Amiseginae* sp01		1	1
Colletidae	sp01		1	1
Dryinidae	sp01		1	1
Halictidae	*Halictinae* sp01	1		1
Halictidae	sp01		1	1

Hymenoptera-Aculeata of Llactahuaman (1710 m)

Family	species	LL02	LL04	Total
Halictidae	sp02		9	9
Halictidae	sp03		1	1
Halictidae	sp04		1	1
Halictidae	sp05		8	8
Halictidae	sp06		2	2
Halictidae	sp07		2	2
Halictidae	sp08		1	1
Halictidae	sp09		2	2
Halictidae	sp10		1	1
Halictidae	sp11		1	1
Halictidae	sp12		2	2
Halictidae	sp13		1	1
Pompilidae	sp01	1		1
Pompilidae	sp03		2	2
Pompilidae	sp04		1	1
Pompilidae	sp05		2	2
Pompilidae	sp06		3	3
Pompilidae	sp07		1	1
Pompilidae	sp08		1	1
Sclerogibbidae	sp01	1		1
Sphecidae	Crabroninae 01		1	1
Sphecidae	*Liris* sp01	1		1
Sphecidae	*Liris* sp02	1		1
Sphecidae	*Liris* sp03		1	1
Sphecidae	*Liris* sp04		2	2
Sphecidae	*Pseninae* sp01		1	1
Vespidae	*Agelaia* sp01	4	1	5
Vespidae	*Polistes* sp01		2	2
Vespidae	*Polybia* sp01	1		1
Vespidae	*Polybia* sp02		1	1
Vespidae	*Polybia* sp03		1	1
Vespidae	*Polybia* sp04		1	1
Vespidae	*Polybia* sp05		3	3
Vespidae	*Polybia* sp06		4	4
Vespidae	*Polybia* sp07		2	2
Vespidae	*Polybia* sp08		1	1
Vespidae	*Polybia* sp09		1	1
Vespidae	*Polybia* sp11		1	1
Vespidae	*Polybia* sp12		3	3
Vespoidea	N.D. 01	2		2
TOTAL		**65**	**428**	**493**

Hymenoptera-Aculeata sampled in three study plots at Wayrapata (2445 m), Southern Cordillera de Vilcabamba, Peru.

APPENDIX 28

Jose Santisteban, Roberto Polo, Gorky Valencia, Saida Córdova, Manuel Laime, and Alicia de la Cruz

WA04, WA05, WA06 = Wayrapata sampling plots 4, 5 and 6

See Chapter 15 for plot descriptions.

Family	species	WA04	WA05	WA06	Total
Apidae	*Apis mellifera*	5	7	3	15
Apidae	*Bombus* sp01	3	1		4
Apidae	*Melipona* sp01		1		1
Apidae	*Melipona* sp02			1	1
Apidae	Meliponini sp04			3	3
Apidae	Partamona sp07	1	2		3
Apidae	Partamona sp08	2	3		5
Apidae	sp09	2			2
Bethylidae	sp01	1	1		2
Bethylidae	sp02	6			6
Bethylidae	sp08		2		2
Bethylidae	sp11		1		1
Bethylidae	sp20	4	7		11
Bethylidae	sp21	1			1
Bethylidae	sp29	4			4
Bethylidae	sp35	1			1
Bethylidae	sp36	1			1
Bethylidae	sp37	1			1
Bethylidae	sp38	1			1
Chrysididae	Amiseginae sp02		1		1
Colletidae	Hylaeinae sp01		1		1
Halictidae	sp05		2		2
Halictidae	sp08			2	2
Halictidae	sp09		2	1	3
Halictidae	sp10		2		2
Halictidae	sp15	2			2
Halictidae	sp16	1			1
Halictidae	sp17	4			4
Halictidae	sp18		9	2	11
Halictidae	sp19		5		5
Halictidae	sp20		8	1	9

Family	species	WA04	WA05	WA06	Total
Halictidae	sp21		2	2	4
Halictidae	sp22		2		2
Halictidae	sp23		10		10
Halictidae	sp24		1	5	6
Halictidae	sp25		1		1
Halictidae	sp26		2		2
Halictidae	sp27		1		1
Halictidae	sp28		1		1
Halictidae	sp29		1		1
Halictidae	sp30		2	2	4
Halictidae	sp31		1		1
Halictidae	sp32		4		4
Pompilidae	sp01	1			1
Pompilidae	sp09	3			3
Pompilidae	sp10		2		2
Pompilidae	sp11	1			1
Pompilidae	sp12	1			1
Pompilidae	sp13		1		1
Sphecidae	Crabronini sp02	1			1
Sphecidae	Crabronini sp03	1			1
Sphecidae	Crabronini sp04	1			1
Sphecidae	Crabronini sp05	1			1
Sphecidae	Crabronini sp06		3		3
Sphecidae	*Trypoxylon* sp01	1			1
Sphecidae	*Trypoxylon* sp02			2	2
Sphecidae	*Trypoxylon* sp03			1	1
Sphecidae	*Trypoxylon* sp04			2	2
Tiphiidae	sp01		1		1
Vespidae	Eumeninae sp01		1		1
Vespidae	*Polybia* sp05		1		1
Vespidae	*Polybia* sp09	1			1
Vespidae	*Polybia* sp13	4			4
Vespidae	*Polybia* sp14	3	13	2	18
Vespidae	*Polybia* sp15	3			3
Vespidae	*Polybia* sp16	9			9
TOTAL		**71**	**105**	**29**	**205**

Crops in "Daniel's" gardens

APPENDIX 29

Glenn Shepard, Jr. and Avecita Chicchón

Adapted from Johnson 1983.

Species	Common name	Matsigenka name	Asháninka name	Use
Alocasia	*Taro*		*Shimpiri*	
Acmella ciliata	*Botoncillo* (daisy relative)	*Koviriki*		Medicinal
Ananas comosus	Pineapple	*Tsirianti*	*Tsivana*	Edible fruit
Anthurium spp.	(Arrow root relative)	*Pankuro*		Root crop
Arachis hypogaea	Peanut	*Iinke*	*Inki*	Food
Bactris gasipaes	Peach palm (*pijuayo*)	*Kuiri*		Palm heart, fruit, wood
Bactris sp.	(Peach palm relative)	*Manataroki*		Edible fruit
Bixa orellana	Annato (*achiote*)	*Potsoti*		Face paint, dye
Bromelia serra	*Fibra* (Bromeliad)	*Tivana*		Fiber
Brugmansia sp.	*Toé* (nightshade relative)	*Saaro*		Medicinal, narcotic
Calathea allouia	*Dale-dale* (Marantaceae)	*Shonaki*	*Shoñaki*	Root crop
Calathea pearcei	(Marantaceae)	*Shirina*		Root crop
Capsicum spp.	Chile pepper	*Tsitikana*	*Tsitsikana*	Condiment
Carica papaya	Papaya	*Tinti*	*Mapocha*	Edible fruit
Coffea arabica	Coffee	*Café*		Cash crop
Coix sp.	(Grass with bead-like seeds)	*Sarioki*		Beads
Crescentia sp.	Tree gourd	*Pamoko*	*Pajo*	Serving bowls
Cucurbita sp.	Squash	*Kemi*	*Kemi*	Food
Cyclanthaceae	*Tamshi*	*Kepia*		Fibers, manufacture
Cyperus sp.	Sedge (*piri-piri*)	*Shigovana*	*Ivenki*	Medicinal
Cyperus spp.	Sedge (*piri-piri*)	*Ivenkiki*		Medicinal
Dioscorea sp.	Yam	*Magona*		Root crop
Eleutherine plicata	*Yawar piri-piri* (Iris relative)	*Tsamiriventaki*		Medicinal
Eryngium foetidum	*Sacha culantro* (coriander relative)	*Sankonka*		Medicinal, condiment
Gossypium barbadense	Cotton	*Ampei*	*Ampeji*	Manufacture
Inga sp.	Guaba (*pacai*)	*Intsipa*	*Intsipa*	Edible fruit pulp
Inga sp.	(Guaba relative)	*Vivingoki*		Edible fruit pulp
Ipomoea batatas	Sweet potato	*Koriti*	*Koriti*	Root crop
Lagenaria sp.	Bottle gourd	*Piarintsi*	*Pachaka*	Water jugs, bowls
Lonchocarpus sp.	*Barbasco*	*Kogi*	*Koñapi*	Fish poison
Manihot esculenta	Manioc	*Sekatsi*	*Kaniri*	Major root crop
Martinella obovata	*Yuquilla* (bignoniaceae)	*Pocharoki*		Medicinal
Mauritia flexuosa	*Aguaje* (Palm)	*Toturoki*		Edible fruit
Musa paradisiaca	Plantain	*Parianti*		Edible fruit
Musa sp.	Banana	*Posuro*	*Parenti*	Edible fruit
Nicotiana tabacum	Tobacco	*Seri*	*Shiri*	Medicinal, narcotic

Species	Common name	Matsigenka name	Asháninka name	Use
Oenocarpus mapora	*Ungurahuillo* (Palm)	*Chorina*		Palm heart, edible fruit
Pachyrhizus tuberosus	*Jicama*	*Poi*		Sweet tuber
Persea americana	Avocado	*Tsivi*	*Akapa*	Edible fruit
Phaseolus vulgaris	Kidney bean	*Mentani*		Legume
Pourouma	*Uvilla*	*Sevantoki*		Edible fruit
Pouteria sp.	Zapote (*caimito*)	*Caimito, segorikashi*		Edible fruit
Psidium guajava	Guava	*Komashiki*	*Tomashiki*	Edible fruit
Renealmia sp.	(Turmeric relative)	*Porenki*		Condiment, medicine
Saccharum officinarum	Sugar cane	*Shanko, impogo*	*Sanko*	Sweet cane, juice
Tephrosia toxicaria	Barbasco	*Kogi*		Fish poison
Theobroma cacao	Cacao	*Cacao, sarigemini*	*Kemito*	Edible fruit, cash crop
Xanthosoma nigra	Cocoyam	*Onko*	*Pitoka*	Root crop
Xanthosoma?	(Arrow root relative)	*Makato*	*Onko*	Root crop
Zea mays	Maize	*Shinki*	*Shinki*	Major seed crop
Zingiber officinale	Ginger	*Katsikonakiri*		Medicinal
Camaralia sp.	Bean	*Poroto*	*Machaki*	Edible
Oryza	Rice	*Aroshi*	*Arooso*	Edible
Nicotana tabacum	Tobacco	*Seri*	*Sheri*	Medicinal
Allium cepa	Onion	*Cebolla, sampera*	*Sheroyo*	Edible

Game and other large animals of the Río Picha

APPENDIX 30

Glenn Shepard, Jr. and Avecita Chicchón

Information on animal encounters based on approximately 30 hours of observation time, during six forest walks.

Animal type	Species	Common name	Matsigenka name	Encounters (number)
Birds (Cracids)	*Aburria pipile*	Common piping guan	*Kanari*	seen (1)
	Crax tuberosa	Razor-billed currasow	*Tsamiri*	--
	Ortalis motmot	Variable chacalaca	*Marati*	heard (2)
	Penelope jacquacu	Spix's guan	*Sangati*	seen (3), killed (1)
Birds (Psittacids)	*Ara ararauna*	Blue-and-yellow macaw	*Kasanto*	seen (group of 3)
	Ara chloroptera	Red-and-green macaw	*Kimaro*	--
	Ara macao	Scarlet macaw	*Kiteri kimaro*	--
	Ara militaris	Military macaw	*Meganto*	(high elevations only)
	Ara severa	Chestnut-fronted macaw	*Eroti*	--
	Pionus menstruus	Blue-headed parrot	*Saveto*	--
	Amazonas spp.	Parrots	*Tsorito*	--
Birds (Other)	*Aramides cajanea*	Wood-rail	*Puvanti*	seen (1)
	Cacicus (C. cela?)	Cacique	*Katsari*	seen (1)
	Columba spp.	Pigeons	*Sheromega*	--
	Crypturellus, Tinamus spp.	Tinamous	*Kentsori, shonkiri*	heard frequently
	Leptotila (Geotrygon?)	Dove	*Sampakiti*	--
	Odontophorus stellatus	Wood-quail	*Kontona*	seen (2)
	Ramphastos sp.	Toucan	*Yotoni*	seen (1)
Ungulates	*Mazama americana*	Red brocket deer	*Maniro*	tracks (1), seen (3), killed (1)
	Mazama gouazoubira	Grey brocket deer	*Maniro potsitari*	--
	Tapirus terrestris	Brazilian tapir	*Kemari*	tracks (3)
	Tayassu pecari	White-lipped peccary	*Santaviri, imarapage*	tracks (3 herds)
	Tayassu tajaca	Collared peccary	*Shintori*	tracks (1 herd), killed (1)
Primates	*Alouatta seniculus*	Red howler monkey	*Yaniri*	seen (group of 4)
	Aotus sp.	Night monkey	*Pitoni*	(nocturnal)
	Aotus sp.	Night monkey	*Marampitoni*	(nocturnal)
	Ateles paniscus	Spider monkey	*Osheto*	--
	Callicebus moloch	Dusky titi monkey	*Togari*	seen (1 troop)
	Cebuella pygmaea	Pygmy marmoset	*Tampianiro tsigeri*	--
	Cebus albifrons	White-fronted capuchin	*Koakoa*	--
	Cebus apella	Brown capuchin	*Koshiri*	seen (2 troops)
	Lagothrix lagotricha	Woolly monkey	*Komaginaro*	--
	Pithecia monachus	Monk saki	*Maramponi*	--
	Saguinus emperator	Emperor tamarin	*Tsintsipoti*	--
	Saguinus fuscicollis	Saddleback tamarin	*Tsintsipoti potsitari*	--
	Saimiri sciureus	Squirrel monkey	*Tsigeri*	seen (2 troops)
Rodents	*Agouti paca*	Paca	*Samani*	--
	Dasyprocta variegata	Agouti	*Sharoni*	seen (1)
	Hydrochaeris hydrochaeris	Capybara	*Iveto*	--
	Sciurus sp.	Squirrel	*Megiri*	--

Animal type	Species	Common name	Matsigenka name	Encounters (number)
Carnivores (Felids)	*Felis concolor*	Puma	*Maniti, matsontsori kiraari*	--
	Felis pardalis	Ocelot	*Matsontsori*	tracks (2), feces (1)
	Felis tigrina	Oncilla	*Pamoko*	--
	Felis wiedii	Margay	*Vamporoshi*	(nocturnal)
	Felis yagouarundi	Jaguarundi	*Potsitari matsontsori*	--
	Panthera onca	Jaguar	*Matsontsori sankenari*	tracks (1)
Carnivores (Others)	*Atelacynus microtis*	Short-eared dog	*Matsiti*	tracks (1)
	Eira barbara	Tayra	*Oati*	--
	Lutra longicaudis	Nutria	*Parari*	--
	Nasua nasua	Coatimundi	*Kapeshi*	seen (2 together)
	Potus flavus	Kinkajou	*Kutsani*	--
	Procyon sp.	Raccoon	*Koventsiri*	--
	Pteronura brasilensis	Giant otter	*Chavaropana*	--
Edentata	*Choloepus* sp.	Two-toed sloth	*Soroni*	--
	Dasypus novemcinctus	Nine-banded armadillo	*Etini*	--
	Myrmecophaga tridactyla	Giant anteater	*Shiani*	--
	Priodontes maximus	Giant armadillo	*Kinteroni*	--

Useful palms of the Río Picha area

APPENDIX 31

Glenn Shepard, Jr. and Avecita Chicchón

Species	Common name	Matsigenka name	Uses
Astrocaryum murumuru	*Huicungo*	*Tiroti*	Edible endocarp
Attalea butyracea	*Shevon*	*Shevoshi*	Roof thatch
Attalea phalerata ('Scheelea')	*Shapaja*	*Tsigaro*	Edible endocarp, heart
Attalea tesmanii	*Conta*	*Konta*	Hard seed for pipe bowls
Bactris concinna	*Pijuayo de tigre*	*Shivarona*	Hard seed for pipe bowls
Bactris gasipaes	*Pijuayo* (peach palm)	*Kuri*	Edible mesocarp, endocarp, heart; wood for bows, arrow points (domesticated)
Bactris sp.	*Chontilla*	*Manataroki*	Edible mesocarp, endocarp (also cultivated)
Euterpe precatoria	*Huasaí*	*Tsireri*	Edible mesocarp, heart
Geonoma sp.	*Crisneja*	*Chogeroshi*	Roof thatch
Geonoma sp.	*Crisneja*	*Metakishi*	Roof thatch
Hyospathe/Chamaedora sp.	*Palmiche, crisneja*	*Kapashi*	Roof thatch
Iriartea deltoidea	*Pona*	*Kamona*	Edible heart; outer bark for floors, walls of houses; trunk for beer kegs; stumps attract beetle larvae
Mauritia flexuosa	*Aguaje*	*Toturoki, koshi*	Edible mesocarp; stumps attract beetle larvae (cultivated in picha, does not grow naturally)
Mauritiella sp.	*Aguajillo*	*Kiniri*	Edible mesocarp
Oenocarpus (Jessenia) bataua	*Ungurahui*	*Sega*	Edible mesocarp, heart
Oenocarpus mapora	*Ungurahuillo*	*Chorina*	Edible mesocarp, heart (also cultivated)
Phytelephas macrocarpa	*Yarina*	*Kompiro*	Edible endocarp
Socratea salazarii	*Pona*	*Kontiri*	Spiny stilt roots used as graters; leaf hatch for temporary housing
Socratea exorrhiza	*Pona*	*Kopapari*	Spiny stilt roots used as graters; leaf thatch for temporary housing
Wettinia augusta	*Ponilla*	*Kepito*	Thatch for temporary shelter

APPENDIX 32 Posanti Inchatoshipage: The diversity of forest types

Glenn Shepard, Jr. and Avecita Chicchón

Term	Translation	Description
A. Floodplain (Riparian) Forest Types		
Ovogeshi	'River bend' forest	Lowland forest, general term, characterized by seasonal or past inundation by river flooding
Imparage	River bed or beach	Sparse vegetation: *impomere* (*Baccharis*), *pochokiroshi* (*Cassia*), *kentakorishi* (Graminae)
Savoroshi	Wild cane vegetation	Beach successional zone: *savoro* (*Gynerium*)
Chakopishi	Arrow cane vegetation	Beach vegetation along upper river course, *chakopi* (*Gynerium*) harvested in December for making arrows
Inkonashi	*Cecropia* forest	Successional zone, seasonally inundated: *inkona* (*Cecropia*), *makota* (*Tachigali*)
Potogoshi	*Ficus* forest	Early successional forest, *Ficus-Cedrela* zone: *potogo* (*Ficus*), *santari* (*Cedrela*), though *Cedrela* greatly reduced due to timber extraction
Tairishi	*Erythrina* forest	Successional forest: *tairi* (*Erythrina*), *songaare* (*Erythrina*), *koshirite* (*Luehea*), *shimashiri* (*Senna*)
Tirotishi, *Nigankivoge*	*Astrocaryum* forest 'middle of bend'	Mature or late successional lowland forest, characterized by herb *sagonto* (Musaceae), palms *tiroti* (*Astrocaryum*), tsigaro (*Scheelea*) *kamona* (*Iriartea*) and emergent trees: *pasaro* (*Ceiba*), *tarovenki* (*Sloanea*), others
B. Upland (Interfluvial) Forest Types		
Nigankipatsa	'Middle earth,' i.e. interfluvial zone	Upland forest, general terms, out of reach of present or past floodplain of river
Kepitoshi	*Wettinia* forest	Most characteristic interfluvial forest; dense understory of *kepito* (*Wettinia augusta*) and other palms; sandy soils, high turnover of emergent trees: *paria* (*Cedrelinga*), *sampoa* (*Parkia*), *konori* (*Hevea*), *sumpa, yoricha, tsivaki* (Burseraceae spp.), *tsirotonaki* (*Cariniana*), *inchoviki* (Lauraceae), *etsiki* (*Helicostylis*), *pochariki* (*Pseudolmedia*), etc.
Pariashi	*Cedrelinga* forest	High canopy, mature upland forest, terraces with high moisture content, supporting the growth of large stands of *paria* (*Cedrelinga*), an important timber resource
Kovuvapishi	*Sagotia* forest	Sloping terrace forest, low canopy with understory dominated by *kovuvapini* (*Sagotia*)
Konorishi	*Hevea* forest	Upland forest with significant stands of *konori* (*Hevea*)
Segashi	*Oenocarpus (Jessenia)* forest	On sandy soils, typically near moist areas, stands of *Oenocarpus*, an important palm for fruits, heart and fibers
Tsirerishi	*Euterpe* forest	Near moist areas, stands of *Euterpe*, important palm for fruits and hearts
Chogeroshi	*Geonoma* forest	Upland forest with dense stands of *Geonoma* in understory, important roof material, in understory
Metakishi	*Geonoma* ? *Chamaedora* ?	Upland forest at higher elevations, near stream headwaters, understory characterized by *metakishi*, important palm thatch material (*Geonoma* group)
Kapashi	*Hyospathe* forest	Upland forest at higher elevations, near stream headwaters, understory characterized by *kapashi* (*Hyospathe*), important palm thatch material; also contains *tegarintsipini* (*Tovitoma weddeliana*) and *kachopitoki* (*Chrysochlamys*)
Songarentsishi	*Aulonemia* forest	On ridges or steep slopes, scandent bamboo *Aulonemia* in understory, used to make pan-pipes
Karororoempeshi	'high branching forest,' high canopy forest	Along, on top of steep ridges by small streams, emergent trees such as *inchoviki* (Lauraceae), *kumpe* (*Copaifera*), *yopo* (*Swietenia*), *tarovenki* (*Sloanea*?), *yomenta* (*Huberodendron*?); high, spreading crowns, difficult to hunt arboreal animals because of height of branches

Term	Translation	Description
C. High-Elevation Vegetation Types		
Otishi	Ridges, mountains, high elevation and cloud forest	General term for ridges and mountains; also used to refer to high elevation vegetational zones
Otyomiageni inchato	'Small trees'	Mixed dwarf forest, general term, characterized by epiphytes such as *tsiriantiniro* 'mother of pineapple' (Spanish moss, in fact a Bromeliad), *tsigiri* (lichens), *keshi* (Bromeliads), *ananta* (Orchids) and *evanaro* (Cyclanthaceae); prominent tree species include *oevaroshi*, fragrant tree, white leaf, many seeds (prob. Asteraceae), *kurikiipinishi,* medicinal shrub w/ spiny serrate leaves (*Ilex?*), *savotaroki* (Melastomataceae), *sangavantoshi* (?); understory includes *tinganari* (*Cyathea*), *tsirompi* (misc. ferns), *kamu* (*Selaginella*)
Katarompanakishi	*Clusia* forest	Mid-high elevation dwarf forest as above characterized by *katarompanaki*, shrub w/ paddle-shaped coriaceous leaves, latex, (probably *Clusia);* dried latex formerly traded as incense
Kasangari koka	Fragrant *Erythroxylum*	Middle elevation mixed forests, *kasangari koka*, 'fragrant coca' (*Erythroxylum*) in the understory; said to be former coca plantations during the Inca Empire
Segashi	*Oenocarpus* forest	Middle elevation terraces characterized by *sega* (*Oenocarpus*) stands
Tipeshi, *Omakaramangaitira*	*Sphagnum* moss, 'soft, spongy ground'	High elevation pajonal vegetation with spongy *tipeshi* (*Sphagnum*) covering soil, said to crunch and crumble underfoot in dry season
Songarentsishi, *Tiposhi,* *Shimpenashi,* *Yaviro*	*Chusque,* *Pariana?,* *Paspalum?,* *Puya?*	Other elements of pajonal vegetation: various Graminae spp.; also *yaviro*, spine-tipped terrestrial herb that 'looks like pineapple' (probably *Puya*) and is used medicinally
Kashikarishi	*Polylepis?* Forest	High elevation dwarf forest characterized by *kashikarishi*, reddish shrub w/ long, narrow coriaceous leaves (possibly *Polylepis*?)
Imperita, *Tsirompishi*	Rocky outcrops, Fern vegetation	Rocky outcrops characterized by *tsirompi* (misc. ferns)
Katsingari, *Terira ontime inchato*	'cold,' 'where no trees are'	Highest elevations, above tree line
D. Disturbed or Inundated Forest Types		
Kapiroshi	*Guadua* bamboo forest	Low canopy forest dominated by spiny *Guadua weberbaueri*; sometimes mixed with lowland trees, sometimes entirely bamboo; very dense along Picha
Shivitsasemai	Liana forest	Low canopy lowland forest with vines in understory, areas of past inundations
Tsamairentsi	'Place of work,' garden	Garden in primary production phase, while it is still being weeded
Pugoroshi	*Vernonia* growth	Secondary growth characterized by *Vernonia* in recent garden fallows
Shintishi	*Guazuma* growth	Secondary growth characterized by *Guazuma crinita* in garden fallows
Yaaroshi	*Cecropia* sp. forest	Stands of *Cecropia* sp. indicating past disturbance, typically older garden fallows or windstorms
Pairani magashipogo	'Old garden clearing'	Mature secondary forest regrown from former garden clearings (25 years +), recognized by historical knowledge and the presence of hardy cultigens like ayahuasca (*Banisteriopsis*)
Inkaarepatsa, *jampovatsa*	'Swampy earth,' 'muddy earth'	Areas of topographical inundation, muddy soils, characterized by dense growth of lianas, *kontiri* (*Socratea*), *maempa* (*Euterpe* sp.), etc.; bad for agriculture
Imereshi	*Diplasia* growth	In moist to swampy areas, understory dominated by *imere* (*Diplasia*), also known as *saviripini*, 'sword plant' due to its sharp edges
Kinirishi	*Mauritiella* forest	Swampy area characterized by kiniri, 'aguajillo' (*Mauritiella*), also contains *kontiri* (*Socratea*), *tsireri* (*Euterpe*), etc.
Koshishi	*Mauritiella* forest	Swamps dominated by the palm *koshi* or *toturoki* (*Mauritia*); does not occur in Picha but is known to occur elsewhere